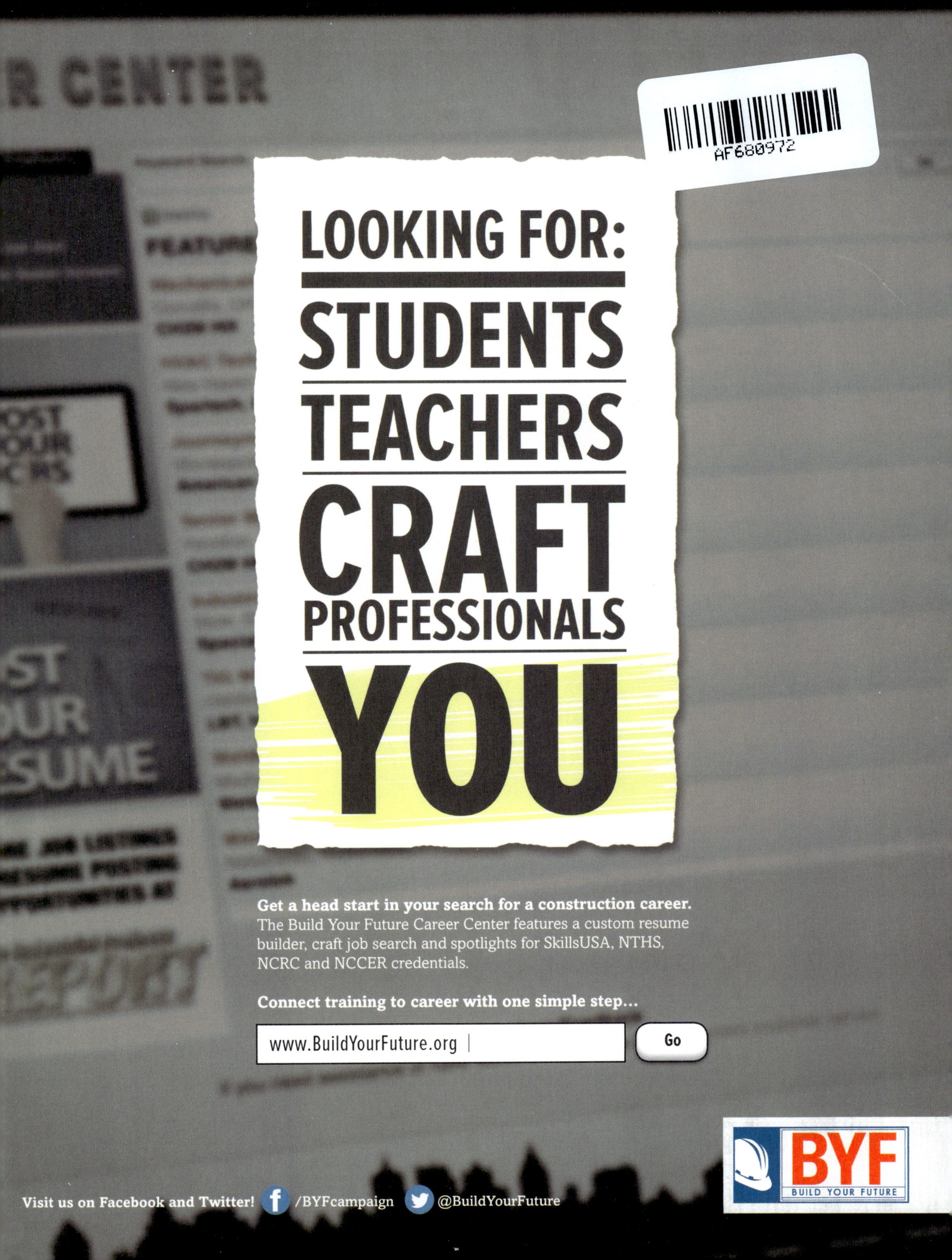
LOOKING FOR:
STUDENTS
TEACHERS
CRAFT
PROFESSIONALS
YOU
Get a head start in your search for a construction career.
The Build Your Future Career Center features a custom resume builder, craft job search and spotlights for SkillsUSA, NTHS, NCRC and NCCER credentials.
Connect training to career with one simple step...
www.BuildYourFuture.org
Go
BYF
BUILD YOUR FUTURE
Visit us on Facebook and Twitter! /BYFcampaign @BuildYourFuture

Masonry

Level One

Trainee Guide
Fourth Edition

PEARSON

Boston Columbus Indianapolis New York San Francisco Upper Saddle River
Amsterdam Cape Town Dubai London Madrid Milan Munich Paris Montreal Toronto
Delhi Mexico City São Paulo Sydney Hong Kong Seoul Singapore Taipei Tokyo

NCCER
President: Don Whyte
Director of Product Development: Daniele Dixon
Masonry Project Manager: Rob Richardson
Senior Manager: Tim Davis
Quality Assurance Coordinator: Debie Ness
Desktop Publishing Coordinator: James McKay
Permissions Specialist: Amanda Werts
Production Specialist: Megan Casey
Editor: Chris Wilson

Writing and development services provided by S4Carlisle Publishing Services, Dubuque, IA
Lead Writer/Project Manager: Michael B. Kopf
Writer: Paul Lagassse, Jack Klasey
Art Development: S4Carlisle Publishing Services
Permissions Specialist: Kim Schmidt, Karyn Morrison, Katherine Benzer
Media Specialist: Genevieve Brand
Copy Editor: Michael H. Toporek

Pearson Education, Inc.
Editorial Director: Vernon R. Anthony
Executive Editor: Alli Gentile
Editorial Assistant: Douglas Greive
Program Manager: Alexandrina B. Wolf
Operations Supervisor: Deidra M. Skahill
Art Director: Jayne Conte
Director of Marketing: David Gesell
Executive Marketing Manager: Derril Trakalo
Marketing Manager: Brian Hoehl
Marketing Coordinator: Crystal Gonzalez
Composition: NCCER
Printer/Binder: LSC Communications
Cover Printer: LSC Communications
Text Fonts: Palatino and Univers

Credits and acknowledgments for content borrowed from other sources and reproduced, with permission, in this textbook appear at the end of each module.

---

21 2022

Case bound ISBN-13: 978-0-13-375402-5
ISBN-10: 0-13-375402-2

Perfect bound ISBN-13: 978-0-13-340248-3
ISBN-10: 0-13-340248-7

PEARSON

Preface

To the Trainee

Masons are recognized as premier craftworkers on any construction site. Although masonry is one of the world's oldest crafts, masons also use 21st-century technology on the job. Using brick, block, or stone, and bound with mortar, masons build durable structures with optimized energy performance.

With the support of the Mason Contractor Association of America (MCAA), NCCER's program has been designed and revised by subject matter experts from across the nation and industry to update the curriculum with modern techniques. Our three levels present an apprentice approach to the masonry field and helps keep you knowledgeable, safe, and effective on the job.

We wish you the best as you begin an exciting and promising career. This newly revised masonry curriculum will help you enter the workforce with the knowledge and skills needed to perform productively in either the residential or commercial market.

New with *Masonry Level One*

NCCER is proud to release the newest edition of *Masonry Level One* in full color with updates to the curriculum that will engage you and give you the best training possible. In this edition, you will find that the layout has changed to better align with the learning objectives. There are also new end-of-section review questions to compliment the module review. The text, graphics, and special features were enhanced to reflect advancements in masonry technology and techniques.

We invite you to visit the NCCER website at **www.nccer.org** for information on the latest product releases and training, as well as online versions of the *Cornerstone* magazine and Pearson's NCCER product catalog.

Your feedback is welcome. You may email your comments to **curriculum@nccer.org** or send general comments and inquiries to **info@nccer.org**.

NCCER Standardized Curricula

NCCER is a not-for-profit 501(c)(3) education foundation established in 1996 by the world's largest and most progressive construction companies and national construction associations. It was founded to address the severe workforce shortage facing the industry and to develop a standardized training process and curricula. Today, NCCER is supported by hundreds of leading construction and maintenance companies, manufacturers, and national associations. The NCCER Standardized Curricula was developed by NCCER in partnership with Pearson, the world's largest educational publisher.

Some features of the NCCER Standardized Curricula are as follows:

- An industry-proven record of success
- Curricula developed by the industry for the industry
- National standardization providing portability of learned job skills and educational credits
- Compliance with the Office of Apprenticeship requirements for related classroom training *(CFR 29:29)*
- Well-illustrated, up-to-date, and practical information

NCCER also maintains a National Registry that provides transcripts, certificates, and wallet cards to individuals who have successfully completed a level of training within a craft in NCCER's Curricula. *Training programs must be delivered by an NCCER Accredited Training Sponsor in order to receive these credentials.*

Special Features

In an effort to provide a comprehensive, user-friendly training resource, we have incorporated many different features for your use. Whether you are a visual or hands-on learner, this book will provide you with the proper tools and information to orient you to the important skills and techniques of the Masonry trade.

> **28103-13**
> **MEASUREMENTS, DRAWINGS, AND SPECIFICATIONS**
>
> **Code Note**
>
> Codes vary among jurisdictions. Because of the variations in code, consult the applicable code whenever regulations are in question. Referring to an incorrect set of codes can cause as much trouble as failing to reference codes altogether. Obtain, review, and familiarize yourself with your local adopted code.
>
> **Objectives**
>
> When you have completed this module, you will be able to do the following:
>
> 1. Recognize the mathematical concepts used in masonry.
>   a. Explain how to read a six-foot rule.
>   b. Explain how to read other measuring devices.
>   c. Explain how to read mason's rules.
>   d. Recognize modular increments.
>   e. Describe how to determine areas and circumferences.
>   f. Explain how to use the 3-4-5 ratio to square a corner.
> 2. Identify the basic parts of a set of drawings and list the information found on each type.
>   a. Identify lines, symbols, and abbreviations used on drawings.
>   b. Identify scales and dimensions used on drawings.
>   c. Identify types of construction drawings.
> 3. Identify the purpose of specifications, standards, and codes used in the building industry and the sections that pertain to masonry.
>   a. Explain the purpose of specifications, standards, and codes.
>   b. Describe the purpose of inspections and testing.
>
> **Performance Tasks**
>
> Under the supervision of your instructor, you should be able to do the following:
>
> 1. Use a mason's rule to measure a space and verify its squareness.
> 2. Use a rule to measure fractional dimensions.
> 3. Locate information on construction drawings.
>
> **Trade Terms**
>
> Converting
> Legend
> MasterFormat™
> Nominal dimension
>
> **Industry-Recognized Credentials**
>
> If you're training through an NCCER-accredited sponsor, you may be eligible for credentials from NCCER's Registry. The ID number for this module is 28103-13. Note that this module may have been used in other NCCER curricula and may apply to other level completions. Contact NCCER's Registry at 888.622.3720 or go to **www.nccer.org** for more information.

Introduction

This page is found at the beginning of each module and lists the Objectives, Performance Tasks, Trade Terms, and Required Trainee Materials for that module. The Objectives list the skills and knowledge you will need in order to complete the module successfully. The Performance Tasks give you an opportunity to apply your knowledge to real-world tasks you will undertake as a mason. The list of Trade Terms identifies important terms you will need to know by the end of the module. Required Trainee Materials list the materials and supplies needed for the module.

Special Features

Features provide a head start for those learning masonry by presenting technical tips and professional practices. These features often include real-life scenarios similar to those you might encounter on the job site.

> **Fall Protection**
>
> Most workers who die from falls were wearing harnesses but had failed to tie off properly. Always follow the manufacturer's instructions when wearing a harness. Know and follow your company's safety procedures when working on roofs, ladders, and other elevated locations.

Color Illustrations and Photographs

Full-color illustrations and photographs are used throughout each module to provide vivid detail. These figures highlight important concepts from the text and provide clarity for complex instructions. Each figure reference is denoted in the text in *italic type* for easy reference.

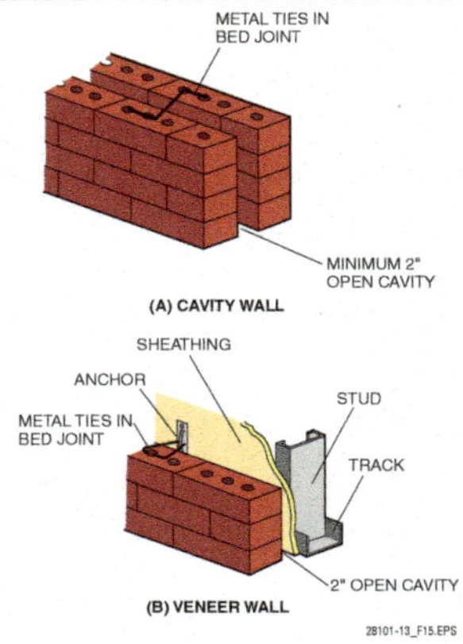

Figure 15 Types of masonry construction.

Notes, Cautions, and Warnings

Safety features are set off from the main text in highlighted boxes and are organized into three categories based on the potential danger of the issue being addressed. Notes simply provide additional information on the topic area. Cautions alert you of a danger that does not present potential injury but may cause damage to equipment. Warnings stress a potentially dangerous situation that may cause injury to you or a co-worker.

> **NOTE**
> Remember to include the units, such as feet or inches, in your calculations.

> **CAUTION**
> Only a competent person has the authority to supervise setting up, moving, and taking down scaffold. Only a competent person can approve the use of scaffold on the job site after inspecting the scaffold.

> **WARNING!**
> Never unlock the wheel brakes of a rolling scaffold while anyone is on it. People on a moving scaffold can lose their balance and fall.

Going Green

Going Green looks at ways to preserve the environment, save energy, and make good choices regarding the health of the planet. Through the introduction of new construction practices and products, you will see how the greening of America has already taken root.

> GOING GREEN
>
> **The ENERGY STAR® Program**
>
> ENERGY STAR® is a joint program developed and administered by the US Environmental Protection Agency (EPA) and the US Department of Energy (DOE). The goal is save energy and protect the environment through the use of energy efficient products and practices. Products identified with the ENERGY STAR® logo meet strict energy efficiency guidelines established by the program. Additional resources available through the ENERGY STAR® program provide energy-efficient guidelines for home construction. The appearance of the ENERGY STAR® logo assures consumers that a product meets the specifications of the program.

Did You Know?

The *Did You Know?* features offer hints, tips, and other helpful bits of information.

> **Did you know?**
>
> **Safety Training**
>
> Safe working conditions on mine sites fall under the Mine Safety and Health Administration (MSHA), and every other job site is regulated by the Occupational Safety and Health Administration (OSHA). However, except for a few industry-specific requirements, the regulations are the same. Safety training is required for all activities. Never operate tools, machinery, or equipment without prior training.

Step-by-Step Instructions

Step-by-step instructions are used throughout to guide you through technical procedures and tasks from start to finish. These steps show you not only how to perform a task but also how to do it safely and efficiently.

> The mason needs to determine whether the brick is too dry for a good bond with the mortar. The following test can be used to measure the absorption rate of brick:
>
> *Step 1* Draw a circle about the size of a quarter on the surface of the brick with a crayon or wax marker.
>
> *Step 2* With a medicine dropper, place 20 drops of water inside the circle.
>
> *Step 3* Using a watch with a second hand, note the time required for the water to be absorbed.
>
> If the time for absorption exceeds 1½ minutes,

Trade Terms

Each module presents a list of Trade Terms that are discussed within the text and defined in the Glossary at the end of the module. These terms are denoted in the text with **bold, blue type** upon their first occurrence. To make searches for key information easier, a comprehensive Glossary of Trade Terms from all modules is located at the back of this book.

> **M**asons are recognized as premier craftworkers at any construction site. Although masonry is one of the world's oldest crafts, masons also use 21st-century technology on the job. Masons build structures out of **masonry units**. The two main types of masonry units manufactured today are made of clay and concrete. Clay products are commonly known as brick and tile; concrete products are commonly known as **concrete masonry units (CMUs)** or block. Masonry units are also made from **ashlar**, glass, **adobe**, and other materials. In the most common forms of masonry, a mason assembles walls and other structures of clay brick or CMUs using **mortar** to bond the units together.

Review Questions

Review Questions are provided to reinforce the knowledge you have gained. This makes them a useful tool for measuring what you have learned.

> **Review Questions**
>
> 1. Exit signs typically have _____.
>   a. red letters on a white field
>   b. white letters on a green field
>   c. white letters on a red field
>   d. red letters on a yellow field
> 2. Safety tags are used to _____.
>   a. identify no-smoking areas
>   b. provide temporary warnings of hazards
>   c. warn against high voltage
>   d. locate emergency eye wash stations
> 3. The construction industry annually suffers millions of dollars in accidents, lost time and productivity because of worker _____.
>   a. waste of mortar
>   b. vehicle accidents
>   c. tardiness
>   d. substance abuse
> 4. Accidents caused by lack of skill can be prevented by _____.
>   a. proper training
>   b. observing warning signs
>   c. working more carefully
>   d. practice
> 5. Making excuses for taking risks on the job is known as _____.
>   a. cutting corners
>   b. rationalizing
>   c. hazardous behavior
>   d. carelessness
> 6. Poor housekeeping on the job site is an example of a(n)_____.
>   a. unsafe act
>   b. management system failure
>   c. unsafe condition
>   d. communication failure
> 7. When dumping mortar bags into a mixer, whenever possible you should _____.
>   a. stand upwind
>   b. warn others to stay away
>   c. stand downwind
>   d. shake the bag vigorously
> 8. Flammable liquids should be stored only in _____.
>   a. outdoor locations
>   b. red plastic gas cans
>   c. glass containers with tight-fitting lids
>   d. approved safety containers
> 9. If safety goggles are not available, the best form of eye protection is _____.
>   a. regular prescription lenses
>   b. safety lenses with side shields
>   c. contact lenses
>   d. tinted lenses
> 10. Large, padded covers for the entire ear are called _____.
>   a. earplugs
>   b. ear shields
>   c. earmuffs
>   d. ear guards
> 11. To protect your hands from abrasion, chemical burns, and other masonry hazards, you should wear gloves made from _____.
>   a. heavy canvas
>   b. nylon and nitrile
>   c. leather
>   d. reinforced polyester
> 12. One of the leading causes of fatalities among construction workers is _____.
>   a. falls from elevated areas
>   b. inhalation of silicon dust
>   c. chemical burns from wet concrete
>   d. trowel-related injuries
> 13. The number of groups that falls are classified into is _____.
>   a. two
>   b. three
>   c. four
>   d. five
> 14. When working at an elevation of 6 feet or more, workers are required to wear _____.
>   a. safety shoes
>   b. suspenders

NCCER Standardized Curricula

NCCER's training programs comprise more than 80 construction, maintenance, pipeline, and utility areas and include skills assessments, safety training, and management education.

Boilermaking
Cabinetmaking
Carpentry
Concrete Finishing
Construction Craft Laborer
Construction Technology
Core Curriculum:
  Introductory Craft Skills
Drywall
Electrical
Electronic Systems Technician
Heating, Ventilating, and
  Air Conditioning
Heavy Equipment Operations
Highway/Heavy Construction
Hydroblasting
Industrial Coating and Lining
  Application Specialist
Industrial Maintenance Electrical
  and Instrumentation Technician
Industrial Maintenance
  Mechanic
Instrumentation
Insulating
Ironworking
Masonry
Millwright
Mobile Crane Operations
Painting
Painting, Industrial
Pipefitting
Pipelayer
Plumbing
Reinforcing Ironwork
Rigging
Scaffolding
Sheet Metal
Signal Person
Site Layout
Sprinkler Fitting
Tower Crane Operator
Welding

Maritime

Maritime Industry Fundamentals
Maritime Pipefitting

Green/Sustainable Construction

Building Auditor
Fundamentals of Weatherization
Introduction to Weatherization
Sustainable Construction
  Supervisor
Weatherization Crew Chief
Weatherization Technician
Your Role in the Green
  Environment

Energy

Alternative Energy
Introduction to the Power Industry
Introduction to Solar Photovoltaics
Introduction to Wind Energy
Power Industry Fundamentals
Power Generation Maintenance
  Electrician
Power Generation I&C
  Maintenance Technician
Power Generation Maintenance
  Mechanic
Power Line Worker
Power Line Worker: Distribution
Power Line Worker: Substation
Power Line Worker: Transmission
Solar Photovoltaic Systems
  Installer
Wind Turbine Maintenance
  Technician

Pipeline

Control Center Operations, Liquid
Corrosion Control
Electrical and Instrumentation
Field Operations, Liquid
Field Operations, Gas
Maintenance
Mechanical

Safety

Field Safety
Safety Orientation
Safety Technology

Supplemental Titles

Applied Construction Math
Careers in Construction
Tools for Success

Management

Fundamentals of Crew Leadership
Project Management
Project Supervision

Spanish Titles

Acabado de concreto: nivel uno,
  nivel dos
Aislamiento: nivel uno, nivel dos
Albañilería: nivel uno
Andamios
Aparejamiento básico
Aparajamiento intermedio
Aparajamiento avanzado
Carpintería:
  Introducción a la carpintería,
  nivel uno; Formas para
  carpintería, nivel tres
Currículo básico: habilidades
  introductorias del oficio
Electricidad: nivel uno, nivel dos,
  nivel tres, nivel cuatro
Encargado de señales
Especialista en aplicación de
  revestimientos industriales:
  nivel uno, nivel dos
Herrería: nivel uno, nivel dos,
  nivel tres
Herrería) de refuerzo: nivel uno
Instalación de rociadores: nivel
  uno
Instalación de tuberías: nivel uno,
  nivel dos, nivel tres, nivel cuatro
Instrumentación: nivel uno, nivel
  dos, nivel tres, nivel cuatro
Orientación de seguridad
Mecánico industrial: nivel uno,
  nivel dos, nivel tres, nivel
  cuatro, nivel cinco
Paneles de yeso: nivel uno
Seguridad de campo
Soldadura: nivel uno, nivel dos,
  nivel tres

Portuguese Titles

Currículo essencial: Habilidades
  básicas para o trabalho
Instalação de encanamento
  industrial: nível um, nível dois,
  nível três, nível quatro

Acknowledgments

This curriculum was revised as a result of the farsightedness and leadership of the following sponsors:

- Arizona Masonry Contractors Association
- Brick Industry Association
- Central Cabarrus High School
- Florida Masonry Apprentice & Educational Foundation, Inc.
- Mason Contractors Association of America
- Mortar Net Solutions
- Mortenson Construction
- Pyramid Masonry
- Rhino Masonry, Inc.
- Rocky Mountain Masonry Institute
- Samuell High School
- Skyline High School

This curriculum would not exist were it not for the dedication and unselfish energy of those volunteers who served on the Authoring Team. A sincere thanks is extended to the following:

- Jeff Buczkiewicz
- Kenneth Cook
- Steve Fechino
- John Foley
- Todd Hartsell
- Lawrence Johnson
- Merritt Johnson
- Bryan Light
- Moroni Mejia
- Dennis Neal

NCCER Partners

- American Fire Sprinkler Association
- Associated Builders and Contractors, Inc.
- Associated General Contractors of America
- Association for Career and Technical Education
- Association for Skilled and Technical Sciences
- Carolinas AGC, Inc.
- Carolinas Electrical Contractors Association
- Center for the Improvement of Construction Management and Processes
- Construction Industry Institute
- Construction Users Roundtable
- Construction Workforce Development Center
- Design Build Institute of America
- GSSC – Gulf States Shipbuilders Consortium
- Manufacturing Institute
- Mason Contractors Association of America
- Merit Contractors Association of Canada
- NACE International
- National Association of Minority Contractors
- National Association of Women in Construction
- National Insulation Association
- National Ready Mixed Concrete Association
- National Technical Honor Society
- National Utility Contractors Association
- NAWIC Education Foundation
- North American Technician Excellence
- Painting & Decorating Contractors of America
- Portland Cement Association
- SkillsUSA®
- Steel Erectors Association of America
- U.S. Army Corps of Engineers
- University of Florida, M. E. Rinker School of Building Construction
- Women Construction Owners & Executives, USA

NCCER Business Partners

Contents

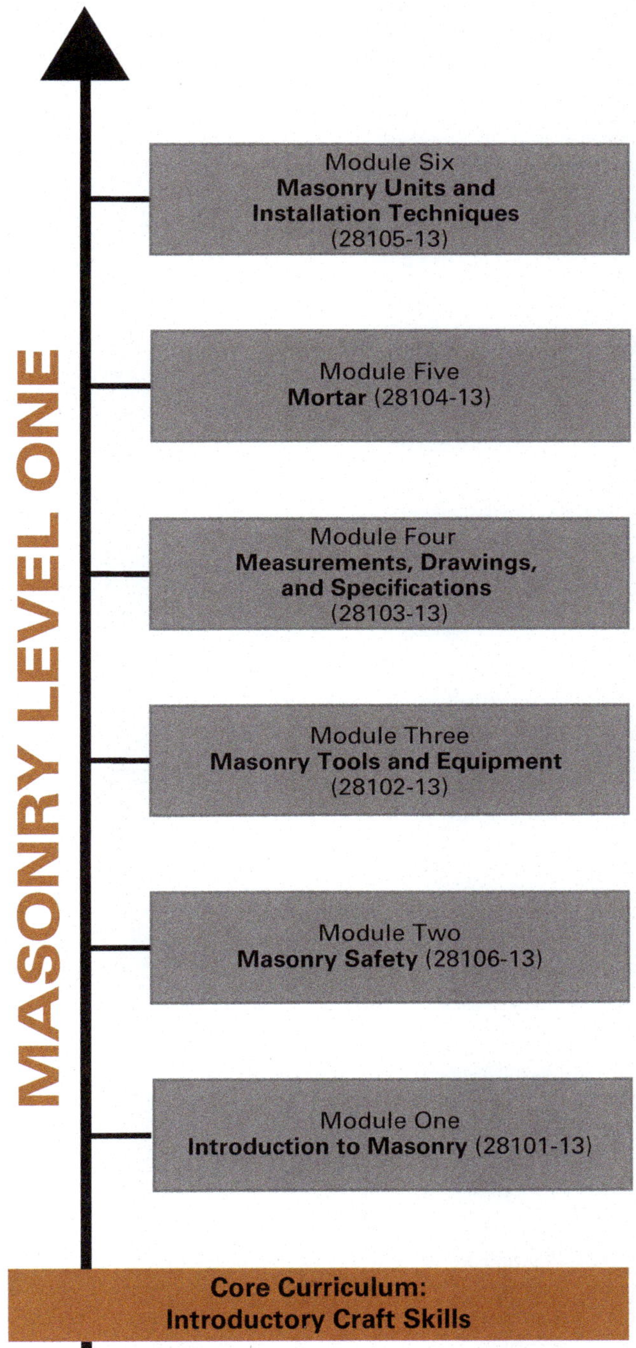

This course map shows all of the modules in *Masonry Level One*. The suggested training order begins at the bottom and proceeds up. Skill levels increase as you advance on the course map. The local Training Program Sponsor may adjust the training order.

28101-13

Introduction to Masonry

Overview

In this module, you will learn about the basic materials, tools, and techniques used by masons, as well as basic safety precautions and the skills, attitudes, and abilities exhibited by successful masons. With the guidance of your instructor, you will learn to mix mortar and lay brick. The challenge is to lay the units perfectly straight and level, and to complete the work to specification and on time, with a high degree of quality.

Module One

Trainees with successful module completions may be eligible for credentialing through NCCER's National Registry. To learn more, go to **www.nccer.org** or contact us at **1.888.622.3720**. Our website has information on the latest product releases and training, as well as online versions of our *Cornerstone* magazine and Pearson's product catalog.

Your feedback is welcome. You may email your comments to **curriculum@nccer.org**, send general comments and inquiries to **info@nccer.org**, or fill in the User Update form at the back of this module.

This information is general in nature and intended for training purposes only. Actual performance of activities described in this manual requires compliance with all applicable operating, service, maintenance, and safety procedures under the direction of qualified personnel. References in this manual to patented or proprietary devices do not constitute a recommendation of their use.

28101-13
INTRODUCTION TO MASONRY

Objectives

When you have completed this module, you will be able to do the following:

1. Describe modern masonry materials and techniques.
   a. Explain how concrete masonry units (CMUs or block) are used in construction.
   b. Explain how clay masonry units (brick) are used in construction.
   c. Explain how stone is used in construction.
   d. Describe how mortar and grout are used in masonry construction.
   e. Describe how wall structures are created using masonry units.
2. Recognize the basic safety precautions when working with masonry materials.
   a. List basic safety practices.
   b. Describe personal protective equipment used in masonry.
3. Explain how to mix mortar and lay masonry units.
   a. Explain how to mix mortar.
   b. Describe how to lay masonry units.
4. Describe the skills, attitudes, and abilities needed to be a successful mason.
   a. Identify the skills of a successful mason.
   b. Identify the attitudes of a successful mason.
   c. Identify the abilities of a successful mason.
   d. Explore career ladders and advancement possibilities in masonry.
5. Summarize how to be connected to the industry through an organization like SkillsUSA.
   a. Understand the program, curriculum, and SkillsUSA Championships.
   b. Understand SkillsUSA membership.
   c. Understand the National Program of Work Standards.

Performance Tasks

Under the supervision of your instructor, you should be able to do the following:

1. Put on eye protection and respiratory protection.
2. Properly mix mortar by hand.
3. Properly spread mortar using a trowel.

Trade Terms

Admixture
Adobe
Aggregate
American Society for Testing and Materials (ASTM) International
Ashlar
Bed joint
Butter
Capital
Competent person
Concrete masonry unit (CMU)
Cored
Cornice
Course
Cube
Facing
Footing
Grout
Head joint
Joints
Manufactured stone veneer
Mason
Masonry unit
Mortar
Nonstructural
Occupational Safety and Health Administration (OSHA)
Parapet
Personal protective equipment (PPE)
Pilaster
Spread
Stringing
Structural
Tuckpointing
Uncored
Weephole
Wythe

Industry-Recognized Credentials

If you're training through an NCCER-accredited sponsor, you may be eligible for credentials from NCCER's Registry. The ID number for this module is 28101-13. Note that this module may have been used in other NCCER curricula and may apply to other level completions. Contact NCCER's Registry at 888.622.3720 or go to **www.nccer.org** for more information.

Code Note

Codes vary among jurisdictions. Because of the variations in code, consult the applicable code whenever regulations are in question. Referring to an incorrect set of codes can cause as much trouble as failing to reference codes altogether. Obtain, review, and familiarize yourself with your local adopted code.

Contents

Topics to be presented in this module include:

Contents (*continued*)

Figures and Tables

Section One

1.0.0 Modern Masonry Materials and Techniques

Objective

Describe modern masonry materials and techniques.

a. Explain how concrete masonry units (CMUs or block) are used in construction.
b. Explain how clay masonry units (brick) are used in construction.
c. Explain how stone is used in construction.
d. Describe how mortar and grout are used in masonry construction.
e. Describe how wall structures are created using masonry units.

Trade Terms

Admixture: A chemical or mineral other than water, cement, or aggregate added to mortar immediately before or during mixing to change its setting time or curing time; to reduce water; or to change the overall properties of the mortar.

Adobe: Sun-dried, molded clay brick.

Aggregate: Materials such as crushed stone or gravel used as a filler in concrete and concrete block.

American Society for Testing and Materials (ASTM) International: The publisher of masonry standards.

Ashlar: A squared or rectangular cut stone masonry unit; or, a flat-faced surface having sawed or dressed bed and joint surfaces.

Capital: The top part of an architectural column.

Concrete masonry unit (CMU): A hollow or solid block made from portland cement and aggregates.

Cored: Brick that has holes extending through it to reduce weight.

Cornice: The horizontal projection crowning the wall of a building.

Course: A row or horizontal layer of masonry units.

Cube: A strapped bundle of approximately 500 standard brick, or 90 standard block. The number of units in a cube will vary according to the manufacturer.

Facing: That part of a masonry unit or wall that shows after construction; the finished side of a masonry unit.

Grout: A mixture of portland cement, lime, and water, with or without fine aggregate, with a high-enough water content that it can be poured into spaces between masonry units and voids in a wall.

Joints: The area between each brick or block that is filled with mortar.

Manufactured stone veneer: A premade veneer consisting of cast cementitious material with pigments and other added materials to give the appearance of natural stone. Also called adhered concrete masonry veneer (ACMV).

Mason: A person who assembles masonry units by hand, using mortar, dry stacking, or mechanical connectors.

Masonry unit: Any building block made of brick, cement, ashlar, clay, adobe, rubble, glass, tile, or any other material that can be assembled into a structural unit.

Mortar: A mixture of portland cement, lime, fine aggregate, and water, plastic or stiff enough to hold its shape between masonry units.

Nonstructural: Not bearing weight other than its own.

Parapet: A low wall or railing.

Structural: Bearing weight in addition to its own.

Uncored: Brick that has no holes extending through it.

Weephole: A small opening in mortar joints or faces to allow the escape of moisture.

Wythe: A continuous section of masonry wall, one masonry unit in thickness, or that part of a wall that is one masonry unit in thickness.

Masons are recognized as premier craftworkers at any construction site. Although masonry is one of the world's oldest crafts, masons also use 21st-century technology on the job. Masons build structures out of masonry units. The two main types of masonry units manufactured today are made of clay and concrete. Clay products are commonly known as brick and tile; concrete products are commonly known as concrete masonry units (CMUs) or block. Masonry units are also made from ashlar, glass, adobe, and other materials. In the most common forms of masonry, a mason assembles walls and other structures of clay brick or CMUs using mortar to bond the units together.

1.1.0 Learning about Concrete Masonry Units (CMUs or Block)

CMUs have not been around as long as brick. The first CMUs were developed in 1850. Block is made of water added to portland cement, aggregates (sand and gravel), and admixtures. Admixtures affect the color and other properties of the cement, such as freeze resistance, weight, and speed of setting.

The block is machine-molded into shape. It is compacted in the molds and cured, typically using live steam. After curing, the block is dried and aged. The moisture content is checked. It must be a specified minimum amount before the block can be shipped for use. *Figure 1* shows commonly used sizes and shapes of concrete block.

While CMUs are generally referred to as block, it's important to note that not all types of block are CMUs. Concrete units are classified according to their intended use, size, and appearance. American Society for Testing and Materials (ASTM) International standards exist for the following types of masonry units:

- Loadbearing and nonbearing concrete block
- Concrete brick
- Calcium silicate face brick
- Pre-faced or prefinished facing units
- Manholes and catch basin units

A concrete block is a large unit, typically with actual dimensions of 7⅝" × 7⅝" × 15⅝", with a hollow core. (Actual dimensions are the dimensions of the block itself; nominal dimensions include the thickness of the mortar in the measurement.) Block comes in modular sizes, and in colors determined by the cement ingredients, the aggregates, or any admixtures. A variety of surface and mixing treatments can give block varied and attractive surfaces. Newer finishing techniques can give block the appearance of brick, rough stone, or cut stone. Like clay masonry units, block can be laid in structural pattern bonds. *Figure 2* shows the names of the parts of a block.

Block takes up more space than other building units, so fewer are needed. Block bed joints usually need mortar only on the shells and webs, so there is less mortaring as well.

Concrete block comes in three weights: normal, lightweight, and aerated. Lightweight block is made with lightweight aggregates. The loadbearing and appearance qualities of the first two weights are similar; the major difference is that lightweight block is easier and faster to lay. Normal-weight block can be made of concrete with regular, high, and extra-high strengths. The last two are made with different aggregates and curing times. They are used to limit wall thickness in buildings over 10 floors high. Aerated block is made with an admixture that generates gas bubbles inside the concrete for a lighter block.

Concrete block is classified as hollow or solid. Like clay products, less than 75 percent of the surface area of a hollow unit is solid. Common hollow

Ancient Wonders

The famous Lion Gate in Istanbul, Turkey, is a fine example of ancient brick artistry. It is made of fired and glazed brick sculpted into high relief before firing. The kiln-firing techniques refined in ancient India, Babylonia, and Rome are still used today.

Brick in Construction

Brick is found in all types of construction, from tract houses to stately mansions. Brick is also used in the construction of banks, schools, churches, and office buildings. Brick not only provides an attractive appearance, it creates a sense of permanence.

28101-13_SA01.EPS

28101-13_SA02.EPS

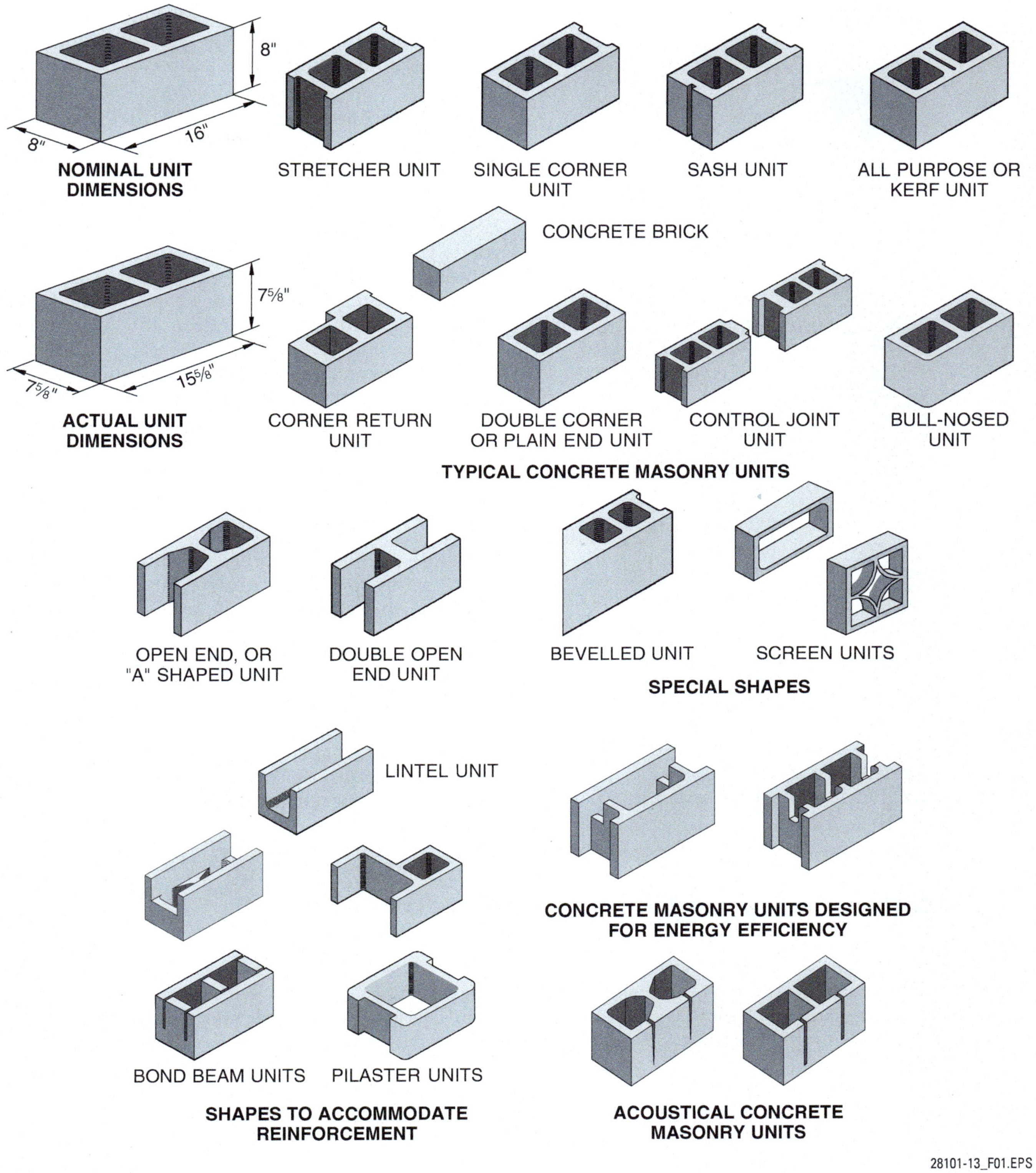

Figure 1 Common concrete block.

units have two or three cores. The hollow cores make it easy to reinforce concrete block walls. **Grout** alone, or steel reinforcing rods combined with grout, can be used to fill the hollow cores. Reinforcement increases loadbearing strength, rigidity, and wind resistance. Less than 25 percent of the surface area of a solid block is hollow. Normal and lightweight solid units are intended for special needs, such as structures with unusually high loads, drainage catch basins, manholes, and firewalls. Aerated block is made in an oversize solid unit used for buildings.

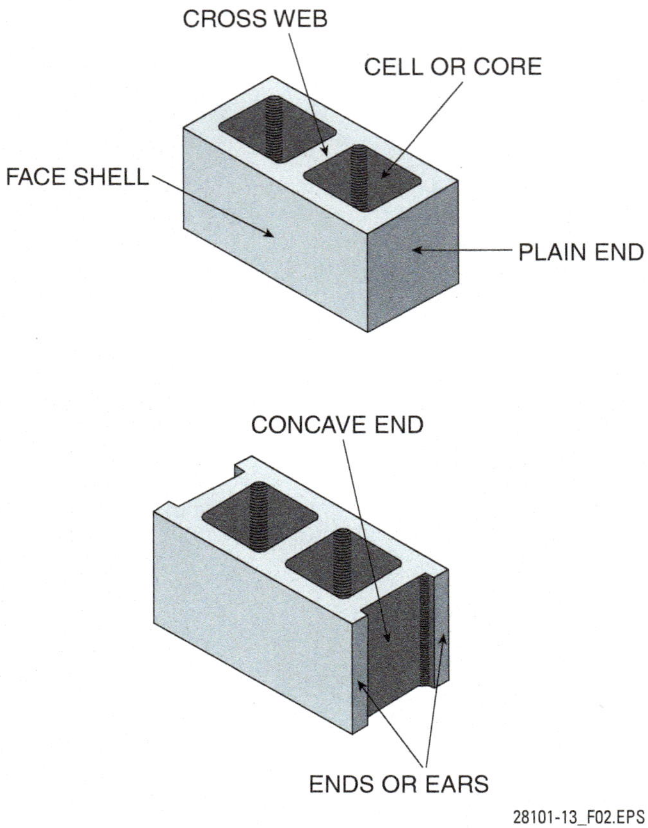

Figure 2 Parts of a block.

Loadbearing block is used as backing for veneer walls, bearing walls, and all structural uses. Both regular and specially shaped block is used for paving, retaining walls, and slope protection. Nonstructural block is used for screening, partition walls, and as a veneer wall for wood, steel, or other backing. Both kinds of block come in a variety of shapes and modular sizes.

1.1.1 Concrete Brick

The length and height dimensions of regular concrete brick are the same as those of standard clay brick (*Figure 3*). The thickness is an additional ⅛-inch. A popular type is slump brick, which is made from very wet concrete. When the mold is removed, the brick bulges because it is not dry enough to completely hold its shape. Slump brick looks like ashlar and adds a decorative element to a wall.

Concrete brick is produced in a wide range of textures and finishes. It is available in specialized shapes for copings, sills, and stairs, just as clay brick is. Concrete brick is more popular in some areas of the country because it is less expensive.

1.1.2 Other Concrete Units

Concrete pre-faced or precoated units are coated with colors, patterns, and textures on one or two face shells. The facings are made of resins, portland cement, ceramic glazes, porcelainized glazes, or mineral glazes. The slick facing is easily cleaned. These units are popular for use in gyms, hospital and school hallways, swimming pools, and food processing plants. They come in a variety of sizes and special-purpose shapes, such as ribbing and bullnose corners. *Figure 4* shows commonly used concrete pre-faced units.

A Short History of Concrete Masonry Units

The first CMUs were invented in 1850 by Joseph Gibbs, who was trying to develop a better way to build masonry cavity walls, which are made of two wythes of masonry units, with a 2- to 4-inch gap between them. Water can get through the outside wall and run down inside the cavity without wetting the inside wall. In his search for a faster way to build a cavity wall, Gibbs developed a block with air cells in it. This idea was refined and patented by several other people. In 1882, someone took advantage of new materials developments to make a hollow block of portland cement.

In 1900, Harmon Palmer patented a machine that made hollow concrete block. Palmer's block was 30 inches long, 10 inches high, and 8 inches wide. Even though Palmer's block was heavy and very hard to lift, cavity walls became cheaper and faster to build. Over time, other machines were developed to produce smaller block. This block was easier to handle, but was still very heavy for its size.

In 1917, Francis Straub patented a block made with the cinders left over from the burning of coal. He used the cinders to replace the sand and small aggregates in the concrete mix. This new cinderblock was lighter, cheaper, and easier to handle. Straub's block made it possible to build a one-wythe wall with a built-in cavity very quickly and inexpensively. Faster machinery was developed to keep up with the demand for this new masonry material.

The demand for block increased with the rise of engineered masonry in the United States. The production of concrete block surpassed that of clay brick in the 1950s. Since the 1970s, there have been more walls built of concrete block in the United States than those of clay brick and all other masonry materials together.

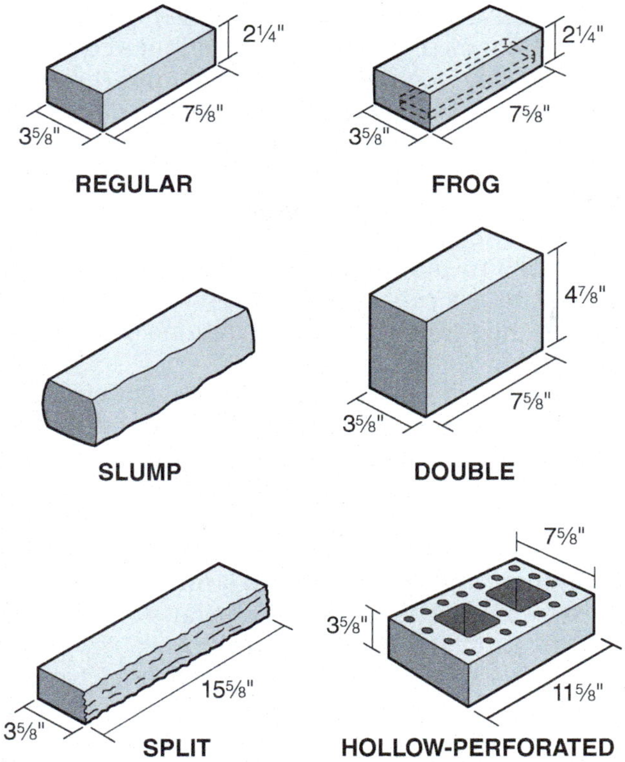

Figure 3 Concrete brick.

Concrete manhole and vault units are specially made with high-strength aggregates (*Figure 5*). They must be able to resist the internal pressure generated by the liquid in the completed compartment. Specially shaped block is manufactured for use on the top of a catchment vault. This type of block is engineered to fit the vault shape and is cast to specification. It is made with interlocking ends for increased strength.

Block Construction

Concrete block is often used in commercial construction. In some parts of the country, it is also used in the walls of residential construction in place of wood framing. The block can be painted on the outside or faced with brick, stucco, or other finish material.

28101-13_SA03.EPS

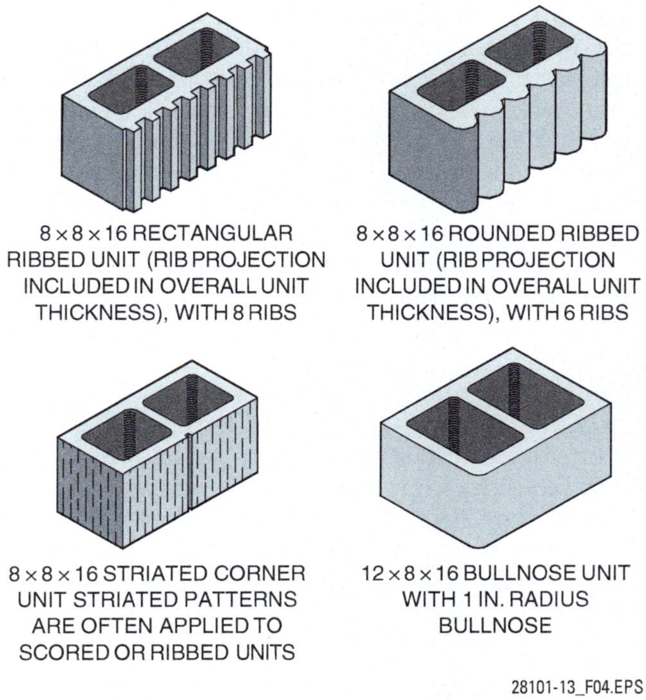

Figure 4 Common pre-faced concrete units.

Insulated block uses foam insulation inserts to provide better energy efficiency and soundproofing than standard block (*Figure 6*). The block is designed to provide a continuous and uninterrupted thermal barrier that absorbs and stores energy, allowing buildings to stay cooler in warm weather and warmer in cool weather. This allows designers

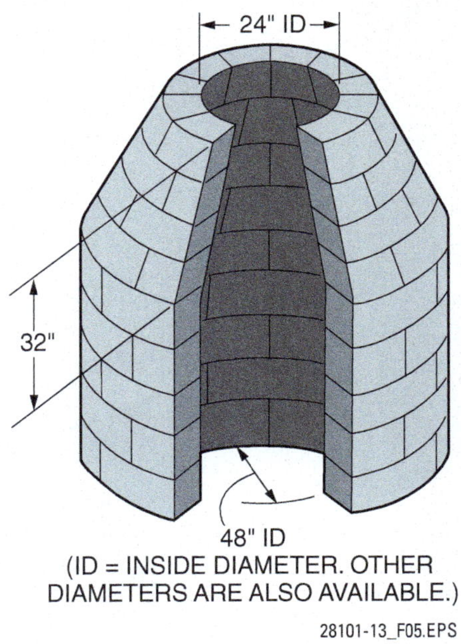

Figure 5 Manhole and vault unit.

to install smaller HVAC (heating, ventilating, and air conditioning) systems, which cost less to operate over time. It is available in standard dimensions in a variety of colors and finishes, and can be installed like conventional block. Because additional wall insulation is not required, the use of insulated block can reduce construction time. The foam inserts are nontoxic and moisture resistant.

Architectural block has a variety of surface finishes that affect the unit's texture. The surface finish allows the block to be used for both structural and finish purposes without the need for veneer (*Figure 7*). Architectural block can be used on interior and exterior walls. The block may be finished on one or both faces. Common types of architectural block facings include split, scored, ribbed, ground, sandblasted, striated (raked), glazed, offset, and slump block. Architectural block will be covered in more depth in the module *Masonry Units and Installation Techniques*.

1.2.0 Learning about Clay Masonry Units (Brick)

Brick has been developed and improved upon for centuries. The process of making brick has not changed much since the advent of the first modern brick-making machines over two centuries ago. The clay is mined, pulverized, and screened. It is mixed with water, formed, and cut into shape. Some plants extrude the clay, punch holes into it, and then cut it into shape. Any coating or glazing is applied before the units are air dried. After drying, the brick is fired in a kiln. Because of small variations in materials and firing temperatures, not all brick is exactly alike. Even brick made and fired in the same batch has variations in color and shading.

The brick is slowly cooled to prevent cracking. It is then bundled into **cubes** and shipped. A cube traditionally holds 500 standard brick, although manufacturers today make cubes of varying sizes.

Today, there are over 100 commonly manufactured structural clay products. ASTM has published standards for masonry design and construction. The standards cover performance specifications for manufactured masonry units. ASTM has also specified standard sizes for various kinds of brick. *Figure 8* shows the standard sizes for today's most commonly used brick. The sizes shown are actual dimensions. Brick is also identified by nominal sizes, which include the thickness of the mortar joint. Some brick is specified by a name, while other brick is specified by its actual dimensions (width × height × length).

Structural clay products include the following:

- Solid masonry units, or brick
- Hollow masonry units, or tile
- Architectural terra-cotta units

The next sections provide more information about these products.

1.2.1 Solid Brick

Brick is classified as solid if no more than 25 percent of its surface is **cored**, which means that the brick has holes extending through it to reduce weight. If the brick has no holes at all, it is called **uncored**. Brick is further divided into the following classifications: building, facing, hollow, paving, ceramic glazed, thin veneer, sewer, and manhole. ASTM standards exist for all of these types of brick. Fire brick has its own standard, but is not considered a major type classification.

Brick comes in modular and nonmodular sizes, in colors determined by the minerals in the clay or by additives. There is also a variety of face textures and a rainbow of glazes. The variety is dazzling. Brick can be laid in structural bonds to create patterns in the face of a wall or walkway.

28101-13_F06.EPS

Figure 6 Insulated block.

Figure 7 Varieties of architectural block.

Figure 9 shows several examples of commonly used bond patterns. Some bond patterns are traditional in some parts of the country. The five most common basic structural bonds are the running bond, common (or American) bond, Flemish bond, English bond, and stack (or block) bond. By varying brick color and texture and joint types and color, many patterns can be created using these bonds.

Brick is also made in special shapes to form arches, sills, copings, columns, and stair treads. Custom shapes can be made to order for architectural or artistic use. *Figure 10* shows some commonly manufactured special shapes of brick.

1.2.2 Hollow Masonry Units

Hollow masonry units, or tile, are machine-made clay tiles extruded through a die and cut to the desired size. A masonry unit is classified as hollow if more than 25 percent of its surface is cored. Hollow units are classified as either structural clay tile or structural clay facing tile.

Clay tile comes in many shapes, sizes, and colors. It is divided into loadbearing and nonbearing types. Structural tile can be used for loadbearing on its side or on its end. In some applications, structural tile is used as a backing **wythe** behind brick. Nonbearing tile is designed for use as fireproofing, furring, or ventilating partitions.

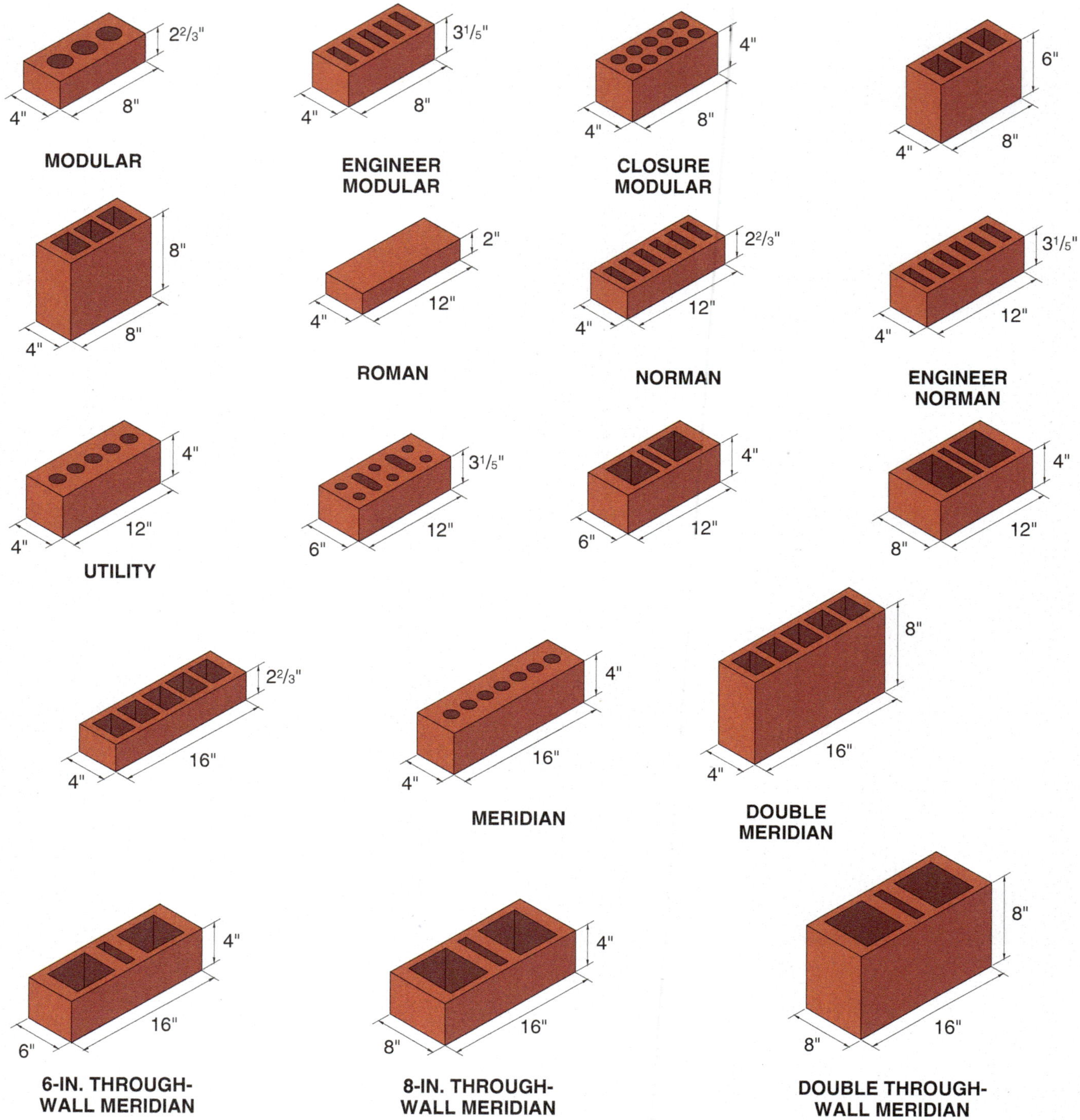

Figure 8 Standard brick.

Structural clay facing tile comes in modular sizes as either glazed or unglazed tile. It is designed for interior uses where precise tolerances are required. A special application of clay facing tile is as an acoustic barrier. The acoustic tiles have a holed face surface to absorb sound. Clay facing tile can also be patterned by shaping the surface face.

1.2.3 Architectural Terra-Cotta

Architectural terra-cotta is a made-to-order product with an unlimited color range. High-temperature-fired ceramic glazes are available in an unlimited color range and unlimited arrangements of parts, shapes, and sizes.

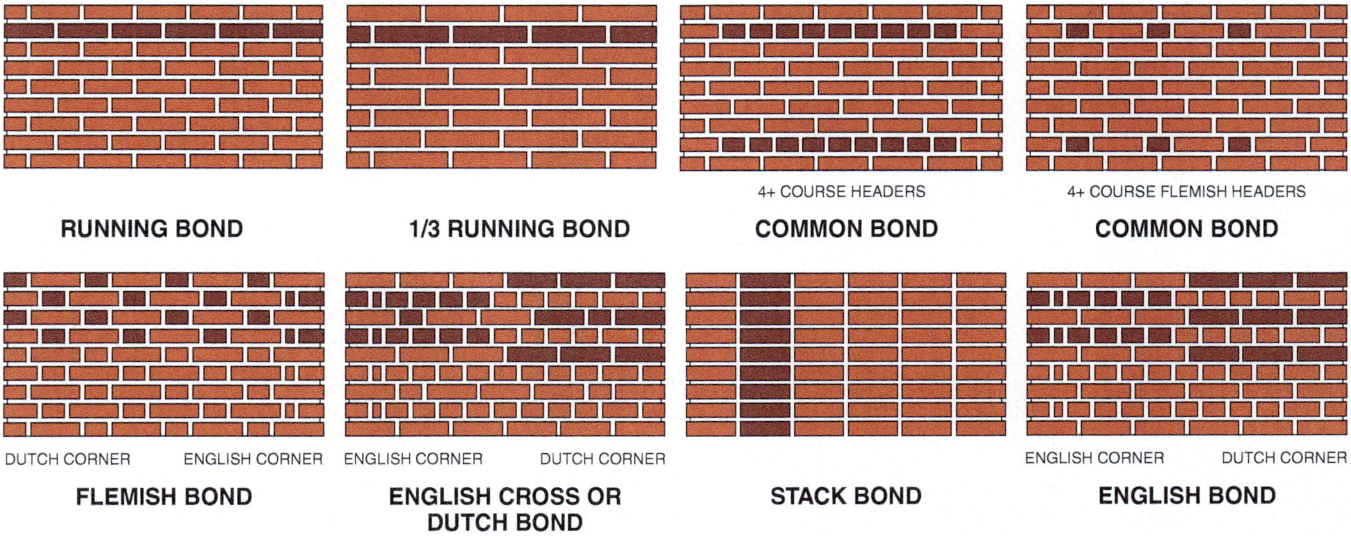

Figure 9 Common bond patterns.

Architectural terra-cotta is classified into anchored ceramic veneer, adhered ceramic veneer, and ornamental or sculptured terra-cotta. Anchored ceramic veneer is thicker than 1 inch, held in place by grout and wire anchors. Adhered ceramic veneer is no more than 1-inch thick, held in place by mortar. Ornamental terra-cotta is frequently used for cornices and column capitals on large buildings.

1.2.4 Brick Classifications

As previously stated, the three general types of brick masonry units are solid, hollow, and architectural terra-cotta. They may serve a structural function, a decorative function, or a combination of both. The three types differ in their formation and composition, and are specific in their use. Brick commonly used in construction includes the following:

- *Building brick* – Also called common, hard, or kiln-run brick, this brick is made from ordinary clays or shales and is fired in kilns. It has no special scoring, markings, surface texture, or color. Building brick is generally used as the backing wythes in solid and cavity brick walls because it does not have a finished face.
- *Face brick* – This is a better-quality brick and has better durability and appearance than building brick, so it is used on exposed wall faces. The most common face brick colors are various shades of brown, red, gray, yellow, and white.
- *Clinker brick* – This brick is oven burnt in the kiln. It is usually rough, hard, durable, and sometimes irregular in shape.
- *Pressed brick* – This brick is made by the dry-press process before kiln firing. It has regular smooth faces, sharp edges, and perfectly square corners. Ordinarily, it is used as face brick.
- *Glazed brick* – This has one surface coated with a white or other color of ceramic glazing. The glazing or texture forms when mineral ingredients fuse together in a glass-like coating during burning. Glazed brick is particularly suited to walls or partitions in hospitals, dairies, laboratories, and other structures requiring sanitary conditions and easy cleaning.
- *Fire brick* – This is made from a special type of fire clay to withstand the high temperatures of fireplaces, boilers, and similar constructions without cracking or decomposing. Fire brick is generally larger than other modular brick, and often is hand molded.
- *Cored brick* – This brick has 3, 5, or 10 holes extending through the brick to reduce weight. Three holes are most common. Walls built entirely from cored brick are not much different in strength than walls built entirely from uncored brick. Both have about the same resistance to moisture penetration. Whether cored or uncored, use the more easily available brick that meets building requirements.

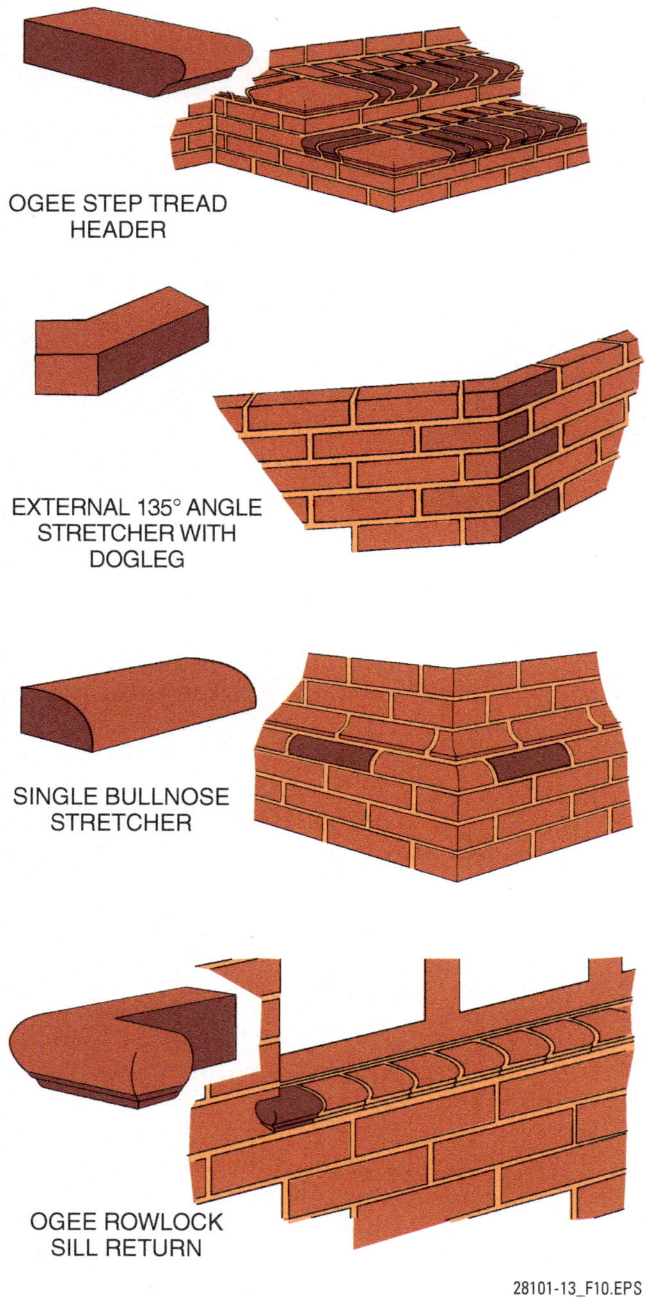

Figure 10 Special brick shapes.

1.2.5 Brick Masonry Terms

You need to know the specific terms that describe the position of brick and mortar joints in a wall (examples are shown in *Figure 11*). These terms include the following:

- **Course** – One of several continuous, horizontal layers (or rows) of masonry units bonded together
- *Wythe* – A vertical wall section that is the width of one masonry unit

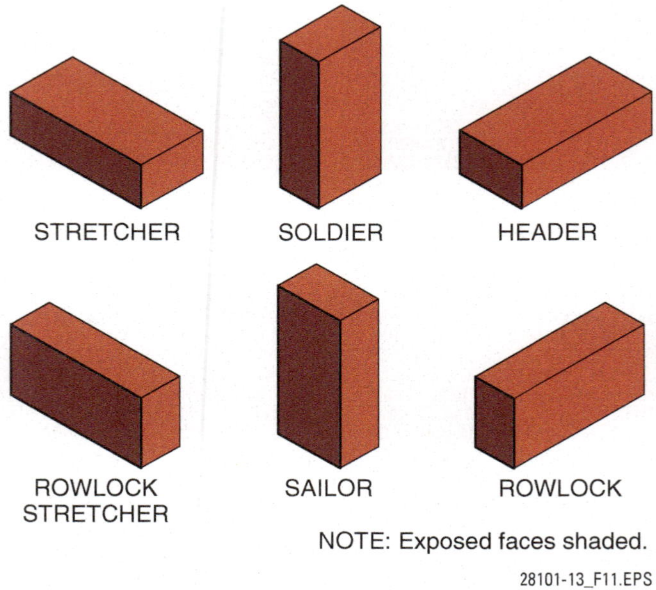

Figure 11 Wall brick positions.

- *Stretcher* – A masonry unit laid flat on its bed along the length of a wall with its face parallel to the face of the wall
- *Header* – A masonry unit laid flat on its bed across the width of a wall with its face perpendicular to the face of the wall; generally used to bond two wythes
- *Rowlock* – A header laid on its face or edge across the width of a wall
- *Rowlock stretcher* – A rowlock brick laid with its bed parallel to the face of the wall
- *Soldier* – A brick laid on in a vertical position with its face perpendicular to the courses in the wall

1.3.0 Learning about Stone

Stone was once used in the construction of all types of buildings, especially churches, schools, and government buildings. Today, it is more commonly used as a decorative material, such as the stone trim shown on the home in *Figure 12*.

1.3.1 Rubble and Ashlar

Rubble and ashlar are used for dry stone walls, mortared stone walls, retaining walls, facing walls, slope protection, paving, fireplaces, patios, and walkways. Rubble stone is irregular in size and shape. Stone collected in a field is rubble. Rubble from quarries is left where shaped block has been removed. It is also irregular with sharp edges. Rubble can be roughly squared with a brick hammer to make it fit more easily.

Ashlar stone is cut at the quarry. It has smooth bedding surfaces that stack easily. Ashlar is usu-

28101-13_F12.EPS

Figure 12 Stone facing used as decorative trim.

ally granite, limestone, marble, sandstone, or slate. Other stone may be common in different parts of the country.

Flagstone is used for paving or floors. It is 2 inches thick or less and cut into flat slabs. Flagstone is usually quarried slate, although other stone may be popular in different areas of the country.

Stone is often used as a veneer over brick or block. The wall shown in *Figure 13* is an example. The brownstone buildings in New York and the gray stone buildings of Paris are veneer over brick. Many of the government buildings and monuments in Washington, DC, are of stone veneer construction.

Stone, including flagstone, can be laid in a variety of decorative patterns. Concrete masonry units are made in shapes and colorings to mimic every kind of ashlar. These units are called cast stone and are more regular in shape and finish than natural stone. ASTM specifications cover cast stone and natural stone.

1.3.2 Manufactured Stone Veneer

Manufactured stone veneer, also called adhered concrete masonry veneer (ACMV) is a pre-made veneer consisting of cast cementitious material with pigments and other added materials that give the veneer the appearance of natural stone (*Figure 14*). Manufactured stone can also be designed to be loadbearing as well as veneer, but if manufactured stone is used as veneer without backing, it is not structural. It can be installed using mortar on wood frame walls with rigid sheathing, as well as on other types of walls including masonry walls, poured-in-place concrete walls, concrete tilt-up panels, and even existing masonry surfaces, provided that the walls have been prepared in accordance with the manufacturer's instructions. The techniques used for installation are similar to those used for adhered ceramic veneer.

CAUTION

Do not install manufactured stone veneer where it will come into frequent water contact from sources such as lawn sprinklers, downspouts, and drainage pipes. Deicing materials, salt, and harsh proprietary cleaners should not be used on manufactured stone veneer. Prolonged exposure to them can discolor and damage the surface.

28101-13_F13.EPS

Figure 13 A block wall faced with stone.

28101-13_F14.EPS

Figure 14 Manufactured stone veneer.

1.4.0 Learning about Mortar and Grout

Portland cement is made of ground earth and rocks burned in a kiln to make clinker. The clinker is ground to become the cement powder. Mixed with water, lime, and rocks, the cement becomes concrete. Mortar is somewhat different from concrete in consistency and use. The components and performance specifications are different also.

Modern mortar is mixed from portland cement or other cementitious material (something that has the properties of cement), along with lime, water, sand, and admixtures. The proportions of these elements determine the characteristics of the mortar.

The three main types of mortar are as follows:

- Cement-lime mortars are made of portland cement, hydrated lime, sand, and water. These ingredients are often mixed at the job site by the mason.
- Masonry cement mortars are premixed with additives. The mason only adds sand and water. The additives affect flexibility, drying time, and other properties.
- Preblended mortars are a dry mix of portland cement, hydrated lime, and dried masonry sand that have been weighed and mixed to ensure consistent proportions. All the mason has to do on the job site is to add the required amount of water and mix the mortar. Preblended mortars with color pigments are also available.

Mortar is mixed to meet four sets of performance specifications, as listed in *Table 1.*

Table 1 Mortar Composition

PROPERTY SPECIFICATIONS FOR LABORATORY-PREPARED MORTAR*			
Mortar Type	Minimum Compressive Strength, PSI at 28 days	Minimum Water Retention,%	Maximum Air Content,%**
M	2,500	75	12***
S	1,800	75	12***
N	750	75	14***
O	350	75	14***

* Adapted from *ASTM C270.*
** Cement-lime mortar only (except where noted).
***When structural reinforcement is incorporated in cement-lime or masonry cement mortar, the maximum air content shall be 12% or 18%, respectively.

Note: The total aggregate shall be not less than 2¼ and not more than 3½ times the sum of the volumes of the cement and lime.

28101-13_T01.EPS

- *Type M* – With high compressive strength, Type M mortar is typically used in contact with earth for foundations, sewers, and walks. This varies with geographic location.
- *Type S* – With medium strength, high bonding, and flex, Type S mortar is used for reinforced masonry and veneer walls.
- *Type N* – With heavy weather resistance, Type N mortar is used in chimneys, **parapets**, and exterior walls. Type N is the preferred mortar for most masonry veneer.
- *Type O* – With low strength, Type O mortar is used in nonbearing applications. It is not recommended for professional use.

Another type of mortar, Type K, has no cement materials but only lime, sand, and water. This type of mortar is used for the preservation or restoration of historic buildings.

Grout is a mixture of cement, water, and sand or other aggregate. Wet enough to be pumped or poured, it is used in reinforcement to bond masonry and steel together. It gives added strength to a structure when it is used to fill the cores of block walls.

NOTE

Even though mortar is no longer made of mud, masons still often call it *mud.*

1.5.0 Learning about Wall Structures

Masonry structures today take many forms in residential, commercial, and industrial construction. Modern engineering has added loadbearing strength so masonry can carry great weight without bulk. In addition to its loadbearing strength, masonry offers these advantages:

- Durability
- Ease of maintenance
- Design flexibility
- Attractive appearance
- Weather and moisture resistance

Another Name for Manufactured Stone Veneer

On the job, you may hear people refer to "lick and stick" or "lick 'em and stick 'em." This is a common way to refer to manufactured stone veneer (also called adhered concrete masonry veneer, or ACMV).

- Sustainability
- Competitive cost

Modern engineering and ASTM standards have been applied directly to everyday masonry work. There are several common classifications of structural wall built with masonry. Masonry walls can fit into more than one classification.

Cavity walls, as shown in *Figure 15A*, have two wythes with a 2- to 4½-inch space between them. Sometimes insulation is put in the cavity. The wythes are tied together with metal ties. Both wythes are loadbearing. Veneer walls, as shown in *Figure 15B*, are not loadbearing. A masonry veneer is usually built 1 to 2 inches away from a loadbearing stud wall or block wall. Veneer walls are used in high-rise and residential construction.

Composite walls have different materials in the facing (outer) and backing (inner) wythes. In a residential installation, the wythes are set with an air space between them of between 1 and 4½ inches and are tied together by metal ties. Unlike a veneer wall, both wythes of a composite wall are loadbearing.

Reinforced walls (*Figure 16*) have steel reinforcing embedded in the cores of block units or between two wythes. The steel is surrounded with grout to hold it in place. This very strong wall is used in high-rise construction and in areas subject to earthquake and high winds. Sometimes, grout is used alone for reinforcement. The grout is pumped into the cores of the block or into the cavity between the wythes.

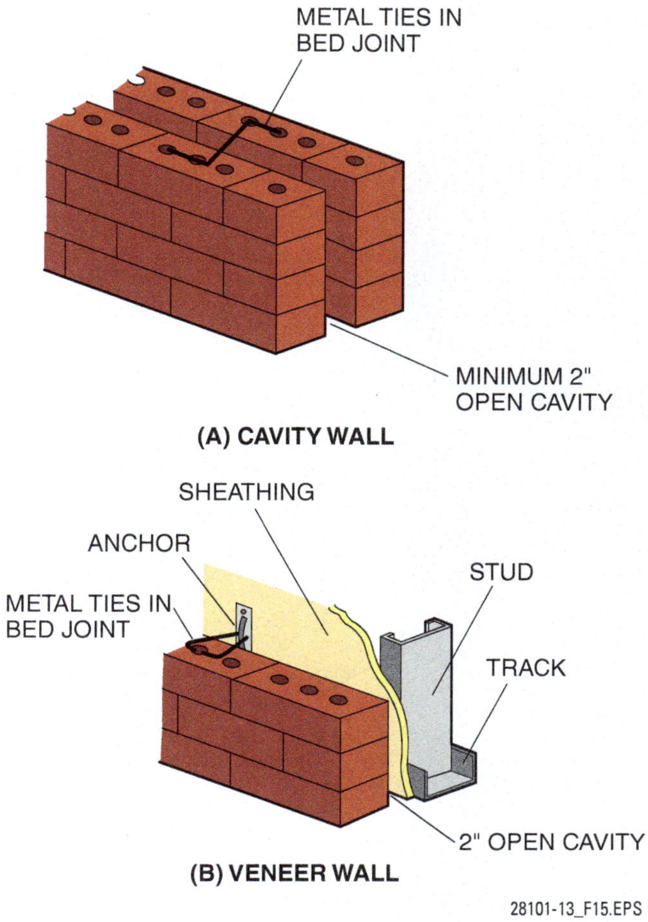

Figure 15 Types of masonry construction.

The Dome and Arch

The development of the dome and arch (in *Figure SA05* [below, left]) were important developments for brick architecture. With domes and arches, early masons could build larger and higher structures, with more open space inside.

The Romans refined arches and domes and built large-scale brickyards. They covered the Roman Empire with roads that spread Roman brick, mortar, and Roman designs for arches (*Figure SA06* [below, right]) and domes, along with Roman civilization, across Europe and North Africa.

28101-13_SA05.EPS

28101-13_SA06.EPS

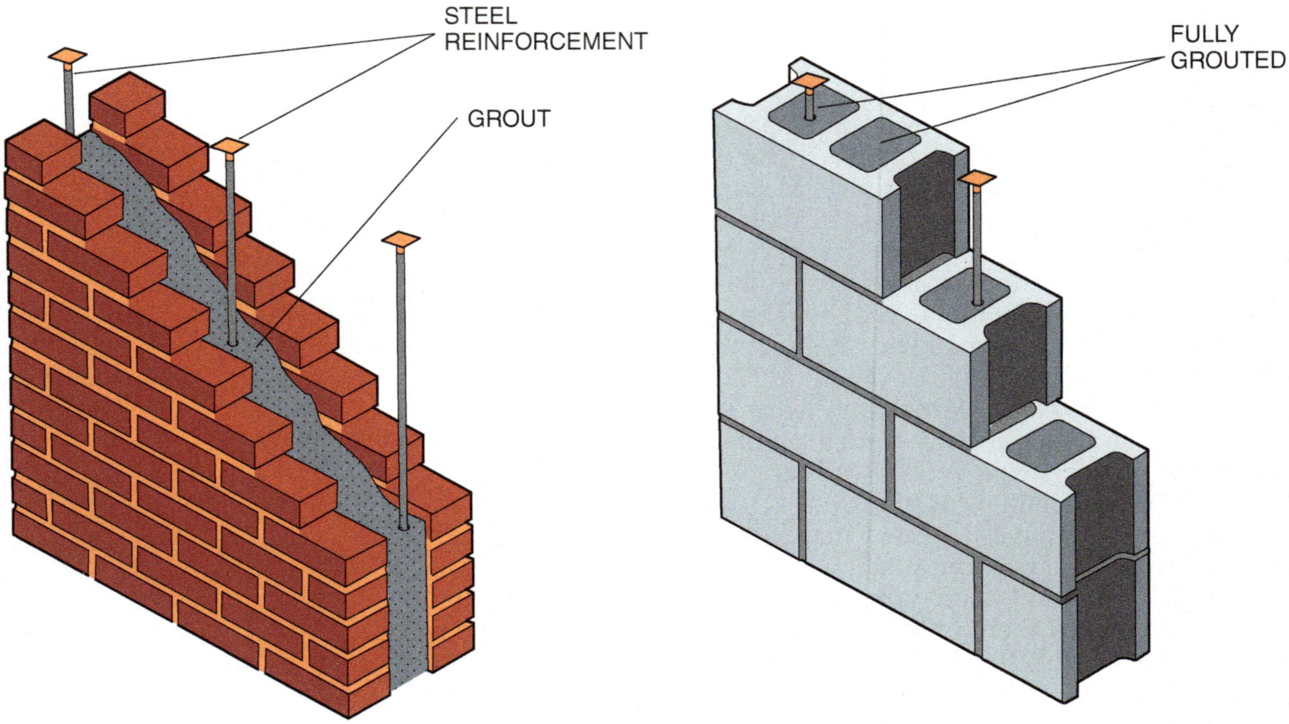

Figure 16 Reinforced walls.

Types of Mortar

You can remember the five types of mortar (M, S, N, O, and K) by taking every other letter from the words *MaSoN wOrK*.

Concrete, Not Cement

Though many people often use the terms *concrete* and *cement* interchangeably, they are not the same thing. Cement is one of the three ingredients of concrete, along with water and an aggregate such as sand. It's a sign of professionalism to use the terms correctly on the job, so take the time to learn the difference.

Contemporary masonry systems are designed not as barriers to water, but as drainage walls. Penetrated moisture is collected on flashing and expelled through **weepholes**. Design, workmanship, and materials are all important to the performance of masonry drainage walls.

A Short History of Masonry

28101-13_SA04.EPS

Masonry is one of the world's oldest and most respected crafts. Masonry construction has been around for thousands of years. The remains of stone buildings date back 15,000 years, and the earliest manufactured brick unearthed by archaeologists is more than 10,000 years old. This brick was made of hand-shaped, dried mud. Among the most well known works of masons are the pyramids of ancient Egypt, the Hanging Gardens of Babylon, and Notre Dame Cathedral in Paris.

Brick is the oldest manufactured building material, invented thousands of years ago. The original hand-formed mud brick was reinforced with straw and dried in the sun, and stacked with wet mud between them—the first mortar. Sometimes they were covered with another coat of mud, which was decorated. This was a common and effective building technique for centuries.

At some point, someone had the idea of laying brick in different patterns instead of simply stacking them. The figure shows a herringbone pattern, seen in ancient walls still standing today. Not only were the Babylonians the first to fire and glaze brick, they also developed two new types of mortar by mixing lime or pitch (asphalt) with the mud.

The Romans refined the Babylonian lime mortar by developing a form of cement that was a waterproof mortar. This mortar was useful for both brick and stone construction. It was also applied as a finish coat to the exterior of the surface as an early form of stucco. Romans produced highly ornate brick architecture using specialized brick shapes and varied brick colors and glazes. The Romans standardized the sizes of their brick, and the standards they developed are recognized even today.

By the Roman period, sand was a common additive in mortar. Burnt limestone, or quicklime, was added to mortar as an ingredient around the first century BC. Experimenting with ways to waterproof mortar, the Romans added volcanic ash and clay. The resulting cement made a strong, waterproof mortar. This made it possible to build aqueducts, water tanks, water channels, and baths that are still in use today. Unfortunately, some of the Roman formula was lost over time.

When the Normans conquered England in 1066, they built many castles there using brick and masons to lay them, who they brought with them from Europe. This construction boom boosted the trade economy of Europe and, along with it, the status of masons. Not only that, but today the size of brick used by the Normans is still widely recognized and used, as are those used by the Romans as mentioned earlier.

As the demand for more elaborate construction grew, so did the need for skilled workers. By the middle 1300s, masons had organized into early unions, known as guilds, across most of Europe. Guilds controlled the practice of the craft by monitoring the skill level of the craft worker. The local guilds controlled the practice of the masons' craft for centuries, monitoring training, judging disputes, and sharing knowledge among members. They collected dues, provided some support to widows and the ill, and celebrated special masons' holidays.

In the early 1700s, Masonic Temples and Orders of Freemasonry were founded in Europe. These organizations were political and spiritual, but based on many of the ideas of the masons' guilds. The pyramid seal on the back of the American dollar bill is a legacy of the masons' guilds.

The modern age of brick manufacture began with the invention of the first brick-making machine. It was powered by a steam engine and patented in 1800. In 1824, portland cement was patented by Joseph Aspdin, a mason. He was trying to recreate the waterproof mortar of the Romans. By 1880, portland cement had become the major ingredient in mortar. The new, waterproof portland cement mortar began to replace the old lime and sand mixture.

Additional Resources

ASTM C270, Standard Specification for Mortar for Unit Masonry. 2012. West Consohocken, PA: ASTM International.

Concrete Masonry Shapes and Sizes Manual CD-ROM. Herndon, VA: National Concrete Masonry Association.

Installation Guide for Adhered Concrete Masonry Veneer, Third Edition. 2012. Washington, DC: Masonry Veneer Manufacturers Association.

1.0.0 Section Review

1. Concrete units are classified according to their _____.
   a. length, width, and height
   b. intended use, size, and appearance
   c. composition, appearance, and weight
   d. thickness, composition, and size

2. The traditional number of brick in a cube is _____.
   a. 90
   b. 100
   c. 500
   d. 750

3. Stone collected in a field is called _____.
   a. ashlar
   b. veneer
   c. clinker
   d. rubble

4. Mortar that is typically used in contact with earth for foundations, sewers, and walks is classified as Type _____.
   a. N
   b. O
   c. M
   d. S

5. In a residential installation, the minimum air space between wythes in a composite wall is set at _____.
   a. 1 inch
   b. 1½ inches
   c. 2 inches
   d. 2½ inches

Section Two

2.0.0 Introduction to Masonry Safety

Objective

Recognize the basic safety precautions when working with masonry materials.

a. List basic safety practices.
b. Describe personal protective equipment used in masonry.

Performance Task 1

Put on eye protection and respiratory protection.

Trade Terms

Competent person: An individual who is capable of identifying existing and predictable hazards or working conditions that are hazardous, unsanitary, or dangerous to employees, and who has authorization to take prompt, corrective measures to eliminate or control these hazards and conditions.

Occupational Safety and Health Administration (OSHA): The division of the US Department of Labor mandated to ensure a safe and healthy environment in the workplace.

Personal protective equipment (PPE): Equipment or clothing designed to prevent or reduce injuries.

Safety is a particularly important issue for you as a masonry worker. You will often work at heights where falls and falling objects are major hazards. You must also be careful when working with mortar and concrete products because they can cause skin irritations and lung ailments if not handled properly. You will often work on sites where heavy equipment is used. For that reason, you must be especially vigilant. In this section, you will be introduced to basic safety practices and the various pieces of **personal protective equipment (PPE)** used in masonry. Safety and personal protective equipment will be covered in more detail in the module titled *Masonry Safety*.

2.1.0 Understanding Basic Safety Practices

Most accidents and injuries on a construction site are caused by worker carelessness, poor safety planning, lack of training, or failure of the employer or employee to follow safety regulations. Accidents do not only affect masons and their employers; they can also affect the health and safety of the public.

To help prevent accidents, your company must have a safety program. This program will provide you with the rules and safeguards you need to work safely. Safety must be part of all phases of the job and must involve employees at every level, including management. The United States Department of Labor's **Occupational Safety and Health Administration (OSHA)** requires that a company-appointed **competent person** be on site before you start any job. OSHA's regulation CFR (*Code of Federal Regulations*) 1926.32 defines a competent person as one who is capable of identifying existing and predictable hazards in the surroundings or working conditions that are unsanitary or dangerous to employees and who is authorized to take prompt, corrective measures to eliminate them. A competent person has experience and training for the job and knows the job's hazards, as well as the rules and regulations associated with the job.

In addition to its own rules, your company must comply with many local, state, and national regulations. For example, OSHA requires the following:

- An employer "shall furnish to each of his employees employment and a place of employment which are free from recognized hazards that are causing or are likely to cause death or serious physical harm to his employees."
- An employer "shall comply with occupational safety and health standards" established in the federal Occupational Safety and Health Act.
- An employee "shall comply with occupational safety and health standards and all rules, regulations, and orders issued pursuant to this Act which are applicable to his own actions and conduct."

This is called OSHA's General Duty Clause. OSHA also says that employees have a duty to fix a recognizable hazard when they see one. Safety regulations like these are intended to make work sites safe and accident free. Safety policies and procedures are available to help you and your company comply with OSHA regulations. Remember that there is a good reason for each regulation. Following good safety practices helps to save lives.

Unsafe acts often lead to serious injury and sometimes death. You can prevent unsafe acts by changing your behavior. It is your responsibility to recognize unsafe acts and stop them immediately. This can mean telling your co-workers to stop what they are doing. If your co-workers do not stop acting in an unsafe manner, stop what you are doing and move as far away from them as possible. In some cases, you may need to inform your supervisor of the problem. Here are some examples of the most common unsafe acts:

- Operating equipment at improper speeds
- Operating equipment without authority
- Using defective equipment
- Disabling a safety device
- Servicing equipment while it is in motion or energized
- Using equipment improperly
- Failing to use PPE
- Failing to warn co-workers of a dangerous or potentially dangerous situation
- Working in an improper position
- Working while impaired by alcohol or illegal drugs
- Operating tools or equipment when taking certain types of prescription drugs
- Lifting loads improperly
- Loading or placing equipment or supplies improperly
- Using cell phones or texting while working
- Working when tired or without enough sleep
- Wearing ear buds or earphones
- Horsing around

An unsafe act can also be defined as work that is not done correctly. Workers who fail to use proper PPE, follow safety procedures, or warn co-workers of potentially hazardous conditions can cause or worsen accidents. Keep yourself safe, and look out for the safety of others. Always follow safety rules, and use the right equipment for the job.

2.2.0 Recognizing Personal Protective Equipment Used in Masonry

PPE is designed to protect you from injury. You won't see all the potentially dangerous conditions on a job site just by looking around. Before doing any job, stop and consider what type of accidents could happen. Using common sense and PPE will greatly reduce your chances of getting hurt. The best PPE is of no use unless you do the following four things:

- Inspect it regularly, and replace any PPE that is damaged or worn.
- Care for it properly.
- Use it properly when it is needed.
- Avoid altering or modifying it in any way.

As a mason, you will most commonly use the following types of PPE:

- Hard hats
- Eye and face protection
- Gloves
- Safety shoes
- Hearing protection
- Fall protection
- Respiratory protection
- Proper clothing

Each of these items will be covered in more detail in the module titled *Masonry Safety*. The following sections review the basic personal protective equipment you will need to use to complete the basic masonry installation requirements in the section titled "Introduction to Masonry Installation".

2.2.1 Gloves

Gloves protect your hands from cuts, abrasion, dust and moisture, burns caused by mortar and portland cement, and other hazards (see *Figure 17*). Gloves used in masonry work often have a tightly knit fabric or nylon shell with palms and fingertips made from latex or synthetic rubber. This allows you to keep your sense of touch while wearing the gloves but still be able to grip effectively and protect your hands.

WARNING!

Wet portland cement can cause caustic burns that can cause blisters, dead skin, and discolored skin. Severe burns from portland cement can even reach the bone, causing scars and even disability. Exposure to portland cement can cause an allergic reaction that results in inflammation from repeated exposure.

Make sure that gloves fit snugly without being too loose or too tight. Always wash and dry your hands before wearing gloves. If the gloves are reusable, clean them after each use. If they are worn, damaged, or contaminated, exchange them for a fresh pair.

2.2.2 Eye, Ear, and Face Protection

Wear safety goggles, eyewear, and face shields (*Figure 18*) to prevent eye injury. Areas where there are potential eye hazards from falling or flying objects are usually identified, but you should always be on the lookout for other possible haz-

28101-13_F17.EPS

Figure 17 Gloves for masonry work.

ards, such as pressurized water. Safety goggles give your eyes the best protection from all directions. Regular safety glasses will protect you from objects flying at you from the front, such as large chips, particles, sand, or dust. You can add side shields for further protection.

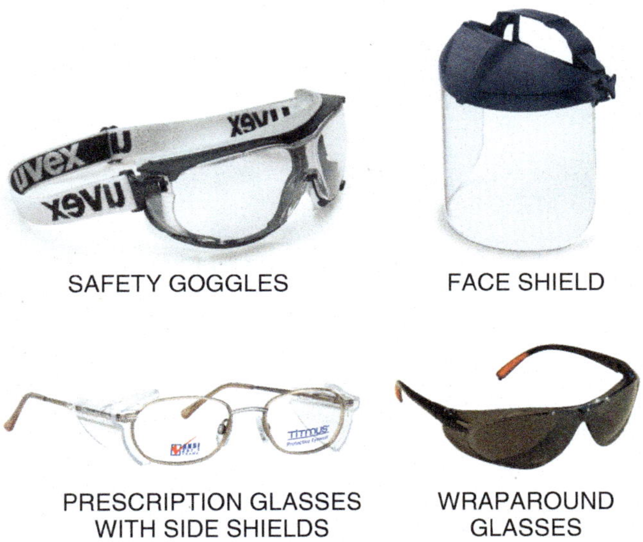

28101-13_F18.EPS

Figure 18 Typical safety goggles, glasses, and face shield.

To protect yourself against the loss of hearing caused by loud noises, wear earmuffs or specially designed disposable earplugs that fit into your ears and filter out noise. Make sure that earmuffs fit snugly against your head to provide maximum protection.

2.2.3 Respiratory Protection

Wherever there is danger of suffocation or other breathing hazards, you must use a respirator. You must also wear a respirator when working with or near fire-resistant asbestos or where hazardous molds are growing. Mandatory special training is required for the use of respirators. It is important to do the following:

- Use the proper type of respiratory protection for the hazard
- Pass a cardiopulmonary fitness test before using respiratory protection
- Conduct a fit test with the respiratory protection equipment prior to use to ensure that it fits properly

CAUTION

OSHA regulations require that workers undergo a cardiopulmonary test before using respiratory protection. Refer to your local applicable standards.

WARNING!

EPA's *Renovation, Repair & Painting Final Rule, 40 CFR 745*, requires that home renovations conducted for compensation, must be performed by certified firms using certified renovators if the home was built prior to 1978, if it qualifies as a child-occupied facility, and if the renovation disturbs over 6 square feet of lead paint per interior room area or 20 square feet of exterior area per wall. All masons working on pre-1978 homes are affected by this rule and need to be aware of their surroundings and test for the presence of lead in their work area.

WARNING!

If you encounter asbestos or mold, do not handle it or attempt to remove it. Tell your supervisor about the presence of these materials immediately. Asbestos and mold are linked to long-term illnesses. Only trained personnel who have the proper equipment for handling and disposing of these materials can deal with such situations.

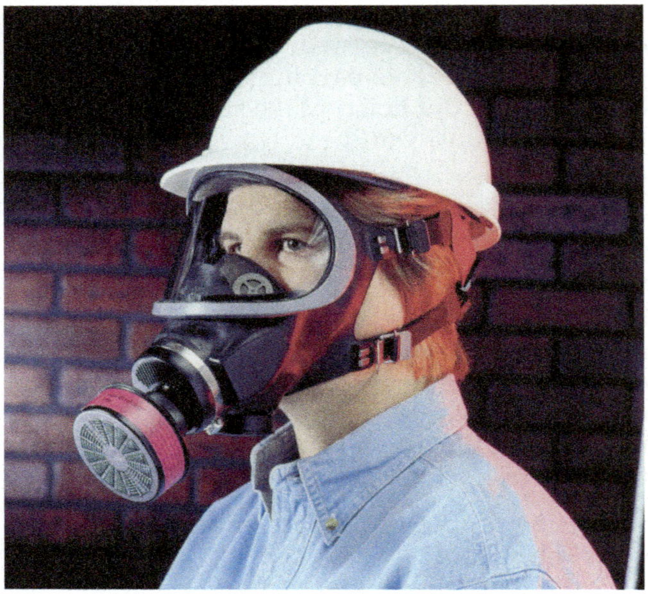

FULL-FACEPIECE RESPIRATOR

HALF-MASK RESPIRATOR

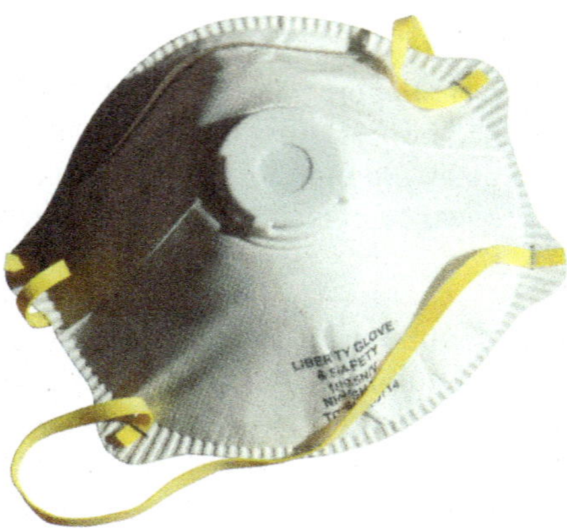

DUST MASK

28101-13_F19.EPS

Figure 19 Full-facepiece respirator, half-mask respirator, and dust mask.

Reusable respirators are made in full-facepiece and half-mask styles; dust masks are also available to provide light-duty respiratory protection (*Figure 19*). These respirators require the replacement of cartridges, filters, and respirator parts. Their use also requires a complete respirator maintenance program. Half-mask air-purifying respirators have several limitations, so be sure to refer to the manufacturer's instructions before using. Dust masks should be discarded after a single use.

2.2.4 Clothing

Figure 20 shows a mason properly dressed and equipped for most masonry jobs. Wear close-fitting clothing that is appropriate for the job. Clothing should be comfortable and should not interfere with the free movement of your body. Clothing or accessories that do not fit tightly, or that are too loose or torn, may get caught in tools, materials, or scaffold. Wear a long-sleeved shirt to provide extra protection for your skin. Wear sturdy work boots or work shoes with thick soles (*Figure 21*). Never show up for work dressed in sneakers, loafers, or sport shoes.

Wear shirts tucked in, with the sleeves tight or buttoned. Do not wear loose-fitting shirts. Always wear your pants up, with the waistband above the hip bone. Some job sites specify that the crotch of the pants cannot be lower than 2 inches from the body.

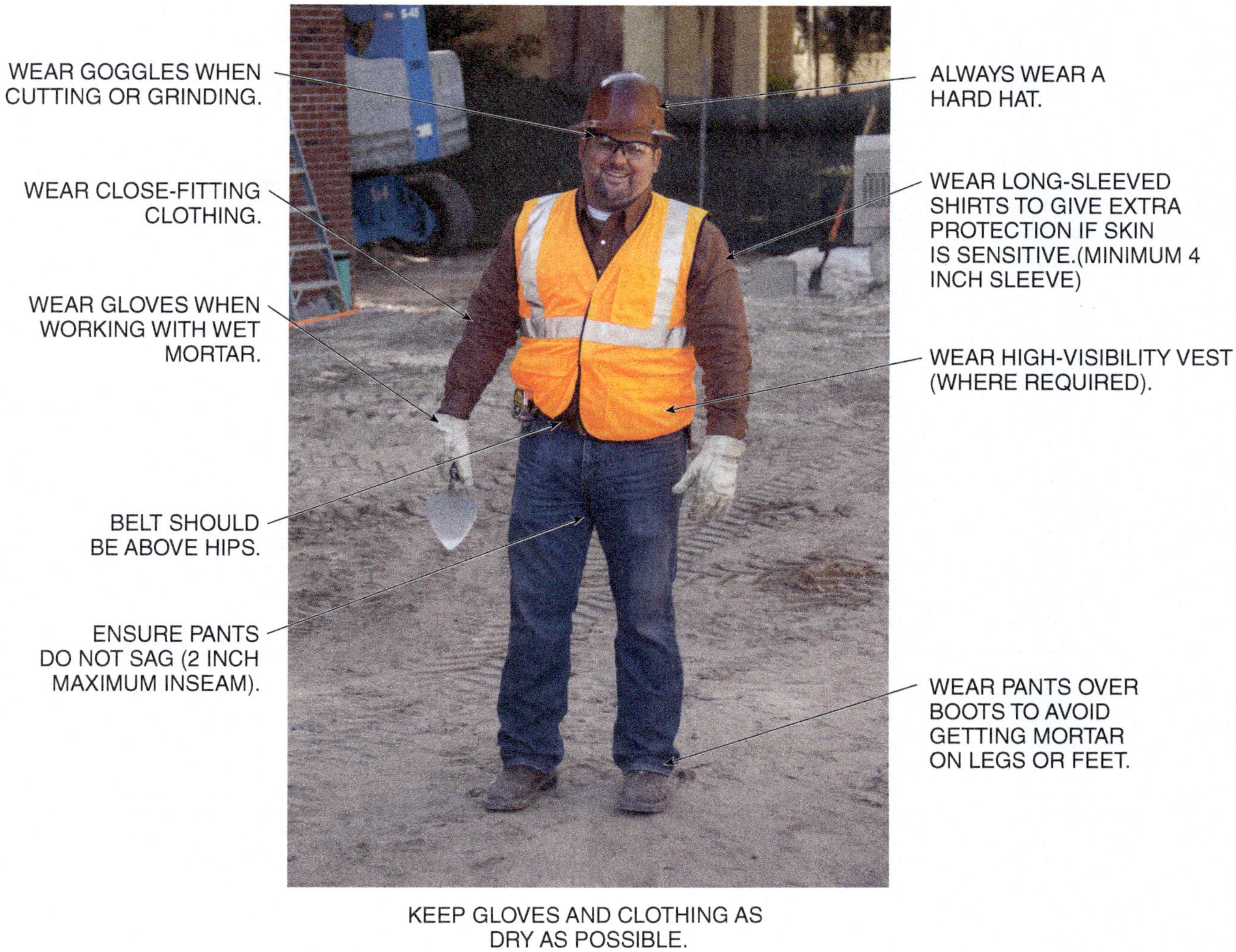

Figure 20 Dressed for masonry work.

Eye Injuries

The average cost of an eye injury is $1,463. That includes both the direct and indirect costs of accidents, not to mention the long-term effects on the health of the worker, which is priceless.

Figure 21 Work boot.

Additional Resources

Building Block Walls: A Basic Guide for Students in Masonry Vocational Training. 1988. Herndon, VA: National Concrete Masonry Association.

Hot & Cold Weather Masonry Construction Manual. 1999. Herndon, VA: National Concrete Masonry Association.

2.0.0 Section Review

1. OSHA's General Duty Clause does *not* require _____.
   a. employers to furnish a place of employment free from recognized hazards
   b. employees to follow the instructions of the employer's designated competent person
   c. employers to comply with occupational safety and health standards
   d. employees to comply with all pursuant rules, regulations, and orders

2. The type of eye protection that provides the best protection from all directions is _____.
   a. safety glasses
   b. regular eyeglasses
   c. face shields
   d. safety goggles

SECTION THREE

3.0.0 INTRODUCTION TO MASONRY INSTALLATION

Objective

Explain how to mix mortar and lay masonry units.

a. Explain how to mix mortar.
b. Describe how to lay masonry units.

Performance Tasks 2 and 3

Properly mix mortar by hand.

Properly spread mortar using a trowel.

Trade Terms

Bed joint: The horizontal joint between two masonry units in separate courses.

Butter: To apply mortar to a masonry unit prior to laying it.

Footing: The base for a masonry unit wall, or concrete foundation, that distributes the weight of the structural member resting on it.

Head joint: The vertical joint between two masonry units.

Pilaster: A square or rectangular pillar projecting from a wall.

Spread: A row of mortar placed into a bed joint.

Stringing: Spreading mortar with a trowel on a wall or footing for a bed joint.

In this section, you will learn the basic elements of bricklaying. When you have completed the section, you should be able to set up a job, mix mortar, and lay brick as directed by your instructor.

Masons use a number of specialized hand and power tools. As you will learn in the module *Masonry Tools and Equipment*, there are many kinds of special trowels, though for this module you will only need to know how to use a basic brick trowel (*Figure 22*). There are also at least six kinds each of hammers, chisels, and steel joint-finishing tools, and seven kinds of measuring and leveling tools. Power tools include several kinds each of saws, grinders, splitters, and powder-actuated tools. Mortar can be mixed by hand, using special equipment, or in a power mixer. Cranes, hoists, and lifts bring the masonry units to the masons working on one of four types of steel scaffold.

While masonry tools have changed over the centuries, one thing has not: the relation between the mason and the masonry unit. The mason uses today's wealth of tools and equipment to perform the following tasks:

- Calculate the number and type of units needed to build a structure
- Estimate the amount of mortar needed
- Assemble the units near the workstation
- Lay out the wall or other architectural structure
- Cut units to fit, as needed
- Mix the appropriate type and amount of mortar
- Place a bed of mortar on the **footing**
- **Butter** the **head joints** and place masonry units on the bed mortar
- Check that each unit is level and true
- Lay courses in the chosen bond pattern or create a new pattern
- Install ties as required for loadbearing
- Install flashing and leave weepholes as required for moisture control
- Clean excess mortar off the units as the work continues
- Finish the joints with jointing tools
- Give the structure a final cleaning
- Complete the work to specification, on time

Masonry work is still very much a craft. The relation between the mason and the masonry unit is personal. The straightness and levelness of each masonry unit in a structure—brick, block, or stone—depend on the hands and the eyes of the mason. These things have not changed in 10,000 years.

The tradition of masonry calls for a bit of art, too. The mason gets trained by work to see the subtle shadings and gradations of color, and learns to create a pattern and to select the right unit to complete the pattern or the shading. The mason grows skilled in building something that is both enduring and attractive.

28101-13_F22.EPS

Figure 22 Brick trowel.

3.1.0 Mixing Mortar

Mortar (*Figure 23*) is a mixture of portland cement, lime, sand, and water. The first three ingredients are determined by the type of mortar being mixed. Water is added until the mix is at the proper consistency. The ability to mix mortar properly, and to produce the same consistency time after time, can only be developed through practice.

You will probably mix (or chop) your first few batches by hand in a wheelbarrow, pan, or mortar box.

For small jobs, manual mixing of the mortar usually makes the most sense. The preliminary steps are the same as those for mechanical or power mixing. Position the material near the mortar box. Ensure that the mortar box is level and stable. Have the mix recipe written down. Make sure that there is enough room at the two ends of the mortar box for you to stand while mixing the mortar. There is no particular height required for the box. It should be at a height that will be convenient for the mason when using the hoe to mix the mortar.

The basic steps for mixing mortar by hand are as follows:

Step 1 Fill a cubic foot box with sand (*Figure 24*). Place half of the sand in the box and spread it out evenly across the bottom of the box.

Step 2 Place the desired amount of cement, lime, or masonry cement over the sand in the box (*Figure 25*). For small batches of mortar, you may wish to use standard shovelsful of material rather than bags as your measuring device. Use a shovel to spread out the remaining sand across the cement and lime layer.

28101-13_F23.EPS

Figure 23 Masonry mortar.

28101-13_F24.EPS

Figure 24 A cubic foot box can be used to measure sand.

28101-13_F25.EPS

Figure 25 Portland cement is added to the sand.

Step 3 Blend the dry ingredients together using the shovel or hoe. When they are thoroughly mixed, push them to one end of the box (*Figure 26*).

28101-13_F26.EPS

Figure 26 Dry ingredients are moved to one end before water is added.

Step 4 Add half the water to the empty end of the box. Begin mixing this water into the dry materials. This can be easily done from the water end of the box, using the hoe with short pull-and-push strokes. Continue in this fashion with the hoe at a 45-degree angle until all the material is well mixed. Add the remaining water to obtain the desired consistency.

Step 5 After the mortar is mixed, pull it to one end of the box to prevent it from drying out.

Step 6 At this time, the mortar can be checked to see if it is of the proper consistency and workability. Pick up a small amount of mortar on a trowel, and set it firmly on the trowel by tapping it once on the side of the box. Turn the trowel upside down. If the mortar is the proper consistency, it will remain on the trowel (*Figure 27*). This is also a measure of the mortar's adhesion, or its ability to stick to a surface.

Step 7 Transport the mortar to the work site with a wheelbarrow or other equipment, but wet the inside surface with water before loading the mortar. This will prevent the mortar from sticking to the sides of the container. When only a small amount of mortar is needed, it can be mixed directly in the wheelbarrow.

28101-13_F27.EPS

Figure 27 Mortar consistency test.

28101-13_F28.EPS

Figure 28 Transferring mortar to mortarboard.

Step 8 Load the mortar from the wheelbarrow to a mortarboard or pan with a shovel. The mortar should be sticky and not runny (*Figure 28*).

Step 9 The final step is to clean the mixing equipment of all mortar as soon as possible after the mixing box or wheelbarrow is no longer needed. A water hose with a spray nozzle and a stiff brush are the best tools to use when cleaning your masonry tools. You can also use a water barrel as a bath to keep the mortar from drying on your tools.

WARNING!

Dry cement and wet concrete are harmful. Dry cement dust can enter open wounds and cause blood poisoning. The cement dust, when it comes in contact with body fluids, can cause chemical burns to the membranes of the eyes, nose, mouth, throat, or lungs. It can also cause a fatal lung disease known as silicosis. Wet cement or concrete can also cause chemical burns to the eyes and skin. Always wear appropriate personal protective equipment when working with dry cement or wet concrete. If wet concrete enters waterproof boots from the top, remove the boots and rinse your legs, feet, boots, and clothing with clear water as soon as possible. Repeated contact with cement or wet concrete can also cause an allergic skin reaction known as cement dermatitis.

3.2.0 Laying Masonry Units

After the mortar is mixed, pick it up on your trowel and spread it. Filling and emptying the trowel is an important skill. Applying the mortar, or spreading it, is the next step. In this section, you will learn how to hold the trowel, pick up the mortar, and lay it down. Then you will learn how to spread and cut mortar, and butter a brick. At this point, you will learn something to be experienced rather than memorized. The techniques in the next sections should be practiced until you feel comfortable using them. Note that the instructions in this section apply to laying brick.

3.2.1 Picking Up Mortar

There are several ways of using the trowel to pick up mortar. This section will introduce you to a general method for picking up mortar from a board and a general method for picking up mortar from a pan. There are many different ways to do these tasks. The instruction here begins with some tips on holding the trowel. All of these instructions are only approximations of the work itself. You can only learn this through watching a skilled mason and practicing until you feel comfortable with these movements.

3.2.2 Holding the Trowel

Pick up your trowel by the handle. Put your thumb along the top of the handle with the tip on the handle, not the shank, as shown in *Figure 29.*

Keep your second, third, and fourth fingers wrapped around the handle of the trowel. Keep the muscles of your wrist, arm, and shoulder relaxed so you can move the trowel freely.

28101-13_F29.EPS

Figure 29 Holding the trowel.

Most of your work with the trowel will require holding the blade flat, parallel to the ground, or rotating the blade so it is perpendicular to the ground.

Rotating the blade gives you a cutting edge. The best edge for cutting is the edge on the side closest to your thumb. It is best this way because you can see what you are cutting. When you turn the trowel edge to cut, rotate your arm so your thumb moves down. This will rotate the trowel so that the bottom of the blade turns away from you. If you rotate only your wrist, after a while you will strain it. Use the larger muscles in your arm and shoulder to rotate the trowel.

Rotating the blade also gives you a scooping motion. Turning your thumb down will give you a forehand scoop. Turning your thumb up will rotate the bottom of the blade toward you and give you a backhand scoop. Using a forehand or backhand movement will depend on the position of the material you are trying to scoop.

3.2.3 Picking Up Mortar from a Board

After putting the mortar on the board, follow these steps:

Step 1 Work the mortar into a pile in the center of the board, and smooth it off with a backhand stroke.

Step 2 Use the trowel edge to cut off a slice of mortar from the edge.

Step 3 Pull and roll the slice of mortar to the edge of the board. Work the mortar into a long, tapered roll, as shown in *Figure 30*.

Step 4 Slide the trowel under the mortar, then lift the mortar up with a light snap of your wrist. Raising the trowel quickly will break the bond between the mortar and the board. If done correctly, the mortar will completely fill the trowel blade.

28101-13_F30.EPS

Figure 30 Picking up mortar from a board.

3.2.4 Picking Up Mortar from a Pan

Try this method when the mortar is in a pan:

Step 1 Cut a slice of mortar, as shown in *Figure 31*.

Step 2 Without removing the trowel from the mortar, slide the trowel under the mortar so the blade becomes parallel to the floor.

Step 3 Firmly push the trowel, with the blade parallel to the floor, toward the middle of the pan. The mortar will pile up on the blade.

Step 4 Lift the trowel from the mortar at the end of the stroke. The trowel should be fully loaded with a tapered section of mortar.

Step 5 To prevent the mortar from falling off the trowel, snap your wrist slightly to set the mortar on the trowel as you lift.

3.2.5 Spreading

Now you will learn about spreading the mortar and shaping its edges. You can practice spreading and cutting the mortar along a 2 × 4 board spread between two cement block or other props. Prac-

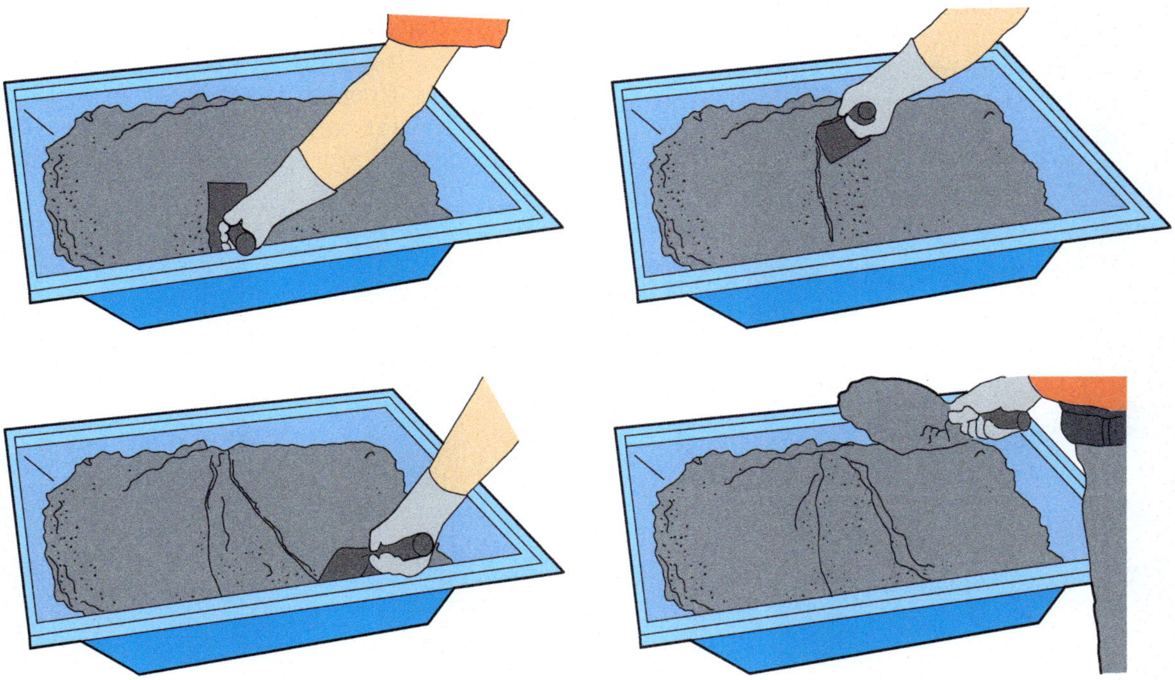

28101-13_F31.EPS

Figure 31 Picking up mortar from a pan.

tice until you feel comfortable with these movements.

Mortar application should adhere to the following guidelines:

- The joints are completely filled with no small voids for water to enter.
- The mortar is still pliable while you level and plumb the unit.
- The finished joint is the specified thickness after you level and plumb the masonry unit.
- The mortar does not smear the face of the masonry unit.

Spreading the mortar means applying it in a desired location at a uniform thickness (see *Figure 32*). Mortar is spread for **bed joints**. The process of spreading the mortar for the bed joint is also called **stringing** the mortar. The spreading motion has two components to it, and they occur at the same time.

The first component of the spreading motion is a horizontal sweep back toward you from the starting point or the point where the last spread of mortar ended. The mortar deposited is called a **spread**. Try to make the spread about two brick long to begin with. If you are working with block, try to string the spread about one block long at first. After practice, you should be able to string the spread three to four brick long, or two block long.

The second component of the motion is a vertical rotation. The trowel starts with its blade horizontal. As you move the trowel back toward you, you are also rotating it. As you rotate it, your thumb moves downward, and the back of the blade moves away from you. As the blade tilts, with the trowel traveling horizontally, the mortar is deposited along the path of the trowel.

Practice spreading until you can deposit a trail rather than a mound of mortar. Keep the trowel in the center of the wall for the length of the spread, so mortar will not get thrown on the face of the masonry. Start with a goal of 16 inches and work up to a spread of 24 to 32 inches.

Full joint spreads are used for all brick but not for all block. Block is usually mortared on its face shells and not its webs. However, block needs a full bed joint when it fits into any of the following categories:

- The first or starting course on a foundation, footing, or other structure
- Part of masonry columns, piers, or **pilasters** designed to carry heavy loads
- In a reinforced masonry structure, where all cores are to be grouted

Check the specifications to be sure. After the first course, the remaining block is mortared on shells, or shells and webs, according to specifications.

Whether you work with block or brick, you will need to know how to spread a full bed joint and cut it.

3.2.6 Cutting or Edging

After each spread, use the edge of the trowel to cut off excess mortar. To cut, hold the edge of the trowel at about a 60-degree angle, perpendicular to the edge of the mortar. Use the edge of the trowel to shave off the edge of the mortar. *Figure 33* shows the correct angle for shaving the edge of the spread.

Keep the edge of the trowel at a flat angle as shown. This will allow you to catch the mortar as you shave the edge. At this stage in your practice, learn how to catch the mortar as you cut it. The excess mortar can be returned to the mortar pan or used to fill any spaces in the bed joint.

28101-13_F32.EPS

Figure 32 Mortar spread at a uniform thickness.

28101-13_F33.EPS

Figure 33 Cutting an edge.

Catching the mortar as you shave it means you do not have to go back and pick it up afterwards. On the job, having mortar stuck to the face of the masonry unit or lying in piles at the foot of a wall is unacceptable. Mortar is hard to remove after it dries, but easy to clean when it is fresh. Learn to clean mortar as you lay it.

3.2.7 Buttering Joints

Buttering the head joint involves applying mortar to a header surface of a masonry unit. Buttering occurs after the bed joint is spread and the first masonry unit is laid in the bed. Buttering techniques are different for block and brick.

Buttering brick is a two-handed job. Begin by spreading the mortar on the bed joint. Keeping the trowel in your hand, pick up the first brick with your other hand. Press this brick into position in the mortar. Cut off the excess mortar on the outside face with the edge of the trowel.

Keeping the trowel in your hand, pick up a second brick in your brick hand. As you hold it, apply mortar to the header end of the brick. *Figure 34* shows a properly buttered head joint.

The buttered mortar should cover all of the header surface but should not extend past the edges of the brick. Hold the trowel at an angle to the header surface to keep the mortar off the sides of the brick.

When the brick is buttered, use your brick hand to press it into position next to the first brick (*Figure 35*). After placing the brick, cut off the excess mortar with the edge of your trowel.

Unlike block, you can easily hold a brick in one hand. Take advantage of this to use both hands for laying brick. Try to develop a rhythmic set of movements. This will make the work faster and easier on you. Remember to use your shoulders and arms, not just your wrists.

28101-13_F34.EPS

Figure 34 A buttered head joint.

28101-13_F35.EPS

Figure 35 Placing the brick.

After you have laid six brick, check them for placement. Use your mason's level to check both plumb and level (*Figure 36*). If a brick is out of line, tap it gently with the handle of your trowel. Do not tap the level. Do not use the point or blade of your trowel or it will lose its edge.

3.2.8 General Rules

The way you work the mortar determines the quality of the joints between the masonry units. The mortar and the joints form a vital part of the structural strength and water resistance of the wall. Learning these general rules and applying them as you spread mortar will help you build good walls:

- Use mortar with the consistency of mud, so it will cling to the masonry unit, creating a good bond.
- Butter the head joints fully for block and brick; butter both ears of the head joints for block.

28101-13_F36.EPS

Figure 36 Checking the level.

- When laying a unit on the bed joint, press down slightly and sideways, so the unit goes against the one next to it.
- If mortar falls off a moving unit, replace the mortar before placing the unit.
- Put down more mortar than the size of the final joint; remember that placing the unit will compress the mortar.
- The length of a spread is determined by the moisture content of the units. Do not string a spread that is longer than what can be laid before water evaporates from the units and the spread becomes too stiff to bond properly.
- Do not move a unit once it is placed, leveled, plumbed, and aligned.
- If a unit must be moved after it is placed, remove all the mortar on it and rebutter it.
- After placing the unit, cut away excess mortar with your trowel and put it back in the pan, or use it to butter the next joint.
- Throw away mortar after 2 hours. At that point, it is beginning to set and will not give a good bond.

Additional Resources

Bricklaying: Brick and Block Masonry. Reston, VA: Brick Industry Association.

Concrete Masonry Handbook. Skokie, IL: Portland Cement Association.

Concrete Masonry Shapes and Sizes Manual CD-ROM. Herndon, VA: National Concrete Masonry Association.

3.0.0 Section Review

1. Adhesion means the ability of mortar to ____.
   a. dry quickly
   b. be mixed to a given consistency
   c. stick to a surface
   d. spread evenly

2. Once you have put the mortar on the board, you should work the mortar ____.
   a. into a pile in the center of the board
   b. into an evenly distributed layer across the board
   c. into a pile at the closest end of the board
   d. into small piles in each of the corners of the board

Section Four

4.0.0 Success in the Masonry Trade

Objective

Describe the skills, attitudes, and abilities needed to be a successful mason.

a. Identify the skills of a successful mason.
b. Identify the attitudes of a successful mason.
c. Identify the abilities of a successful mason.
d. Explore career ladders and advancement possibilities in masonry.

Trade Term

Tuckpointing: Filling fresh mortar into cutout or defective joints in masonry.

Becoming a good mason takes more than the ability to lay a masonry unit and level it. A competent mason is one who can be trusted to perform the required work and meet the project specifications. This mason must have the necessary knowledge, skills, and ability, as well as good attitudes about the work itself, about safety, and about quality. This section explains the skills, attitudes, and abilities of successful masons, and discusses the career opportunities available to them.

4.1.0 Identifying the Skills of a Successful Mason

Masons need to know how to handle all aspects of masonry work, including all of the following:

- Read and interpret drawings and specifications
- Calculate and estimate quantities, lengths, weights, and volumes
- Select the proper materials for the job
- Lay masonry units into structural elements
- Work productively alone or as part of a team
- Assemble and disassemble scaffold
- Keep tools and equipment in good repair and safe condition
- Follow safety precautions to protect themselves and other workers on the job

4.1.1 Job-Site Knowledge

Masons need to be skilled in applying their knowledge to the challenges they face each day on the job. The best way to do the work at a particular job site will depend on the layout of the work, what is happening around the masonry site, and the conditions surrounding the project.

Most masonry work is done outside in temperature and weather variations. You must be able to work under these conditions and not be distracted by them. You must know how to react to changing conditions around you.

Much of this knowledge can be learned as you work, if you will pay attention. Notice what others do and ask questions. Ask your supervisor questions, too. Learn to respond to conditions at the job site.

4.1.2 Learning More

Masons need to keep on learning after they finish their apprenticeships. They need to keep updating their skills all the time. The environment, tools, and expectations about masonry have evolved and will continue to change. Craftworkers and contractors alike will need to change the way they think about their work and how they do it.

National, regional, and local organizations offer continuing education for masons. Technical seminars, training sessions, publications, and classes are often free or low cost. They can bring you the latest information about tools, materials, and methods. To succeed, you must be alert to change and willing to learn new ways.

4.1.3 Quality

Quality in work is not a new idea. Those who work in masonry construction and finishing have been concerned about quality for thousands of years. The quality of masonry depends on many factors. When building a wall, you may have little control over its design or the choice of masonry units. But you do have control over the quality of the completed job. A wall out of level or with poorly finished joints is your responsibility.

The quality of the finished masonry structure depends directly on your knowledge, skill, and ability. Good work is easily recognized. Poor work is seen even more easily. Given the durability of masonry, the quality of the work is a monument to your skill for a very long time. The skilled, proud mason always strives for the highest quality that can be achieved.

4.2.0 Identifying the Attitudes of a Successful Mason

Attitude can build an invisible bridge, or build an invisible wall, between people. No one wants to hang around a grouch or count on someone who is not dependable. No one minds helping someone who can do something in return or working with a friendly, cooperative partner. On top of knowledge, skills, and ability, you need the right attitude. Your attitude comes from how you think and feel about your work and yourself.

4.2.1 Dependability

You must be dependable. Masonry work, like all construction, is a closely timed operation. Once started, it cannot stop without waste of material and money. Employers need workers who report to work on time. An undependable, absent worker will slow or stop masonry work and cost the project time and money. An undependable worker will not be able to depend on having a job for very long.

4.2.2 Responsibility

You must be responsible for doing the assigned work in a proper and safe manner, be responsible enough to work without supervision, and work until the task is complete.

Being responsible for your own work includes admitting your mistakes. It also includes learn-

Smartphone Apps for Construction

Although the craft of masonry has been around for centuries, not everything in masonry is old. Smartphones are becoming an increasingly popular form of communication, and also offer a great deal of versatility for craftworkers. Smartphone cameras can be used to document on-the-job activities or potential safety violations. Best practices can be communicated to crewmembers using video clips. Construction calculators can be downloaded, providing craftworkers with the same (or even greater) capabilities than a handheld calculator.

The smartphone version of Calculated Industries, Inc.'s popular Construction Master® Pro, for example, lets craftworkers calculate and convert dimensions, plot right-angle conversions, find area and volume, and determine measures such as board feet, cost per unit, and even the angles of an equal-sided polygon. Calculated Industries, Inc., also makes an app version of its ConcreteCalc™ Pro app (see the figure), which masons can use to calculate stair dimensions, the length and weight of rebar, and the number of brick loads and mortar bags needed for a job.

28101-13_SA07.EPS

ing from your mistakes. Nobody is expected to be perfect. Everyone is expected to learn and to grow more skilled.

Employers are always in need of workers who are ambitious and want to become leaders. Being responsible for what others do may be your career goal. The path to that goal starts with being responsible for what you do.

4.2.3 Adaptability

On any construction project, a large amount of work must be done in a short time. Planning and teamwork are needed in order to work efficiently and safely. Supervisors sometimes form teams of two or more workers to do specific tasks. You may work in a team to erect a scaffold, then work alone for most of the day, then team with someone else to do a cleanup.

On a job site, you may find yourself teaming with different people at different times. The ability to be a team player is an important part of being a mason. Team players accept instruction and direction. They communicate clearly, keep an eye out for potential problems, and share information. They meet problems squarely with constructive ideas, not criticism.

All team players treat each other with respect. Everyone must be willing to work together. Everyone must be willing to bring their best attitude to the team. Team members need to be able to depend on each other. Team priorities must be more important than individual priorities.

4.2.4 Pride

Pride in what you do comes from doing high-quality work in a timely manner and from knowing you are doing your best. Being proud of what you do can overflow into other areas. Proud workers take pride in their personal appearance. Their work clothes are clean, safe, neat, and suitable. Proud masons take pride in their tools. They have a complete set of well-maintained tools and other special equipment they need to do their jobs. They keep their tools safe and orderly, and know how to use the right tool for the work at hand.

Proud masons work so that they can continue to be proud of what they do and how they do it. Being proud of what you do is an important part of being proud of who you are.

4.2.5 Ethics

Members of the construction trades are expected to observe the following ethical principles:

- *Honesty* – Be honest and truthful in all dealings. Conduct business according to the highest professional standards. Faithfully fulfill all contracts and commitments. Do not deliberately mislead or deceive others.
- *Integrity* – Demonstrate personal integrity and the courage of your convictions by doing what is right even when there is great pressure to do otherwise. Do not sacrifice your principles for expediency, be hypocritical, or act in an unscrupulous manner.
- *Loyalty* – Be worthy of trust. Demonstrate fidelity and loyalty to companies, employers, fellow craftspeople, and trade institutions and organizations.
- *Fairness* – Be fair and just in all dealings. Do not take undue advantage of another's mistakes or difficulties. Fair people display a commitment to justice, equal treatment of individuals, tolerance for and acceptance of diversity, and open-mindedness.
- *Respect for others* – Be courteous and treat all people with equal respect and dignity, regardless of sex, race, or national origin.
- *Law abiding* – Abide by laws, rules, and regulations relating to all personal and business activities.
- *Commitment to excellence* – Pursue excellence in performing your duties, be well informed and prepared, and constantly endeavor to increase your proficiency by gaining new skills and knowledge.
- *Leadership* – By your own conduct, seek to be a positive role model for others.

4.3.0 Identifying the Abilities of a Successful Mason

A mason must be a responsible person with a high degree of concern for the safety of workers and the quality of the work.

4.3.1 Willingness to Take Responsibility

In general, responsibility should be delegated and not assumed; once responsibility has been delegated to a mason, the mason should continue to perform the duties without further direction. Every mason should have the responsibility for working safely. Most contractors expect their masons to see what needs to be done, then go ahead and do it. It is very tiresome to have to ask again and again that a certain job be done.

4.3.2 Willingness to Follow Rules and Regulations

People can work together well only if there is some understanding about what work is to be done, when it will be done, and who will do it. Rules and regulations are a necessity in any work situation.

4.3.3 Willingness to Avoid Tardiness and Absenteeism

Tardiness means being late for work and absenteeism means being off the job for one reason or another. Consistent tardiness and frequent absences are an indication of poor work habits, unprofessional conduct, and a lack of commitment to your contractor.

Work life is governed by the clock. All members of a masonry crew are required to be at work at a specific time. Failure to get to work on time results in confusion, lost time, and resentment on the part of those who do come on time. In addition, frequent tardiness or absenteeism may lead to penalties, including dismissal. When accepting a job with a contractor, you agree to the terms of work. Perhaps it will allow you to see the picture more clearly if viewed from the supervisor's point of view. Supervisors cannot keep track of people if they come in any time they please. It is not fair to others to ignore tardiness. Failure to be on time may hold up the work of other masons and craftworkers. Better planning of your morning routine will often keep you from being delayed and so prevent a breathless, late arrival. In fact, arriving a little early indicates your interest and enthusiasm for your work, which is appreciated by contractors. The habit of being late is another one of those things that stand in the way of promotion.

It is sometimes necessary to take time off from work. No one should be expected to work when sick or when there is a serious issue at home. However, it is possible to get into the habit of letting unimportant and unnecessary matters keep you from the job. This results in lost production and hardship on those who try to carry on the work with less help. The contractor that hires you has a right to expect you to be on the job unless there is some very good reason for staying away. Certainly, do not let some trivial reason keep you home. Do not stay up late at night and be too tired to go to work the next day. If you are ill, spend the time at home to recover quickly. This, after all, is no more than what you would expect of a person you hired, and on whom you depended to do a certain job.

If it is necessary to stay home, then at least notify your supervisor early in the morning so the supervisor can find another worker for the day, if needed. Some workers remain at home without contacting the contractor, which is the worst possible way to handle the matter. It leaves those at work uncertain about what to expect. They have no way of knowing whether you have merely been held up and will be in later, or whether immediate steps should be taken to assign your work to someone else. Courtesy alone demands that you let the supervisor know if you cannot come to work.

The most frequent causes of absenteeism are illness or death in the family, accidents, personal business, and dissatisfaction with the job. Some of the causes are legitimate and unavoidable, while others can be controlled. For most situations, you can carry on most personal business affairs after working hours. Frequent absences will reflect unfavorably on a worker when promotions are being considered.

Contractors sometimes resort to docking pay, demotion, and even dismissal in an effort to control tardiness and absenteeism. No contractor likes to impose restrictions of this kind. However, in fairness to those workers who do come on time and who do not stay away from the job, a contractor is sometimes forced to discipline those who will not follow the rules.

4.4.0 Exploring Career Ladders and Advancement Possibilities in Masonry

Masonry offers a rewarding career for people who want to work with their hands. As masons, they will be skilled workers who understand the principles and practices of masonry construction.

The Customer

When you are on a job site, consider yourself to be working for both your contractor and the customer. If you are honest and maintain a professional attitude when interacting with customers, everyone will benefit. Your contractor will be pleased with your performance, and the customer will be happy with the work that is being done. Try seeing things from a customer's point of view. A good, professional attitude goes a long way toward ensuring repeat business.

They will earn good pay and be rewarded for initiative. They will have opportunity for advancement.

Masons will continue to play an important part in building homes, schools, offices, and commercial structures. They can add artistic elements to their work and create beauty. They can be proud of their skills and the fact that they produce something people need.

Masons work on different projects, so each job is different and never boring. If they like to travel, masons can find good jobs all over the country. They can be independent and creative while working outdoors. Masons will be in demand as long as buildings are being constructed.

Masons can find work on large construction projects for commercial buildings, as well as projects for building homes, patios, sidewalks, or walls. They can also specialize in repair work, cleaning, and **tuckpointing** old buildings. They can specialize in restoring historic brick buildings, which is a recognized craft specialty in some parts of the country. Historically, they have been well paid.

Masonry is more than physical labor. It is a skilled occupation that calls for good hand-eye coordination, balance, and strength. It also requires good mental skills. This means ongoing study, concentration, and continued learning in an environment free from substance abuse.

Because masonry is a highly skilled craft, it takes time to learn. Your learning starts with this course, combined with, or followed by, an apprentice's job. Masons usually work as part of a team, so you will also learn to be a good team player. Masons work outdoors and do a lot of lifting and bending, so you will learn to keep yourself in good shape. Masons work on high scaffolds, so you will learn safety rules and practices. Masons bring their skills and tools wherever they go.

Late for Work

Showing up on time is a basic requirement for just about every job. Your contractor is counting on you to be there at a set time, ready to work. While legitimate emergencies may arise that can cause you to be late for or even miss work, starting a bad habit of consistent tardiness is not something you want to do. What are the possible consequences that you could face as a result of tardiness and absenteeism?

4.4.1 Career Stages

Masons were among the first workers to band together. During the Middle Ages, they formed influential groups that still shape trade practices. Today, as in the past, masons' organizations recognize several stages of skill:

- Tender
- Apprentice
- Journeyman
- Supervisor
- Superintendent
- Contractor

The tender is a laborer, not a mason. The tender carries masonry materials, tools, and mortar and gets things for the mason. The tender mixes mortar, cleans tools, and learns by watching the mason at work. Sometimes, tenders decide they want to become masons. If they do, they may enter an apprenticeship program. Apprentices are at the beginning level of the masonry career path. Their training will lead them to full participation in the mason's trade and the opportunity for higher job levels.

4.4.2 Apprentice

An apprentice is a person who has signed an apprenticeship agreement with a local joint apprenticeship committee. The committee works with local contractors who have agreed to take apprentices. The US Department of Labor regulates the apprenticeship process. In most states, the state department of labor is also involved, as state labor regulations provide guidelines on legal age requirements, pay, hours, and other aspects of apprenticeship.

The length of the apprenticeship will vary. The US Department of Labor program is three years and 4,500 hours with a minimum of 432 hours of classroom instruction. Programs such as the NCCER Standardized Craft Training program are used for classroom instruction or as part of apprenticeship training.

The apprentice is assigned to work with a contractor and to take classes. The apprentice must study and work under supervision. The apprentices agree to do the following:

- Perform the work assigned by the contractor.
- Abide by the rules and regulations of the contractor and the committee.
- Complete the hours of instruction.
- Keep records of work experience, training, and instruction.
- Learn and use safe working habits.

- Work with the assigned contractors for the entire apprenticeship period, unless reassigned by the committee.
- Conduct themselves in an ethical manner, realizing that time, money, and effort are being spent to afford them this opportunity to become a skilled worker.
- Remain free from drug and alcohol abuse.

A typical three-year apprenticeship is divided into six periods of six months each. The first six months is a trial period. The committee reviews the apprentice's performance and may end the agreement.

The apprentice attends classes and works under the supervision of a journeyman mason. As part of the supervised work, the apprentice learns to lay masonry units and perform other craftwork. The apprentice's pay increases for each six-month period as skill and performance increase. At the end of the period, the apprentice receives a certificate of completion from the craft-training program. A certificate from NCCER (*Figure 37*) is known and accepted everywhere in the United States. After completing three years of training, an apprentice can become a journeyman mason.

4.4.3 Journeyman

Unlike an apprentice, a journeyman is a free agent who can work for any contractor. A journeyman can work without close supervision and is skilled in most tasks. The successful journeyman knows that the end of the apprenticeship is not the end of learning. Masons' organizations recognize the journeyman to be the highest stage of skill for a mason.

Journeymen are people with an excellent trade. They earn good wages in a trade that is always in demand. They have the satisfaction of creating and the opportunity to grow as masonry artists. They also have the opportunity to grow as layout persons, trainers, and supervisors.

An experienced and skilled journeyman can work as a layout person. For a pay premium, the layout person lays out the work and lays the leads. Less experienced masons and apprentices work between the leads set by the layout person. Experienced and skilled journeymen also train apprentices and supervise their work. With further experience, journeymen can supervise crews.

Journeymen can continue to learn by studying and handling more complex tasks. They can continue to develop their skills as they work. Further education in masonry innovations and techniques is available, as is training in leadership and supervision.

4.4.4 Supervisors, Superintendents, and Contractors

Supervisors are responsible for managing and supervising a group of workers. This job requires a high degree of knowledge about masonry and leadership skills. Supervisors are typically responsible for training workers in safety measures and keeping work areas safe. They also train workers in new techniques and easier ways of working. They solve daily problems, keep on top of materials and supplies, and make sure workers meet job schedules. They check work to ensure it is done to standards. Supervisors may be called crew leaders or forepersons, depending on the company that hires them.

Superintendents have several supervisors reporting to them. Usually, the superintendent is the lead person on a large job. For a smaller company, the superintendent may be in charge of all the work in the field for the contractor. The superintendent oversees the work of the supervisors and makes major decisions about the job under construction. The superintendent must have strong masonry, leadership, and business skills.

A masonry contractor owns the company. The contractor bids on jobs, organizes the work and the workers, inspects the work, confers with the clients, and runs the business. The contractor needs to be able to plan ahead to keep up with change.

Contractors, along with journeymen, supervisors, and superintendents, need to keep up with the latest materials and methods. Like apprentices, they need to continue learning their trade.

4.4.5 The Role of NCCER

This course is part of a curriculum produced by NCCER. NCCER is an independent, private educational foundation founded and funded by the construction industry to provide quality instruction and instructional materials for a wide variety of crafts. The basic idea of the NCCER is to supplant governmental control and credentialing of the construction workforce with industry-driven training and education programs. NCCER has departed from traditional classroom learning and has adopted a pure competency-based training regimen. Competency-based training means that instead of requiring specific hours of classroom training and set hours of on-the-job training (OJT), you simply have to prove that you know what is required and can demonstrate that you can perform the specific skill. NCCER also uses the latest technology, such as interactive computer-based training, to deliver the classroom

NCCER
The Standard for Developing Craft Professionals

This is to certify that

Earl Ordway

has fulfilled the requirements for

Masonry Level One

in NCCER's standardized training curriculum
this Tenth day of May, 2013

The Standard for Developing Craft Professionals
Founded 1996

Donald E. Whyte
President, NCCER

28101-13_F37.EPS

Figure 37 Example of apprenticeship training recognition.

portions of the training. All completion information for every trainee is sent to NCCER and kept within the National Registry. The National Registry can then confirm training and skills for workers as they move from company to company, state to state, or even within their own company.

The dramatic shortage of skills within the construction workforce, combined with the shortage of new workers coming into the industry, is providing an opportunity for the construction industry to design and implement new training initiatives. When enrolling in an NCCER program, it is critical that you work for a contractor who supports a national, standardized training program that includes credentials to confirm your skill development.

The construction industry knows that the future construction workforce will largely be recruited and trained in the nation's secondary and postsecondary schools. Schools know that to prepare their students for a successful construction career they must use the curriculum that is developed and recognized by the industry. Nationwide, thousands of schools have adopted NCCER's standardized curricula.

The primary goal of NCCER is to standardize construction craft training throughout the country so that both you and your employer will benefit from the training, no matter where you and your job are located. As a trainee in a NCCER-accredited program, you will be listed in the National Registry. You will receive a certificate for each level of training you complete (*Figure 37*), which can then travel with you from job to job as you progress through your training. In addition, many technical schools and colleges use NCCER's programs.

Additional Resources

Hot & Cold Weather Masonry Construction Manual. 1999. Herndon, VA: National Concrete Masonry Association.

4.0.0 Section Review

1. Masons should update their skills _____.
   a. with the permission of their employer
   b. when classes are available
   c. all the time
   d. until they have finished their apprenticeship

2. Being a team player involves communicating clearly, keeping an eye out for potential problems, and _____.
   a. reading and interpreting drawings
   b. sharing information
   c. making decisions on your own
   d. seeking opportunities for advancement

3. A legitimate cause of absenteeism is _____.
   a. car problems
   b. traffic conditions on the way to work
   c. oversleeping
   d. illness

4. A mason who is a free agent who can work for any contractor is called a(n) _____.
   a. supervisor
   b. journeyman
   c. apprentice
   d. superintendent

Section Five

5.0.0 SkillsUSA

Objective

Summarize how to be connected to the industry through an organization like SkillsUSA.

a. Understand the program, curriculum, and SkillsUSA Championships.
b. Understand SkillsUSA membership.
c. Understand the National Program of Work Standards.

SkillsUSA is a partnership of students, teachers, and industry representatives working together to ensure America has a skilled workforce. SkillsUSA is a national organization serving teachers and high school and college students who are preparing for careers in technical, skilled, and service occupations, including masonry and other building trades occupations. More than 320,000 students and advisers, who are organized into more than 17,000 sections and 54 state and territorial associations, join SkillsUSA annually. Combining alumni and lifetime membership, the total number impacted is more than 320,000. SkillsUSA has served more than 10.5 million members in its history.

The mission of SkillsUSA is to assist its members in becoming world-class workers, leaders, and responsible citizens. SkillsUSA is an applied method of instruction for preparing America's high-performance workers in public career and technical programs. It provides quality education experiences for students in leadership, teamwork, citizenship, and character development. SkillsUSA builds and reinforces self-confidence, positive work attitudes, and communications skills. It emphasizes total quality at work: high ethical standards, superior work skills, lifelong education, and pride in the dignity of work. SkillsUSA also promotes understanding of the free-enterprise system and involvement in community service. SkillsUSA helps each student to excel.

5.1.0 Understanding the Program, Curriculum, and SkillsUSA Championships

SkillsUSA programs help to establish industry standards for job skill training in the lab and classroom, and promote community service. SkillsUSA is recognized by the US Department of Education and is cited as a "successful model of employer-driven youth development training program" by the US Department of Labor.

SkillsUSA Championships include local, state, and national competitions in which students demonstrate occupational and leadership skills. The SkillsUSA Championships is the showcase for the best career and technical students in the nation. This is a multimillion-dollar event that occupies a space equivalent to 16 football fields. In 2011, there were more than 5,700 contestants in 94 separate events. Nearly 1,500 judges and contest organizers from labor and management make the national event possible. The philosophy of the Championships is to reward students for excellence, to involve industry in directly evaluating student performance, and to keep training relevant to employers' needs. The national masonry competition at the SkillsUSA Championships is sponsored by NCCER.

> Learn more about the SkillsUSA Championships: **http://www.skillsusa.org/compete/skills.shtml**

5.2.0 Understanding SkillsUSA Membership

In 2011, more than 16,600 teachers and school administrators served as professional SkillsUSA members and advisers. More than 1,100 business, industry, and labor sponsors actively support SkillsUSA at the national level through financial aid, in-kind contributions, and involvement of their people in SkillsUSA activities. Many more work directly with state associations and local chapters. NCCER and SkillsUSA have a long-standing partnership, as both organizations share the goal of a skilled workforce.

5.2.1 The Value for Students

For many students, SkillsUSA is the first professional organization they will join. The experiences and knowledge gained provide an excellent platform for career development and success. SkillsUSA also sets the stage for involvement in other professional and service organizations. Advantages include:

- Teamwork and leadership development
- Reinforcement of employability skills
- Competition in a nationally recognized contest program
- Community service opportunities

- Access to scholarships
- Networking with potential employers

5.2.2 The Value for the Classroom and School

Great instructors are always looking for ways to engage students and build relationships. SkillsUSA provides the tools to do both. As a student-run organization, members feel a sense of empowerment and belonging. SkillsUSA is a motivator for students to put forth their best effort in the classroom, making daily lessons even more relevant to career success. As a SkillsUSA adviser, the activities, projects, and contests provide opportunities for instructors to build stronger relationships with students. Chapter activities and accomplishments can build a positive image for participating schools and their programs. Benefits include:

- Recognition for the school within the community
- Opportunities to meet educational standards
- Development of career technical education (CTE) pathways
- Improved recruitment and enrollment
- More graduates equipped with essential skills

5.3.0 Understanding the National Program of Work Standards

The heart of SkillsUSA is the Program of Work (POW), or what each chapter is going to do during the school year. It is the activities and projects—the plan of action—that a chapter will carry out.

The National Program of Work sets the pace for SkillsUSA nationwide. The expectation is that each chapter will carry out this Program of Work. All of the SkillsUSA programs are in some way related to the following seven major goals: professional development, community service, employment, ways and means, SkillsUSA Championships, public relations, and social activities.

Professional development – The goal of professional development is to prepare each SkillsUSA member for entry into the workforce and provide a foundation for success in a career. Becoming a professional does not stop with acquiring a skill, but involves an increased awareness of the meaning of good citizenship and the importance of labor and management in the world of work.

Community service – The goal of the community service standard is to promote and improve goodwill and understanding among all segments of the community through services donated by SkillsUSA chapters. In addition, SkillsUSA hopes to instill in its members a lifetime commitment to community service.

Employment – The goal of this standard is to increase student awareness of quality job practices and attitudes, and to increase the opportunities for employer contact and eventual employment.

Ways and means – The ways and means goal is to plan and participate in fundraising activities to allow all members to carry out the chapter's projects.

SkillsUSA Championships – The goal of the SkillsUSA Championships is to offer students the opportunity to demonstrate their skills and be recognized for them through competitive activities in occupational areas and leadership.

Public relations – The goal of the public relations standard is to make the general public aware of the good work that students in career and technical education are doing to better themselves and their community, state, nation, and world.

Social activities – The goal of this standard is to increase cooperation in the school and community through activities that allow SkillsUSA members to get to know each other in something other than a business or classroom setting.

Learn more about the Program of Work standards: **http://www.skillsusa.org/educators/chapmanage5.shtml**

http://skillsusa.org/courses/07_Program/player.html

5.3.1 Chapter Activity Planner

Chapter members should discuss and develop their own program of work. Instructors will assist in selecting activities that relate to students' vocational training and will guide them in developing their personal skills in communications, organization, planning, and follow-through.

Chapter activities provide some of the best opportunities for students to learn by doing. A successful program of work creates a positive learning atmosphere in the classroom, and allows students to learn how to accept responsibility, work as a team, manage a budget, and handle success and failure.

5.3.2 Chapter Elections and Training

Effective chapter officers ensure the chapter functions effectively and efficiently. Officers frequently are responsible for routine management tasks, such as organizing meetings, conducting meetings, scheduling work, and leading chapter activities. This helps students learn simple supervisory skills and creates responsible team spirit.

The election of chapter SkillsUSA officers is often one of the highlights of the SkillsUSA year. The outcome of the election affects the entire group's chances for having a successful program. Officers not only spark enthusiasm in the organization, but also carry on the routine business affairs that keep the program moving.

Being elected as a SkillsUSA officer provides an opportunity to hone leadership abilities. The officer selection process is an excellent way to learn valuable, practical lessons in leadership and teamwork.

5.3.3 Chapter Meetings

Valuable skills are learned and practiced through organized activities. Well-run meetings are a good example. Learning how to plan and work cooperatively with others is an important skill set. As individuals, everyone has good ideas, but when people combine their ideas and efforts, great things can occur.

Additional Resources

SkillsUSA Professional Development Resources. **http://www.skillsusa.org/store/curricula.html**

5.0.0 Section Review

1. The national masonry competition at the SkillsUSA Championships is sponsored by _____.
   a. the Mason Contractors Association of America
   b. local contractors
   c. NCCER
   d. chapter officers
2. The number of business, industry, and labor sponsors that actively support SkillsUSA at the national level is more than _____.
   a. 60
   b. 600
   c. 900
   d. 1,100
3. Which of the following is *not* a goal of the SkillsUSA Program of Work?
   a. Community service
   b. Professional development
   c. Social activities
   d. Improved recruitment and enrollment

Summary

Masonry is a craft that has existed for thousands of years. Science and engineering have brought masonry into the modern age. With modern construction techniques, masonry is now used in high-rise buildings as well as residential and commercial projects. This allows for widespread use of masonry construction throughout North America.

Modern clay products are categorized as solid and hollow brick, structural and nonstructural tile, and made-to-order architectural terra-cotta. Clay has been joined by concrete as a modern masonry material. Stone, the oldest recovered building material, is still laid by masons. Mortars and grouts have evolved from mud to special-purpose, high-strength cements.

Masonry offers advancement through the recognized career steps of apprentice, journeyman, layout person, supervisor, superintendent, and masonry contractor. Your success as a mason requires the willingness to keep on learning.

Review Questions

1. Concrete masonry units are usually cured by using _____.
   a. hot air
   b. live steam
   c. sunlight
   d. infrared heaters

2. Drainage openings in masonry walls are called _____.
   a. scuppers
   b. drip vents
   c. cores
   d. weepholes

3. The custom brick shape shown in *Review Question Figure 1* is a(n)_____.
   a. water table rowlock
   b. ogee step tread header
   c. single bullnose stretcher
   d. internal radial

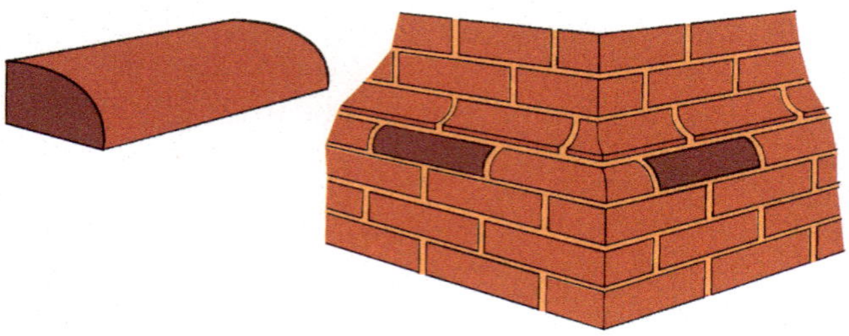

28101-13_RQ01.EPS

Figure 1

4. For ease of cleaning, the type of brick often used in dairies and laboratories is _____.
   a. glazed
   b. faced
   c. fired
   d. pressed

5. A brick positioned to bond two wythes of masonry is a _____.
   a. rowlock
   b. header
   c. soldier
   d. stretcher

6. Stone is often laid over brick or block to form a _____.
   a. screen
   b. face course
   c. moisture barrier
   d. veneer

7. Ashlar stone _____.
   a. is irregular in size and shape
   b. has smooth bedding surfaces
   c. has poor weather resistance
   d. is 2 inches thick or less

8. Cement-lime mortar is a type of mortar in which _____.
   a. all the ingredients (cement, sand, lime, and water) are often mixed on the job site
   b. cement, water, and fine aggregates form a mixture wet enough to be pumped
   c. the mason only adds sand and water to a premixture of cement, lime, and additives
   d. only water must be added to preblended dry ingredients

9. The number of wythes in a cavity wall is _____.
   a. four
   b. three
   c. two
   d. one

10. Loadbearing walls with different materials in the facing and backing wythes are called _____.
    a. veneer walls
    b. composite walls
    c. cavity walls
    d. reinforced walls

11. OSHA's regulation *CFR 1926.32* defines the duties and responsibilities of a(n) _____.
    a. legally liable person
    b. competent person
    c. enforcement person
    d. responsible person

12. To allow masons to keep their sense of touch while still being able to grip effectively and protect their hands, many gloves used in masonry work have palms and fingertips made from _____.
    a. canvas or nylon
    b. felt
    c. latex or synthetic rubber
    d. cotton

13. For safety on the job, a mason should wear appropriate clothing that is _____.
    a. loose fitting
    b. polyester
    c. disposable
    d. close fitting

14. The straightness and levelness of each masonry unit in a structure depends upon the _____.
    a. quality of the supplied materials
    b. hands and eyes of the mason
    c. time available to complete the job
    d. skill and experience of the architect

15. Before using a wheelbarrow to transport mortar to the work site, you should _____.
    a. check whether the wheelbarrow is right handed or left handed
    b. oil the inside surface of the wheelbarrow
    c. wet the inside surface of the wheelbarrow
    d. grease the wheelbarrow's axle

16. In *Review Question Figure 2*, mortar has been placed on the header end of the brick in a process known as _____.
    a. buttering
    b. spreading
    c. smoothing
    d. bedding

28101-13_RQ02.EPS

Figure 2

17. To be sure of a good bond, excess mortar should be thrown away after _____.
    a. 1½ hours
    b. 2 hours
    c. 3 to 4 hours
    d. 8 hours

18. In the following list, the highest stage of skill recognized by masons' organizations is _____.
    a. journeyman
    b. tender
    c. sawman
    d. apprentice

19. Apprentices are trained and supervised by _____.
    a. OSHA
    b. journeyman masons
    c. the US Department of Labor
    d. foremen

20. The activities and projects a SkillsUSA chapter will carry out during a school year are outlined in the _____.
    a. Master Schedule of Events
    b. Activities Planner and Guide
    c. National Program of Work
    d. National Action Plan

Trade Terms Quiz

Fill in the blank with the correct term that you learned from your study of this module.

1. The division of the US Department of Labor mandated to ensure a safe and healthy environment in the workplace is called the ________.
2. ________ is materials such as crushed stone or gravel used as a filler in concrete and concrete block.
3. A square or rectangular pillar projecting from a wall is called a(n) ________.
4. A(n) ________ is any building block made of brick, cement, ashlar, clay, adobe, rubble, glass, tile, or any other material that can be assembled into a structural unit.
5. To apply mortar to the end of a masonry unit prior to laying it is to ________ it.
6. A continuous section of masonry wall, one masonry unit in thickness, or that part of a wall that is one masonry unit in thickness, is called a(n) ________.
7. ________ is spreading mortar with a trowel on a wall or footing for a bed joint.
8. An individual who is capable of identifying existing and predictable hazards or working conditions that are hazardous, unsanitary, or dangerous to employees, and who has authorization to take prompt, corrective measures to eliminate or control these hazards and conditions, is called a ________.
9. A (n) ________ is a chemical or mineral other than water, cement, or aggregate added to mortar immediately before or during mixing to change its setting time or curing time; to reduce water; or to change the overall properties of the mortar.
10. The horizontal projection crowning the wall of a building is called a(n) ________.
11. The ________ is that part of a masonry unit or wall that shows after construction.
12. Pre-made veneer consisting of cast cementitious material with pigments and other added materials to give the appearance of natural stone is called ________.
13. ________ is the publisher of masonry standards.
14. A mixture of portland cement, lime, and water, with or without fine aggregate, with a high-enough water content that it can be poured into spaces between masonry units and voids in a wall, is called ________.
15. The ________ is the top part of an architectural column.
16. A vertical joint between two masonry units is called a(n) ________.
17. ________ are the area between each brick or block that is filled with mortar.
18. A person who assembles masonry units by hand, using mortar, dry stacking, or mechanical connectors is called a(n) ________.
19. ________ brick has holes extending through it to reduce weight.
20. A wall that does not bear weight other than its own is ________.
21. ________ is sun-dried, molded clay brick.
22. Brick that has no holes extending through it is called ________.
23. A(n) ________ is a strapped bundle of approximately 500 standard brick.
24. Equipment or clothing designed to prevent or reduce injuries is called ________.
25. A(n) ________ is a hollow or solid block made from portland cement and aggregates.

26. The base for a masonry unit wall, or concrete foundation, that distributes the weight of the structural member resting on it is called a(n) ________.

27. A(n) ________ is a row or horizontal layer of masonry units.

28. A row of mortar placed into a bed joint is called a(n) ________.

29. A(n) ________ is a small opening in mortar joints or faces to allow the escape of moisture.

30. A wall that bears weight in addition to its own is called ________.

31. ________ is filling fresh mortar into cutout or defective joints in masonry.

32. A mixture of portland cement, lime, fine aggregate, and water, plastic or stiff enough to hold its shape between masonry units is called ________.

33. A(n) ________ is a low wall or railing.

34. A squared or rectangular cut stone masonry unit; or, a flat-faced surface having sawed or dressed bed and joint surfaces is called ________.

35. A(n) ________ is the horizontal joint between two masonry units in separate courses.

Trade Terms

Admixture
Adobe
Aggregate
American Society for Testing and Materials (ASTM) International
Ashlar
Bed joint
Butter
Capital
Competent person
Concrete masonry unit (CMU)
Cored
Cornice
Course
Cube
Facing
Footing
Grout
Head joint
Joints
Manufactured stone veneer
Mason
Masonry unit
Mortar
Nonstructural
Occupational Safety and Health Administration (OSHA)
Parapet
Personal protective equipment (PPE)
Pilaster
Spread
Stringing
Structural
Tuckpointing
Uncored
Weephole
Wythe

Moroni Mejia

Workforce Development Committee Chairman
Mason Contractors Association of America

Ancient architecture inspired Moroni Mejia to pursue a career in construction from the ground up. Today, Moroni inspires young craft professionals to seek the same sense of challenge and opportunity that had also inspired him.

How did you get started in the construction industry?

My first day on a real job came in 1997. I can still remember how amazed I was at the huge machinery and powerful tools that were used. I can also remember how invincible I felt back then and how humbling it was to work through the fear I felt as I tended to a team of bricklayers on a temporary work platform several stories above the ground.

Who or what inspired you to enter the industry? Why?

I'd always admired the magnificence of the stone structures built by our ancestors all around the world. After visiting the Aztec and Mayan ruins in southern Mexico in my teenage years, I knew I wanted to be a part of creating buildings. My original goal was to become an architect, and I felt that by serving an apprenticeship in masonry I would learn about buildings from the ground up. The apprenticeship also allowed me to earn money so I could support myself and pay for college. Not long after enrolling, I fell in love with the trade, realized the tremendous earning potential associated with it, and ran with it.

What do you enjoy most about your career?

I feel a deep sense of satisfaction when I drive around the state and see the projects I've been involved in being enjoyed or put to use. When I do, I remember all the different challenges that had to be overcome to build the projects successfully. Properly designed and built masonry work gives a building character, identity, and leaves a lasting impression on our communities.

Why do you think training and education are important in construction?

First and foremost, finish high school. Enroll in any construction technology courses your school offers, or seek out other places near you that offer them. Once you decide on a trade, complete an accredited, certified apprenticeship program.

Continuing education is critical if you want to keep up with the constantly evolving construction industry and advance your career.

Would you recommend construction as a career to others?

The career opportunities offered by the masonry industry are extremely diverse and rewarding. Whatever type of work you enjoy and challenges you seek, there is a career in the masonry industry that will help you achieve your goals. I absolutely recommend exploring a career in masonry to anyone who is interested. My masonry career has had an unbelievably positive impact on my life.

What does craftsmanship mean to you?

There are quite a few things that contribute to craftsmanship: quality, professionalism, efficiency, and reliability. When you boil it all down, though, it's about respecting yourself and your craft enough to do your very best work on every unit, every time, no matter who is watching.

Trade Terms Introduced in This Module

Admixture: A chemical or mineral other than water, cement, or aggregate added to mortar immediately before or during mixing to change its setting time or curing time; to reduce water; or to change the overall properties of the mortar.

Adobe: Sun-dried, molded clay brick.

Aggregate: Materials such as crushed stone or gravel used as a filler in concrete and concrete block.

American Society for Testing and Materials (ASTM) International: The publisher of masonry standards.

Ashlar: A squared or rectangular cut stone masonry unit; or, a flat-faced surface having sawed or dressed bed and joint surfaces.

Bed joint: The horizontal joint between two masonry units in separate courses.

Butter: To apply mortar to a masonry unit prior to laying it.

Capital: The top part of an architectural column.

Competent person: An individual who is capable of identifying existing and predictable hazards or working conditions that are hazardous, unsanitary, or dangerous to employees, and who has authorization to take prompt, corrective measures to eliminate or control these hazards and conditions.

Concrete masonry unit (CMU): A hollow or solid block made from portland cement and aggregates.

Cored: Brick that has holes extending through it to reduce weight.

Cornice: The horizontal projection crowning the wall of a building.

Course: A row or horizontal layer of masonry units.

Cube: A strapped bundle of approximately 500 standard brick, or 90 standard block. The number of units in a cube will vary according to the manufacturer.

Facing: That part of a masonry unit or wall that shows after construction; the finished side of a masonry unit.

Footing: The base for a masonry unit wall, or concrete foundation, that distributes the weight of the structural member resting on it.

Grout: A mixture of portland cement, lime, and water, with or without fine aggregate, with a high-enough water content that it can be poured into spaces between masonry units and voids in a wall.

Head joint: The vertical joint between two masonry units.

Joints: The area between each brick or block that is filled with mortar.

Manufactured stone veneer: A premade veneer consisting of cast cementitious material with pigments and other added materials to give the appearance of natural stone. Also called adhered concrete masonry veneer (ACMV).

Mason: A person who assembles masonry units by hand, using mortar, dry stacking, or mechanical connectors.

Masonry unit: Any building block made of brick, cement, ashlar, clay, adobe, rubble, glass, tile, or any other material that can be assembled into a structural unit.

Mortar: A mixture of portland cement, lime, fine aggregate, and water, plastic or stiff enough to hold its shape between masonry units.

Nonstructural: Not bearing weight other than its own.

Occupational Safety and Health Administration (OSHA): The division of the US Department of Labor mandated to ensure a safe and healthy environment in the workplace.

Parapet: A low wall or railing.

Personal protective equipment (PPE): Equipment or clothing designed to prevent or reduce injuries.

Pilaster: A square or rectangular pillar projecting from a wall.

Spread: A row of mortar placed into a bed joint.

Stringing: Spreading mortar with a trowel on a wall or footing for a bed joint.

Structural: Bearing weight in addition to its own.

Tuckpointing: Filling fresh mortar into cutout or defective joints in masonry.

Uncored: Brick that has no holes extending through it.

Weephole: A small opening in mortar joints or faces to allow the escape of moisture.

Wythe: A continuous section of masonry wall, one masonry unit in thickness, or that part of a wall that is one masonry unit in thickness.

Additional Resources

This module presents thorough resources for task training. The following reference material suggested for further study.

ASTM C270, *Standard Specification for Mortar for Unit Masonry*. 2012. West Consohocken, PA: ASTM International.

Bricklaying: Brick and Block Masonry. Reston, VA: Brick Industry Association.

Building Block Walls: A Basic Guide for Students in Masonry Vocational Training. 1988. Herndon, VA: National Concrete Masonry Association.

Concrete Masonry Handbook. Skokie, IL: Portland Cement Association.

Concrete Masonry Shapes and Sizes Manual CD-ROM. Herndon, VA: National Concrete Masonry Association.

Hot & Cold Weather Masonry Construction Manual. 1999. Herndon, VA: National Concrete Masonry Association.

Installation Guide for Adhered Concrete Masonry Veneer, Third Edition. 2012. Washington, DC: Masonry Veneer Manufacturers Association.

SkillsUSA Professional Development Resources. **http://www.skillsusa.org/store/curricula.html**

Figure Credits

Courtesy of National Concrete Masonry Association, Figure 1, Figure 2, Figure 4, Figure 7

Provided by Haskell, SA03

Courtesy of Portland Cement Association, Figure 3, Figure 5

Courtesy of NRG Insulated Block, Figure 6 (top)

COURTESY OF RAE PARAVIA, LV, NV, Figure 6 (bottom)

Courtesy of the Brick Industry Association, Figure 8, Figure 9, Figure 11, E01

Hanson Brick, Figure 10, SA05, SA06, RQ01

Courtesy of Cultured Stone® by Boral®, Figure 14

Adapted, with permission, from C270-12a Standard Specification for Mortar for Unit Masonry, copyright ASTM International, 100 Barr Harbor Drive, West Conshohocken, PA 19428. A copy of the complete standard may be obtained from ASTM International, **www.astm.org.**, T01

West Chester Holdings, Inc., Figure 17

Courtesy of Honeywell Safety Products, Figure 18 (top right, bottom left)

Courtesy of MSA, Figure 19 (top, middle), E02

Anna Mead, Figure 21

Bon Tool Co., Figure 22, E03

The QUIKRETE Companies, Figure 23

Courtesy of Calculated Industries, SA07

Section Review Answer Key

Answer	Section Reference	Objective
Section One		
1. b	1.1.0	1a
2. c	1.2.0	1b
3. d	1.3.0	1c
4. c	1.4.0	1d
5. a	1.5.0	1e
Section Two		
1. b	2.1.0	2a
2. d	2.2.0	2b
Section Three		
1. c	3.1.0	3a
2. a	3.2.3	3b
Section Four		
1. c	4.1.2	4a
2. b	4.2.3	4b
3. d	4.3.3	4c
4. b	4.4.3	4d
Section Five		
1. c	5.1.0	5a
2. d	5.2.0	5b
3. d	5.3.0	5c

NCCER CURRICULA — USER UPDATE

NCCER makes every effort to keep its textbooks up-to-date and free of technical errors. We appreciate your help in this process. If you find an error, a typographical mistake, or an inaccuracy in NCCER's curricula, please fill out this form (or a photocopy), or complete the online form at **www.nccer.org/olf**. Be sure to include the exact module ID number, page number, a detailed description, and your recommended correction. Your input will be brought to the attention of the Authoring Team. Thank you for your assistance.

Instructors – If you have an idea for improving this textbook, or have found that additional materials were necessary to teach this module effectively, please let us know so that we may present your suggestions to the Authoring Team.

NCCER Product Development and Revision

13614 Progress Blvd., Alachua, FL 32615

Email: curriculum@nccer.org
Online: www.nccer.org/olf

❑ Trainee Guide ❑ Lesson Plans ❑ Exam ❑ PowerPoints Other ____________

Craft / Level: Copyright Date:

Module ID Number / Title:

Section Number(s):

Description:

Recommended Correction:

Your Name:

Address:

Email: Phone:

28106-13

Masonry Safety

Overview

In this module, you will learn how to ensure your safety and that of the people you work with by following safe work practices and procedures, inspecting safety equipment before use, and using safety equipment properly. Your employer is obligated to maintain a safe workplace for all employees, however, the ultimate responsibility for on-the-job safety rests with you.

Module Two

Trainees with successful module completions may be eligible for credentialing through NCCER's National Registry. To learn more, go to **www.nccer.org** or contact us at **1.888.622.3720**. Our website has information on the latest product releases and training, as well as online versions of our *Cornerstone* magazine and Pearson's product catalog.

Your feedback is welcome. You may email your comments to **curriculum@nccer.org**, send general comments and inquiries to **info@nccer.org**, or fill in the User Update form at the back of this module.

This information is general in nature and intended for training purposes only. Actual performance of activities described in this manual requires compliance with all applicable operating, service, maintenance, and safety procedures under the direction of qualified personnel. References in this manual to patented or proprietary devices do not constitute a recommendation of their use.

28106-13 MASONRY SAFETY

Objectives

When you have completed this module, you will be able to do the following:

1. Understand the importance of safety on a job site.
   a. Identify the costs of job accidents.
   b. Identify the causes of accidents.
   c. Recognize hazards.
   d. Demonstrate proper housekeeping techniques.
   e. Observe mortar and concrete safety.
   f. Observe flammable liquid safety.
2. Recognize the proper use of personal protective equipment in masonry.
   a. Describe how to use protective lenses and face shields.
   b. Describe how to use hearing protection.
   c. Describe how to use gloves.
   d. Describe how to use respirators.
3. Explain how to work safely from elevated surfaces.
   a. Explain fall protection procedures.
   b. Describe personal fall arrest systems.
   c. List basic scaffold safety guidelines.
   d. Explain how to protect against falling objects.
4. Describe how to use tools and equipment safely.
   a. Describe how to use hand tools safely.
   b. Describe how to use saws safely.
   c. Describe how to use mixers safely.
   d. Describe how to use grinders safely.
   e. Describe how to work safely around forklifts.
   f. List basic electrical safety guidelines.
   g. Describe how to use powder-actuated tools safely.
5. Explain how to handle materials properly.
   a. Describe how to store and stockpile masonry materials.
   b. Describe how to stack brick.

Performance Tasks

Under the supervision of your instructor, you should be able to do the following:

1. Properly use the following personal protective equipment:
   - Face shield
   - Protective lenses
   - Hearing protection
   - Respirator
2. Properly use the following personal protective equipment:
   - Safety harness
   - Connector
   - Anchor point

Trade Terms

Arresting force
Bundled
Carabiner
Caustic
Deceleration device
Deceleration distance
Flammable
Free-fall distance
Hygroscopic
Management system
Nitrile
Personal protective equipment (PPE)
Potable
Respirator
Safety Data Sheet (SDS)
Scaffold
Silicosis
Working stack

Industry-Recognized Credentials

If you're training through an NCCER-accredited sponsor, you may be eligible for credentials from NCCER's Registry. The ID number for this module is 28106-13. Note that this module may have been used in other NCCER curricula and may apply to other level completions. Contact NCCER's Registry at 888.622.3720 or go to **www.nccer.org** for more information.

Contents

Topics to be presented in this module include:

Figures and Tables

SECTION ONE

1.0.0 JOB-SITE SAFETY

Objective

Understand the importance of safety on a job site.

a. Identify the costs of job accidents.
b. Identify the causes of accidents.
c. Recognize hazards.
d. Demonstrate proper housekeeping techniques.
e. Observe mortar and concrete safety.
f. Observe flammable liquid safety.

Trade Terms

Caustic: Capable of causing chemical burns.

Flammable: Capable of easily igniting and rapidly burning; used to describe a fuel with a flash point below 100°F.

Hygroscopic: Having a high initial rate of moisture absorption.

Management system: The organization of a company's management, including reporting procedures, supervisory responsibility, and administration.

Personal protective equipment (PPE): Equipment or clothing designed to prevent or reduce injuries.

Potable: Water that is safe for cooking and drinking.

Respirator: A device that provides clean, filtered air for breathing, no matter what is in the surrounding air.

Safety data sheet (SDS): A document that must accompany any hazardous substance. The SDS identifies the substance and gives the exposure limits, the physical and chemical characteristics, the kind of hazard it presents, precautions for safe handling and use, and specific control measures. Formerly known as material safety data sheet (MSDS).

Scaffold: An elevated platform for workers and materials.

Silicosis: A respiratory disease caused by the inhalation of silica dust.

Masons operate in a high-risk environment. All around them are stacks of materials, trucks, and heavy equipment. Work sites have many possibilities for accidents. Workers themselves can cause accidents. They can drop masonry or tools off **scaffold** and onto other workers. They can assemble scaffold so poorly that it collapses under the weight of the load. They can fall off scaffold. They can use damaged or poorly maintained tools that can injure themselves or others.

You must think and practice safety at all times. Your work must be planned so that it is safe as well as efficient. All workers at a construction site must wear appropriate **personal protective equipment (PPE)** to protect their skin and eyes from mortar, grout, and flying masonry chips. They also need to protect themselves by being aware of what is happening around them. Workers need to keep track of the rest of their crew and of other crews. Unusual movements or noises can indicate something is moving that should not be.

Masons need to have the knowledge, skill, and ability to do the following:

- Recognize an unsafe situation.
- Alert fellow workers to the danger.
- Take action to avoid or correct the situation.

1.1.0 Identifying the Costs of Job Accidents

When an accident happens, everyone loses—the injured worker, the employer, and the insurance company. Accidents cost billions of dollars each year and cause much needless suffering. This section examines why accidents happen and how you can help prevent them.

Unsafe working conditions and practices can result in the following:

- Personal injury or death
- Injury or death of other workers
- Damage to equipment
- Damage to the work site

Insurance may cover some of the costs, but there are hidden (indirect or uninsured) costs as well (*Figure 1*). Some examples of indirect costs might include:

- Damage to a wall due to the careless operation of a forklift

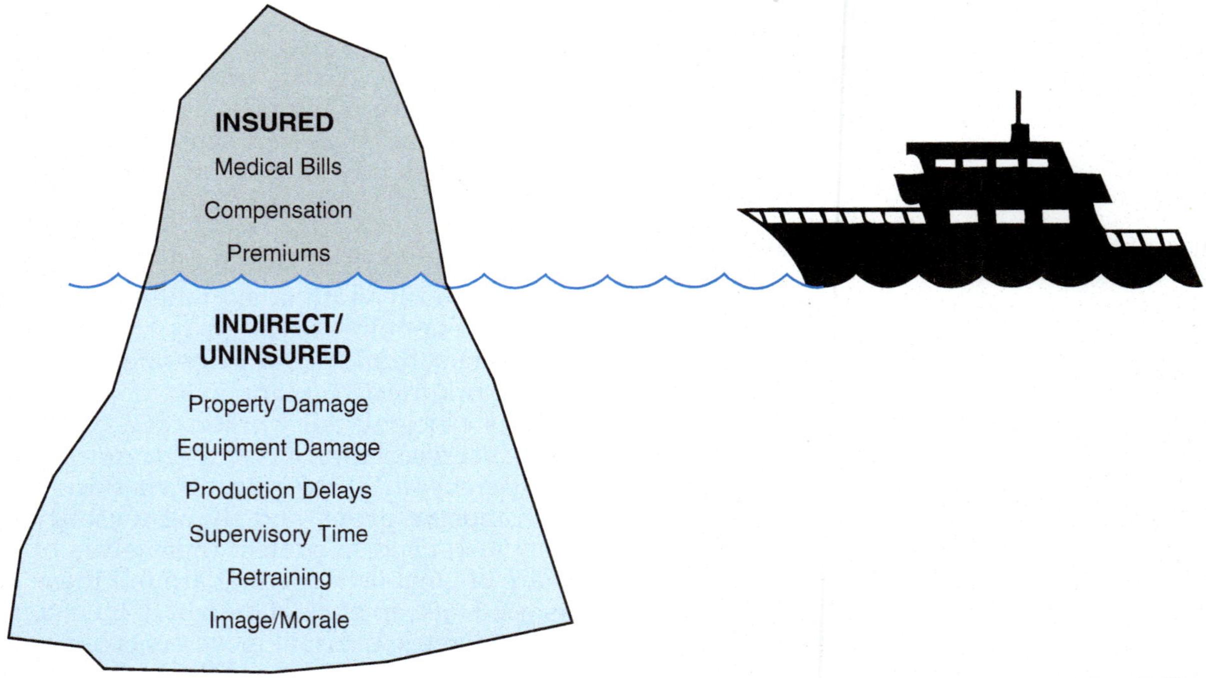

Figure 1 Hidden costs of accidents.

- Burning out a mortar mixer's engine due to improper maintenance
- Delays to a project as a result of injuries sustained due to recklessness
- Excessive supervisory costs incurred by the need to address poor workmanship
- Delays caused by safety training due to repeated improper equipment operation
- Lowered morale from picking up the slack caused by a co-worker who is chronically absent because of an accident
- Loss of family income due to hospitalization due to carelessness on the job
- Damage to a company's reputation because of unprofessional conduct on the job site

In addition to the pain and suffering for the individuals involved and their families, accidents can also affect others For example, accidents can slow down or stop a job, thereby putting the entire operation in jeopardy, and possibly resulting in site-wide layoffs. A high accident rate can cause an employer's insurance rates to rise, making the company less competitive with other construction companies and less likely to secure future work.

1.2.0 Identifying the Causes of Accidents

As you learned in the *Core Curriculum*, some of the main causes of accidents include the following:

- Failure to communicate
- Poor work habits
- Alcohol or drug abuse
- Lack of skill
- Intentional acts
- Unsafe acts
- Rationalizing risks
- Unsafe conditions
- Failure of the **management system**

1.2.1 Failing to Communicate

Many accidents happen because of a lack of communication. For example, you may learn how to

GOING GREEN

Sustainable Practices on Job Sites

Concrete landscape products are naturally environmentally sustainable. They don't require the use of large machinery to deliver or place them, which prevents pollution and damage to the ground, and helps reduce the development footprint. Not only that, but permeable concrete pavers help reduce stormwater runoff and open-cell concrete products permit the growth of grass and plants.

do things one way on one job, but what happens when you go to a new job site? You need to communicate with the people at the new job site to find out whether they do things the way you have learned to do them. If you do not communicate clearly, accidents can happen. Remember that different people, companies, and job sites do things in different ways.

NOTE

Toolbox talks are one way to effectively keep all workers aware and informed of safety issues and guidelines. Toolbox talks are short 5- to 10-minute meetings that review specific health and safety topics. Keep a signed and dated log of each training meeting.

If you think that people know something without talking with them about it, then you are assuming that they know. Assuming that other people know and will do what you think they will do can cause accidents.

CAUTION

Never assume anything! It never hurts to ask questions, but disaster can result if you don't ask. For example, do not assume that an electrical current is turned off. First ask whether the current is turned off, then check it yourself to be completely safe.

All work sites have specific markings and signs to identify hazards and provide emergency information (*Figure 2*). Learn to recognize these types of signs:

- Informational
- Safety
- Caution
- Danger
- Temporary warnings

Informational markings or signs provide general information. These signs are blue. The following are considered informational signs:

- No Admittance
- No Trespassing
- For Employees Only

Safety signs give general instructions and suggestions about safety measures. The background on these signs is white; most have a green panel with white letters. These signs tell you where to find such important areas as the following:

- First-aid stations
- Emergency eye wash stations
- Evacuation routes

Figure 2 Communications signs.

- Safety data sheet (SDS) stations
- Exits (usually have white letters on a red field)

Caution markings or signs tell you about potential hazards or warn against unsafe acts. When you see a caution sign, protect yourself against a possible hazard. Caution signs are yellow and have a black panel with yellow letters. They may give you the following information:

- Hearing and eye protection are required
- Respirators are required
- Smoking is not allowed

Danger markings or signs tell you that an immediate hazard exists and that you must take certain precautions to avoid an accident. Danger signs are red, black, and white. They may indicate the presence of the following:

- Defective equipment
- Flammable liquids and compressed gases

WARNING!

Always use the proper containers for storing flammable liquids and compressed gases. Improper containers may burst or leak, causing contamination, fires, and injury or death.

> **WARNING!**
> Never store acids and flammable materials near each other.

- Safety barriers and barricades
- Emergency stop button
- High voltage

Safety tags are temporary warnings of immediate and potential hazards. They are not designed to replace signs or to serve as permanent means of protection. Learn to recognize the standard accident prevention tags (*Table 1*).

1.2.2 Poor Work Habits

Poor work habits can cause serious accidents. Examples of poor work habits are procrastination, carelessness, and horsing around. Procrastination, or putting things off, is a common cause of accidents. For example, delaying the repair, inspection, or cleaning of equipment and tools can cause accidents. If you try to push machines and equipment beyond their operating capacities, you risk injuring yourself and your co-workers.

Machines, power tools, and even a pair of pliers can hurt you if you don't use them safely. It is your responsibility to be careful. Tools and machines don't know the difference between wood or steel and flesh and bone.

Work habits and work attitudes are closely related. If you resist taking orders, you may also resist listening to warnings. If you let yourself be easily distracted, you won't be able to concentrate. If you aren't concentrating, you could cause an accident.

Your safety is affected not only by how you do your work, but also by how you act on the job site. This is why most companies have strict policies for employee behavior. Horsing around and other inappropriate behavior are forbidden. Workers who engage in foolish and inappropriate behavior on the job site will be fired.

These strict policies are for your protection. There are many hazards on construction sites. Each person's behavior—at work, on a break, or at lunch—must follow the principles of safety.

If you fool around on the job, play pranks, or don't concentrate on what you are doing, you are showing a poor work attitude that can lead to a serious accident.

1.2.3 Alcohol and Drug Abuse

Alcohol and drug abuse costs the construction industry millions of dollars a year in accidents, lost time, and lost productivity. The true cost of alcohol and drug abuse is much more than just money, of course. Substance abuse can cost lives. Just as drunk driving kills thousands of people on our highways every year, alcohol and drug abuse can have deadly results on the construction site.

Using alcohol or drugs creates a risk of injury for everyone on a job site. Many states have laws that prevent workers from collecting insurance benefits if they are injured while under the influence of alcohol or illegal drugs.

Would you trust your life to a crane operator who was high on drugs? Would you bet your life on the responses of a co-worker using alcohol or drugs? Alcohol and drug abuse have no place in the construction industry. A person on a construction site who is under the influence of alcohol or drugs is an accident waiting to happen—possibly a fatal accident.

People who work while using alcohol or drugs are at risk of accident or injury, and their co-workers are at risk as well. That's why your employer probably has a formal substance abuse policy.

Table 1 Tags and Signs

Basic Stock (background)	Safety colors (ink)	Message(s)
White	Red panel with white or gray letters	Do Not Operate Do Not Start
White	Black square with a red oval and white letters	Danger Unsafe Do Not Use
Yellow	Black square with yellow letters	Caution
White	Black square with white letters	Out of Order Do Not Use
Yellow	Red/magenta (purple) panel with black letters and a radiation symbol	Radiation Hazard
White	Fluorescent orange square with black letters and a biohazard symbol	Biological Hazard

28106-13_T01.EPS

You should know that policy and follow it for your own safety.

You don't have to be abusing illegal drugs such as marijuana, cocaine, or heroin to create a job hazard. Many prescription and over-the-counter drugs, taken for legitimate reasons, can affect your ability to work safely. Amphetamines, barbiturates, and antihistamines are only a few of the legal drugs that can affect your ability to work or operate machinery safely.

> **CAUTION**
>
> If your doctor prescribes any medication that you think might affect your job performance, ask about its effects. Your safety and the safety of your co-workers depend on everyone being alert on the job.

Do yourself and the people you work with a big favor. Be aware of and follow your employer's substance abuse policy. Avoid any substances that can affect your job performance. The life you save could be your own.

1.2.4 Lack of Skill

You should learn and practice new skills under careful supervision. Never perform new tasks alone until your supervisor allows it.

Lack of skill can cause accidents quickly. For example, suppose you are told to cut some masonry units with a saw, but you aren't skilled with that tool. A basic rule of power saw operation is never to cut without a properly functioning guard. Because you haven't been trained, you don't know this. You find that the guard on the saw is slowing you down. So you jam the guard open with a small block of wood. The result could be a serious accident. Proper training can prevent this type of accident.

1.2.5 Intentional Acts

When someone purposefully causes an accident, it is called an intentional act. Sometimes an angry or dissatisfied employee may purposefully create a situation that leads to property damage or personal injury. If someone you are working with threatens to get even or pay back someone, let your supervisor know at once.

1.2.6 Unsafe Acts

An unsafe act is a change from an accepted, normal, or correct procedure that usually causes an accident. It can be any conduct that causes unnecessary exposure to a job-site hazard or that makes an activity less safe than usual. Here are examples of unsafe acts:

- Failing to use personal protective equipment (PPE)
- Failing to warn co-workers of hazards
- Lifting improperly
- Loading or placing equipment or supplies improperly
- Making safety devices (such as saw guards) inoperable
- Operating equipment at improper speeds
- Operating equipment without authority
- Servicing equipment in motion
- Taking an improper working position
- Using defective equipment
- Using cell phones or texting while working
- Wearing ear buds or earphones
- Operating equipment while tired or without enough sleep
- Using equipment improperly

Stress

Stress creates a chemical change in your body. Although stress may heighten your hearing, vision, energy, and strength, long-term stress can harm your health.

Not all stress is job related; some stress develops from the pressures of dealing with family and friends and daily living. In the end, your ability to handle and manage your stress determines whether stress hurts or helps you. Use common sense when you are dealing with stressful situations. For example, consider the following:

- Keep daily occurrences in perspective. Not everything is worth getting upset, angry, or anxious about.
- When you have a particularly difficult workday scheduled, get plenty of rest the night before.
- Manage your time. The feeling of always being behind creates a lot of stress. Waiting until the last minute to finish an important task adds unnecessary stress.
- Talk to your supervisor. Your supervisor may understand what is causing your stress and may be able to suggest ways to manage it better.

> **WARNING!**
> Dull blades cause more accidents than sharp ones. If you do not keep your cutting tools sharpened, they won't cut very easily. When you have a hard time cutting, you exert more force on the tool. When that happens, something is bound to slip. And when something slips, you can get cut.

1.2.7 Rationalizing Risk

Everybody takes risks every day. When you get in your car to drive to work, you know there is a risk of being involved in an accident. Yet when you drive using all the safety practices you have learned, you know that there is a good chance that you will arrive at your destination safely. Driving is an appropriate risk because you have some control over your own safety and that of others.

Some risks are not appropriate. On the job, you must never take risks that endanger yourself or others just because you can make an excuse for doing so. This is called rationalizing risk. Rationalizing risk means ignoring safety warnings and practices. For example, because you are late for work, you might decide to run a red light. By trying to save time, you could cause a serious accident.

The following are common examples of rationalized risks on the job:

- Crossing boundaries because no activity is in sight
- Not wearing gloves because it will take only a minute to make a cut
- Removing your hard hat because you are hot and you cannot see anyone working overhead
- Not tying off your fall protection because you only have to lean over by about a foot

Think about the job before you do it. If you think that it is unsafe, then it is unsafe. Stop working until the job can be done safely. Bring your concerns to the attention of your supervisor. Your health and safety, and that of your co-workers, make it worth taking extra care.

1.2.8 Unsafe Conditions

An unsafe condition is a physical state that is different from the acceptable, normal, or correct condition found on the job site. It usually causes an accident. It can be anything that reduces the degree of safety normally present. The following are some examples of unsafe conditions:

- Congested workplace
- Defective tools, equipment, or supplies
- Excessive noise
- Fire and explosive hazards
- Hazardous atmospheric conditions (such as gases, dusts, fumes, and vapors)
- Inadequate supports or guards
- Inadequate warning systems
- Poor housekeeping
- Poor lighting
- Poor ventilation
- Radiation exposure
- Unguarded moving parts such as pulleys, drive chains, and belts

1.2.9 Management System Failure

Sometimes the cause of an accident is failure of the management system. The management system should be designed to prevent or correct the acts and conditions that can cause accidents. If the management system did not do these things, that system failure may have caused the accident.

What traits could mean the difference between a management system that fails and one that succeeds? A company implementing a good management system will:

- Put safety policies and procedures in writing
- Distribute written safety policies and procedures to each employee
- Review safety policies and procedures periodically
- Enforce all safety policies and procedures fairly and consistently
- Evaluate supplies, equipment, and services to see whether they are safe
- Provide weekly, documented safety meetings
- Provide regular, periodic safety training for employees, and provide safety mentoring for inexperienced workers

1.3.0 Recognizing Hazards

Construction sites may contain numerous hazards. You need to walk and work with all due respect for those hazards. Be aware of, and always follow, the local applicable code and all applicable safety guidelines for working on site. The following list includes some, but not all, of the hazardous conditions at a typical job site:

- Improper ventilation
- Inadequate lighting
- Uncapped exposed rebar
- High noise levels
- Slippery floors
- Unmarked low ceilings

- Excavations, holes, and open, unguarded spaces, including open, unbarricaded elevator shafts
- Poorly constructed or poorly rigged scaffolds
- Improperly stacked materials
- Live wires, loose wires, and extension cords
- Unsafe ladders and access points
- Unsafe crane or heavy-equipment operations
- Water and mud
- Unsafe storage of hazardous or flammable materials
- Unlabeled or improperly labeled containers
- Buried power and utility lines
- Overhead power lines
- Defective or unsafe tools and equipment
- Poor housekeeping

Your safety and that of your fellow workers should be a primary consideration in your work life. Some common-sense rules and ways of doing things can make the job site safer for everyone. These safety tips should be a part of your everyday thinking. Develop a positive safety attitude. It will keep you and your co-workers safe and sound.

1.4.0 Demonstrating Proper Housekeeping Techniques

In construction, housekeeping means keeping your work area clean and free of scraps or spills. It also means being orderly and organized. You must store your materials and supplies safely and label them properly. Arranging your tools and equipment to permit safe, efficient work practices and easy cleaning is also important. As veteran craftworkers say, "Plan the work, then work the plan."

Weather Hazards

Masons usually work outdoors. Under certain environmental conditions, such as extreme hot or cold weather, work can become uncomfortable and possibly dangerous. There are specific things to be aware of when working under these adverse conditions.

If you live in a place with cold weather, you will most likely be exposed to it when working. Spending long periods of time in the cold can be dangerous. It's important to know the symptoms of cold-weather exposure and how to treat them. Symptoms of cold exposure include shivering, numbness, low body temperature, and drowsiness. The amount of injury caused by exposure to abnormally cold temperatures depends on wind speed, length of exposure, temperature, and humidity. Freezing is increased by wind, humidity, or a combination of the two factors. Follow these guidelines to prevent injuries during extremely cold weather:

- Always wear the proper clothing.
- Limit your exposure as much as possible.
- Ask permission to take frequent, short rest periods.
- Keep moving. Exercise fingers and toes if necessary, but do not overexert.
- Do not drink alcohol before exposure to cold. Alcohol can dull your sensitivity to cold and make you less aware of overexposure.
- Do not expose yourself to extremely cold weather if any part of your clothing or body is wet.
- Do not smoke before exposure to cold. Breathing can be difficult in extremely cold air. Smoking can worsen the effect.
- Learn how to recognize the symptoms of overexposure.
- Place cold hands under dry clothing against the body, such as in the armpits.

Hot weather can be as dangerous as cold weather. When someone is exposed to excessive amounts of heat, they run the risk of overheating. Conditions associated with overheating include heat cramps, heat exhaustion, and heat stroke. Follow these guidelines when working in hot weather in order to prevent heat exhaustion, cramps, or heat stroke:

- Drink plenty of potable (drinkable) water.
- Do not overexert yourself.
- Wear lightweight clothing.
- Keep your head covered and face shaded.
- Ask to take frequent, short work breaks.
- Rest in the shade whenever possible.

Other hazards that masonry workers can encounter outdoors include lightning, which can strike even when it's not storming, and high winds, which can blow over green walls even if they are properly braced.

If the work site is indoors, make sure it is well lit and ventilated. Don't allow aisles and exits to be blocked by materials and equipment. Make sure that flammable liquids are stored in safety cans. Oily rags must be placed only in approved, self-closing metal containers.

Remember that the primary goal of housekeeping is to prevent accidents. Good housekeeping reduces the chances for slips, fires, explosions, and falling objects. Cleanup is also about looking after yourself and your fellow workers. It is much easier to push a wheelbarrow up a ramp that is clear of debris than one that is bumpy and rough. Walking along a clear path doesn't use up as much energy as it takes to step over and around rubble and rocks, or to slog through mud.

Here are some good housekeeping rules:

- Remove from work areas all scrap material and lumber with nails protruding.
- Clean up and properly dispose of leftover mortar.
- Clean up spills to prevent falls.
- Remove all combustible scrap materials regularly.
- Make sure you have containers for the collection and separation of refuse. Containers for flammable or harmful refuse must be covered.
- Dispose of wastes often.
- Store all tools and equipment when you're finished using them.

Another term for good housekeeping is pride of workmanship. If you take pride in what you are doing, you won't let trash build up around you. The saying "a place for everything and everything in its place" is the right idea on the job site.

1.5.0 Observing Mortar and Concrete Safety

Another hazard encountered by masonry craftworkers is exposure to mortar, grout, and concrete. These cement-based materials have ingredients that can hurt your eyes or skin. The basic ingredient of mortar, portland cement, is **caustic** in nature. It also **hygroscopic**, which means it has a high initial rate of water absorption and it can absorb moisture from your skin. Prolonged contact between the fresh mix and skin can cause skin irritation and chemical burns to hands, feet, and exposed skin areas. It can also saturate a worker's clothes and transmit caustic or hygroscopic effects to the skin. In addition, the sand contained in fresh mortar can cause skin abrasions through prolonged contact. Inhaling dust or sand that contains silica, a natural crystal commonly found in sand, can cause a respiratory condition called **silicosis**. Silicosis causes scarring and inflammation (swelling) of the lungs, and the symptoms include chronic coughing and shortness of breath, and low levels of oxygen in the blood. Silicosis can be caused by short-term exposure to large amounts of silica dust, or by long-term exposure to small amounts of silica dust.

WARNING!

Whether wet or dry, mortar and grout can be harmful. Both contain cement, which is caustic. Dry cement dust can enter open wounds and cause blood poisoning. The cement dust, when it comes in contact with body fluids, can cause chemical burns to the membranes of the eyes, nose, mouth, throat, or lungs. It can also cause a fatal lung disease known as silicosis.

Wet mortar and grout, as well as wet concrete, can also cause chemical burns to the eyes and skin. Always wear appropriate personal protective equipment when working with dry cement or wet concrete. If wet concrete enters waterproof boots from the top, remove the boots and rinse your legs, feet, boots, and clothing with potable water as soon as possible. Repeated contact with cement or wet concrete can also cause an allergic skin reaction known as cement dermatitis.

WARNING!

Silicosis is a serious lung disease that is caused by inhaling sand dust. Silica is a major component of sand and is therefore present in concrete products and mortar. Silica dust is released when cutting brick and cement, especially when dry-cutting with a power saw. Any time you are involved in the cutting or demolition of concrete or masonry materials, be sure to wear approved respiratory equipment.

Avoid injuries by taking the following precautions:

- Hold the trowel so as to keep your thumb out of the mortar.
- Keep cement products off your skin at all times by wearing the proper protective clothing, including boots, gloves, and clothing with snug wristbands, ankle bands, and a cooling neckband when it is appropriate for the local weather conditions. Make sure they are all in good condition.

- Prevent your skin from rubbing against cement products. Rubbing increases the chance of serious injury. If your skin does come in contact with any cement products, wash your skin promptly. If a reaction persists, seek medical attention.
- Wash thoroughly as soon as possible to prevent skin damage or contamination from cement dust.
- Keep cement products out of your eyes by wearing safety lenses when mixing mortar or pumping grout. If any cement or cement mixtures get in your eye, flush it immediately and repeatedly with clean **potable** water, and consult a physician promptly.
- Rinse off any clothing that becomes saturated from contact with fresh mortar. A prompt rinse with potable water will prevent continued contact with skin surfaces.
- Be alert! Watch for trucks backing into position and overhead equipment delivering materials. Listen for the alarms or warning bells on mixers, pavers, and ready-mix trucks.
- Never put your hands, arms, or any tools into rotating mixers.
- Be certain adequate ventilation is provided when using epoxy resins, organic solvents, brick cleaners, and other toxic substances.
- Use good work practices to reduce dust in the air when handling mortar and lime. For example, do not shake out mortar bags unnecessarily and do not use compressed air to blow mortar dust off clothing or a work surface. Whenever possible, stand upwind when dumping mortar bags, so that the dust does not blow back into your face.
- Never use solvents to clean skin.
- Immediately remove epoxy, solvents, and other toxic substances from skin using the appropriate cleansing agents, as outlined in the SDS/MSDS.
- If eye wash stations and emergency showers are provided on the job site, know their locations and how to use them properly.

1.6.0 Observing Flammable Liquid Safety

Flammable liquids are particularly dangerous and require additional safety precautions. Always adhere to the local applicable code, job-site rules, and the following guidelines:

- Carefully read labels on all flammable liquids, and use flammable liquids only in open, well-ventilated areas.
- Do not inhale or ignite fumes from flammable liquids.
- Be sure that all flammable liquids are labeled correctly for storage.
- Store all flammable liquids properly and only in approved safety containers, such as safety cans and safety cabinets.
- Store oily rags in metal containers with self-sealing lids.
- When clothing is soaked by a flammable liquid, immediately change clothing, and cleanse the body with an appropriate cleaner.
- Use flammable liquids only for their intended purposes; for example, never use gasoline as a cleaner.
- Never use flammable liquids near fire or flame.
- Never use starting fluid to start a diesel engine.
- Always be aware of the location of an appropriate fire extinguisher.
- Learn your company's emergency response procedures for fire and explosion.

Additional Resources

OSHA website. **www.osha.gov**

WorkSAFE masonry safety resources. **www.worksafecenter.com/safety/tutorial/masonry/step-1.page**

1.0.0 Section Review

1. The two types of accident costs are _____.
   a. insured and uninsured
   b. covered and uncovered
   c. incidental and accidental
   d. primary and secondary

2. Signs that provide general instructions and suggestions about safety measures are _____.
   a. informational signs
   b. safety signs
   c. caution signs
   d. danger signs

3. On a construction site, exposed rebar, slippery floors, and improperly stacked materials are considered _____.
   a. intentional
   b. acceptable
   c. obstacles
   d. hazards

4. When working indoors, never place oily rags in anything other than approved _____.
   a. open containers
   b. self-closing metal containers
   c. self-sealing plastic bags
   d. buckets or pails

5. The basic ingredient of mortar is _____.
   a. portland cement
   b. silica
   c. grout
   d. ready-mix concrete

6. Flammable liquids should never be used near _____.
   a. potable or unpotable water
   b. block and brick
   c. fire or flame
   d. sand and mortar

SECTION TWO

2.0.0 PERSONAL PROTECTIVE EQUIPMENT

Objective

Recognize the proper use of personal protective equipment in masonry.

a. Describe how to use protective lenses and face shields.
b. Describe how to use hearing protection.
c. Describe how to use gloves.
d. Describe how to use respirators.

Performance Task 1

Properly use the following personal protective equipment:

- Face shield
- Protective lenses
- Hearing protection
- Respirator

Trade Term

Nitrile: A synthetic rubberlike material used in masonry gloves to protect hands while permitting a tactile response.

In general, the employer is responsible to the Occupational Safety and Health Administration (OSHA) for making sure that all employees are wearing appropriate personal protective equipment whenever those employees are exposed to possible hazards to their safety. In turn, you are responsible for wearing the gear and clothing assigned to you. *Figure 3* shows a mason properly dressed and equipped for many common masonry tasks.

It is important to take the following safety precautions when dressing for masonry work:

- Confine long hair in a ponytail or in your hard hat. Flying hair can obscure your view or get caught in machinery.
- Wear appropriate clothing and personal protective equipment.
- Always wear a hard hat.
- Wear goggles when cutting or grinding.
- Wear a high-visibility vest.
- Wear close-fitting clothing, including long-sleeved shirts (minimum 4-inch sleeves) to give extra protection if skin is sensitive.
- Wear gloves when working with wet mortar.
- Wear pants over boots to avoid getting mortar on legs or feet and keep your shirt tucked in. Your belt should be above your hips and your pants should not be sagging.
- Keep gloves and clothing as dry as possible.
- Wear face and eye protection as required, especially if there is a risk from flying particles, debris, or other hazards such as brick dust or proprietary cleaners.
- Wear hearing protection as required.
- Wear respiratory protection as required.
- Protect any exposed skin by applying skin cream, body lotion, or petroleum jelly.
- Wear sturdy work boots or work shoes with thick soles. Never show up for work dressed in sneakers, loafers, or sport shoes.
- Wear fall protection equipment as required.

2.1.0 Using Protective Lenses and Face Shields

Wear eye protection such as safety glasses, goggles, lenses and face shields whenever there is even the slightest chance of an eye injury (*Figure 4*). Areas where there are potential eye hazards from falling or flying objects are usually identified, but you should always be on the lookout for other possible hazards, such as sewage and pressurized water.

There are different types of eye protection. Safety goggles give your eyes the best protection from all directions. Regular safety lenses will protect you from objects flying at you from the front, such as large chips, particles, sand, or dust. You can add side shields for further protection, or wear wraparound lenses. Do not wear contact lenses because they will not protect your eyes. You may, however, substitute prescription safety lenses if they provide the same protection as regular safety lenses.

CAUTION

Never wear tinted lenses indoors.

2.2.0 Using Hearing Protection

Damage to most parts of the body causes pain. Ear damage, however, does not always cause pain. Most companies follow OSHA's rules regarding when workers must use hearing protection. Specially designed earplugs that fit into your ears and filter out noise are one type of hearing protection

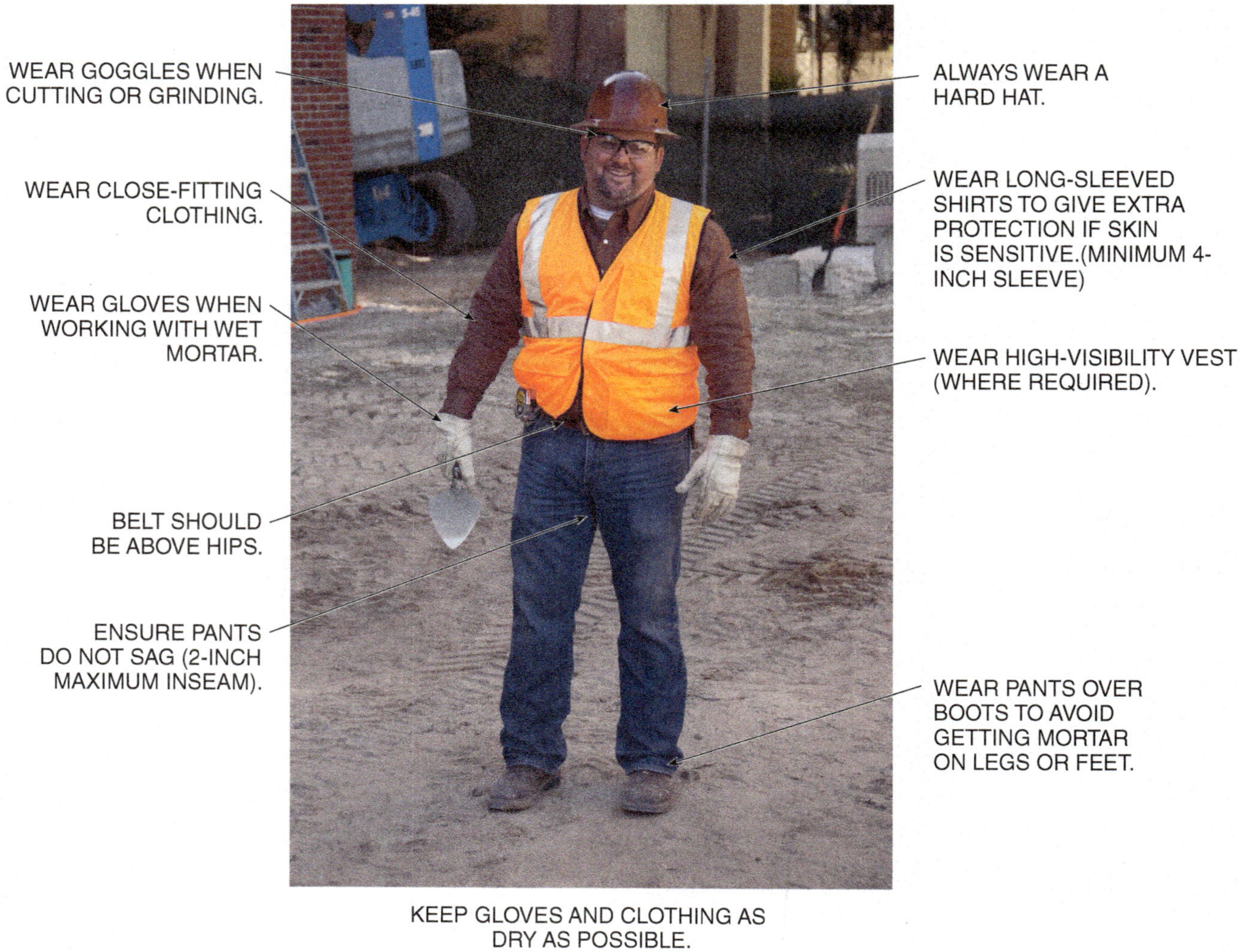

28106-13_F03.EPS

Figure 3 Dressed for masonry work.

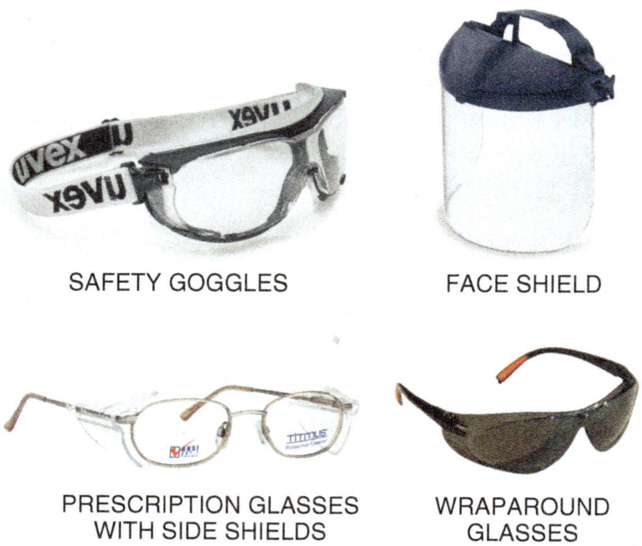

28106-13_F04.EPS

Figure 4 Typical safety goggles and lenses.

(see *Figure 5*). When reusing earplugs, you must clean them regularly to prevent ear infections. Because earplugs are made in different sizes, make sure you use earplugs that fit properly.

WARNING!

Exposure to loud noise over a long period can cause hearing loss, even if the noise is not loud enough to cause pain.

Another type of hearing protection is earmuffs, which are large, padded covers for the entire ear (see *Figure 6*). You must adjust the headband on earmuffs for a snug fit. If the noise level is very high, you may need to wear both earplugs and earmuffs.

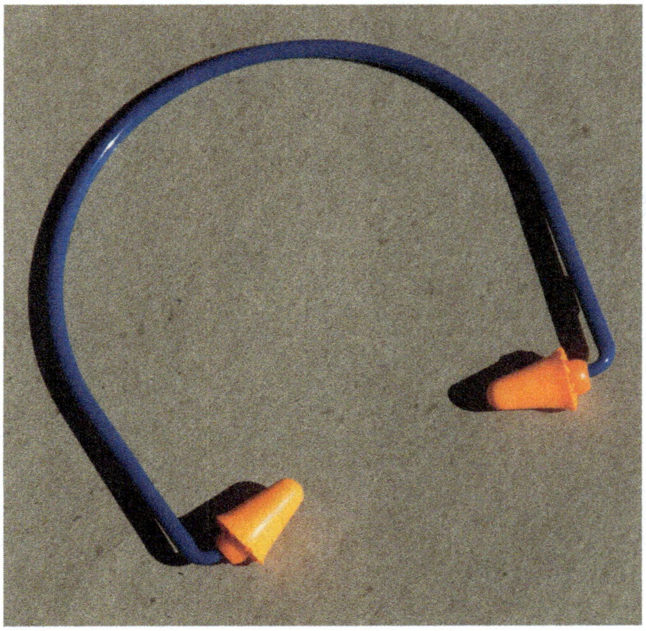

28106-13_F05.EPS

Figure 5 Earplugs.

28106-13_F06.EPS

Figure 6 Earmuffs.

2.3.0 Using Gloves

On many jobs, you must wear gloves to protect your hands from cuts, abrasion, dust and moisture, burns caused by mortar and portland cement, and other hazards (see *Figure 7*). Gloves used in masonry work often have a tightly knit fabric or nylon shell with palms and fingertips made from latex or a synthetic rubber called **nitrile**. Latex and nitrile allow you to keep your sense of touch while wearing the gloves, while still being able to grip effectively and protect your hands.

Occupational Safety and Health Administration (OSHA) regulations (*29 CFR 1926.95*) require workers to wear gloves when working with or around portland cement and other hazardous materials. Inspect your gloves every day to ensure that they are in good condition. When they are worn, damaged, or contaminated on the inside, exchange them for new gloves. Never wear cloth gloves around rotating or moving equipment. Immediately replace gloves that are worn or torn, or that no longer fit.

WARNING!

Wet portland cement can cause caustic burns that can cause blisters, dead skin, and discolored skin. Severe burns from portland cement can even reach the bone, causing scars and even disability. Exposure to portland cement can cause an allergic reaction that results in inflammation from repeated exposure.

When choosing gloves, make sure they fit snugly without being too loose or too tight. Always wash and dry your hands before putting on your gloves. If the gloves are reusable, clean them after each use. If they are disposable, follow the manufacturer's instructions for disposing of them properly.

2.4.0 Using Respirators

As you learned in the module *Introduction to Masonry*, respirators must be worn wherever there is danger of suffocation or other breathing hazards, when working with or near fire-resistant asbestos, or where hazardous molds are growing. Mandatory special training is required for the use of respirators. Always use the proper type of respiratory protection for the hazard. You must pass a cardiopulmonary fitness test before using respiratory protection. Always conduct a fit test with

28106-13_F07.EPS

Figure 7 Gloves for masonry work.

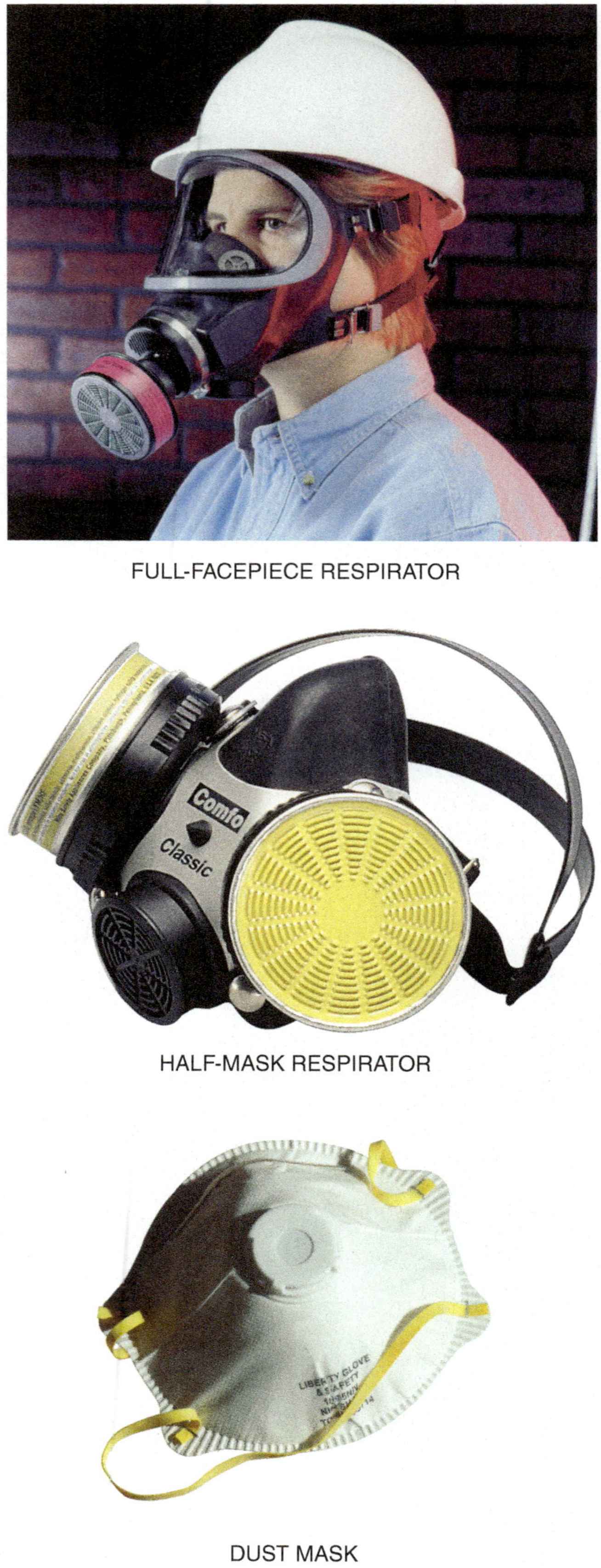

Figure 8 Reusable respirators for masonry work.

the respiratory protection equipment prior to use to ensure that it fits properly.

> **CAUTION**
>
> OSHA regulations require that workers undergo a cardiopulmonary test before using respiratory protection. Refer to your local applicable standards.

Reusable respirators for masonry work are available in three types: full-facepiece, half-mask, and dust masks (*Figure 8*). Full-facepiece and half-face respirators require the replacement of cartridges, filters, and respirator parts. Their use also requires a complete respirator maintenance program. Half-mask air-purifying respirators have several limitations, so be sure to refer to the manufacturer's instructions before using. Dust masks should be discarded after a single use. Dust masks provide light-duty respiratory protection and should be discarded after a single use.

WARNING!

EPA's *Renovation, Repair & Painting Final Rule* (*40 CFR 745*) requires that home renovations conducted for compensation, must be performed by certified firms using certified renovators if the home was built prior to 1978, if it qualifies as a child-occupied facility, and if the renovation disturbs over 6 square feet of lead paint per interior room area or 20 square feet of exterior area per wall. All masons working on pre-1978 homes are affected by this rule and need to be aware of their surroundings and test for the presence of lead in their work area.

WARNING!

If you encounter asbestos or mold, do not handle it or attempt to remove it. Tell your supervisor about the presence of these materials immediately. Asbestos and mold are linked to long-term illnesses. Only trained personnel who have the proper equipment for handling and disposing of these materials can deal with such situations.

WARNING!

Silicosis is a serious lung disease that is caused by inhaling sand dust. Silica is a major component of sand and is therefore present in concrete products and mortar. Silica dust is released when cutting brick and cement, especially when dry-cutting with a power saw. Any time you are involved in the cutting or demolition of concrete or masonry materials, be sure to wear approved respiratory equipment.

Additional Resources

"The Sense of Safety." K. K. Snyder. *Masonry Magazine*, March 2008. **www.masonrymagazine.com/3-08/safety.html**

Personal Protective and Life Saving Equipment. OSHA. 29 CFR 1926 Subpart E.

2.0.0 Section Review

1. The best protection for your eyes from all directions is provided by _____.
   a. safety lenses
   b. safety goggles
   c. face shields
   d. respirators

2. Pain is not always felt when an injury happens to the _____.
   a. feet
   b. hands
   c. eyes
   d. ears

3. Gloves should be exchanged for new ones when they are worn, damaged, or _____.
   a. contaminated on the outside
   b. made from nitrile
   c. contaminated on the inside
   d. stiff

4. An example of a scenario in which respiratory protection is *not* required on a masonry job site is _____.
   a. where hazardous molds are growing
   b. when mixing mortar, grout, or concrete
   c. when working with or near fire-resistant asbestos
   d. wherever there is danger of suffocation or other breathing hazards

SECTION THREE

3.0.0 WORKING SAFELY FROM ELEVATED SURFACES

Objective

Explain how to work safely from elevated surfaces.

a. Explain fall protection procedures.
b. Describe personal fall arrest systems.
c. List basic scaffold safety guidelines.
d. Explain how to protect against falling objects.

Performance Task 2

Properly use the following personal protective equipment:

- Safety harness
- Connector
- Anchor point

Trade Terms

Arresting force: The force needed to stop a person from falling. The greater the free-fall distance, the more force is needed to stop or arrest the fall.

Carabiner: A coupling link fitted with a safety closure.

Deceleration device: A device such as a shock-absorbing lanyard or self-retracting lifeline that brings a falling person to a stop without injury.

Deceleration distance: The distance it takes before a person comes to a stop when falling. The required deceleration distance for a fall arrest system is a maximum of 3½ feet.

Free-fall distance: The vertical distance a worker moves after a fall before a deceleration device is activated.

Falls from elevated areas are one of the leading causes of fatalities among construction workers. Falls from elevated heights account for approximately ⅓ of all deaths in the construction trade. Most workers injured in falls lose time from work; many of them require hospitalization; and some never return to the job.

While the risk of falls is high in construction, there is much you can do to safeguard yourself. Using the appropriate personal protective equipment, following proper safety procedures, practicing good housekeeping habits, and staying alert at all times will help you stay safe when working at an elevation.

Scaffolds provide safe elevated work platforms for people and materials. They are designed and built to comply with high safety standards, but normal wear and tear or accidentally putting too much weight on them can weaken them and make them unsafe. That's why it is important to inspect every part of a scaffold before each use.

3.1.0 Understanding Fall Protection Procedures

Fall protection is required when workers are exposed to falls from work areas with elevations that are 6 feet or higher. The types of work areas that put the worker at risk include the following:

- Scaffolds
- Ladders
- Leading edges
- Ramps or runways
- Wall or floor openings
- Roofs
- Excavations, pits, and wells
- Concrete forms
- Unprotected sides and edges

Falls happen because of the inappropriate use or lack of fall protection systems. They also happen because of worker carelessness (*Figure 9*). It is your responsibility to learn how to set up, inspect, use, and maintain your own fall protection equipment. Not only will this keep you alive and uninjured, it could save the lives of your co-workers.

Falls are classified into two groups: falls from an elevation and falls on the same level. Falls from an elevation can happen when you are doing work from scaffold, work platforms, decking, concrete forms, ladders, or excavations. Falls from elevations are almost always fatal. This is not to say that falls on the same level aren't also extremely dangerous. When a worker falls on the same level, usually from tripping or slipping, head injuries often occur. Sharp edges and pointed objects such as exposed rebar could cut or stab the worker.

The following safe practices can help prevent slips and falls:

- Wear safe, strong work boots that are in good repair.
- Watch where you step. Be sure your footing is secure.
- Install cables, extension cords, and hoses so that they will not become tripping hazards.
- Do not allow yourself to get in an awkward position. Stay in control of your movements at all times.

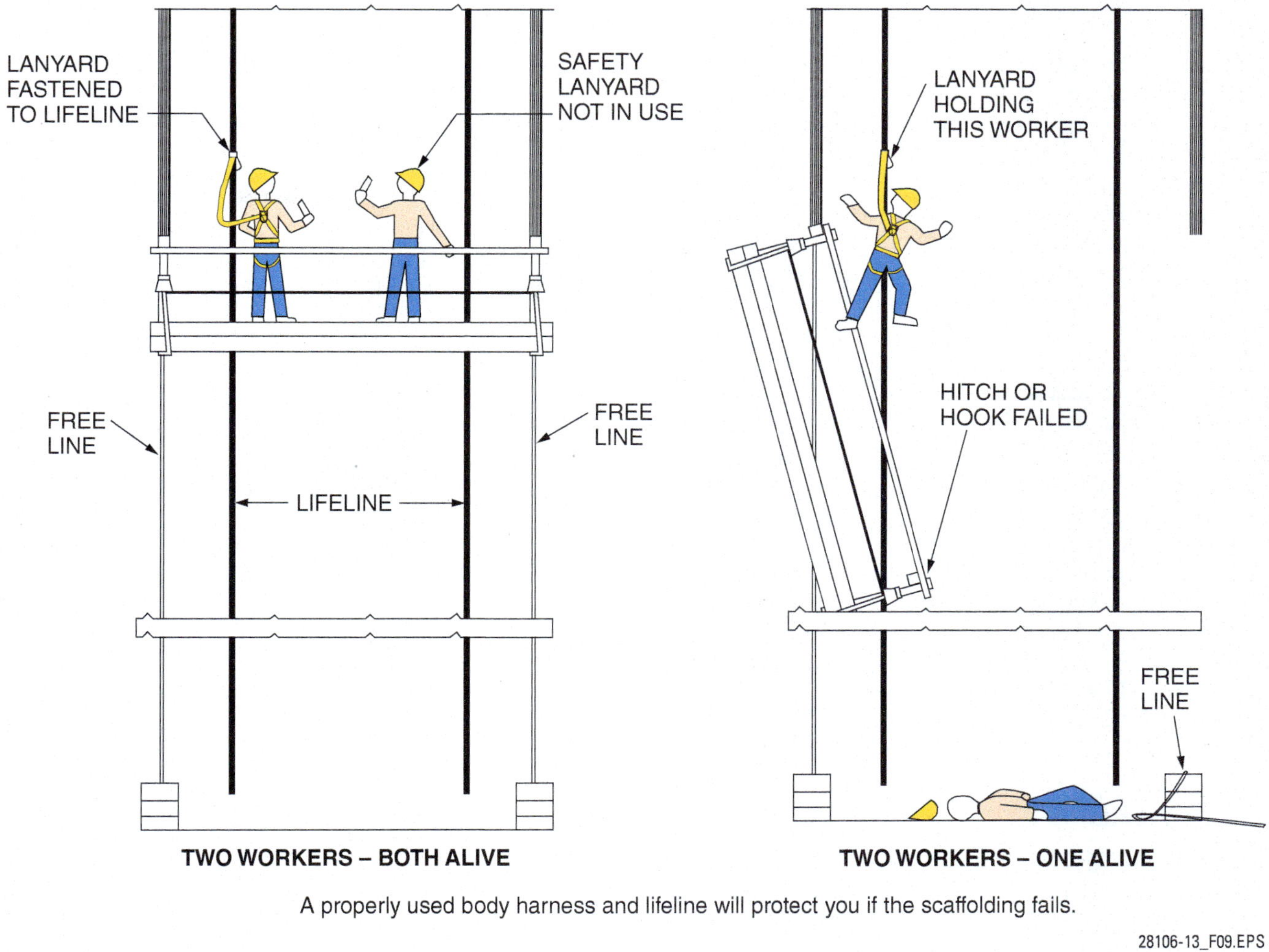

Figure 9 Proper and improper safety harness use.

- Maintain clean, smooth walking and working surfaces. Fill holes, ruts, and cracks. Clean up slippery material and litter.
- Do not run on scaffold, work platforms, decking, roofs, or other elevated work areas.

The best way to survive a fall from an elevation is to use fall protection equipment. The three most common types of fall protection equipment are guardrails, personal fall arrest systems, and safety nets.

3.2.0 Using Personal Fall Arrest Systems

Personal fall arrest systems catch workers after they have fallen. They are designed and rigged to prevent a worker from free falling a distance of more than 6 feet and hitting the ground or a lower work area. When describing personal fall arrest systems, these terms must be understood:

- **Free-fall distance** – The vertical distance a worker moves after a fall before a deceleration device is activated.
- **Deceleration device** – A device such as a shock-absorbing lanyard or self-retracting lifeline that brings a falling person to a stop without injury.
- **Deceleration distance** – The distance it takes before a person comes to a stop when falling. The required deceleration distance for a fall arrest system is a maximum of 3½ feet.
- **Arresting force** – The force needed to stop a person from falling. The greater the free-fall distance, the more force is needed to stop or arrest the fall.

Personal fall arrest systems use specialized equipment. This equipment includes the following:

- Body harnesses
- Lanyards
- Deceleration devices
- Lifelines
- Anchoring devices and equipment connectors

3.2.1 Body Harnesses

Full-body harnesses with a sliding D-ring in the back (*Figure 10*) are used in personal fall arrest systems. They are made of straps that are worn securely around the user's body. This allows the arresting force to be distributed throughout the body, including the shoulders, legs, torso, and buttocks. This distribution decreases the chance of injury. When a fall occurs, the sliding D-ring moves to the nape of the neck. This keeps the worker in an upright position and helps to distribute the arresting force. The worker then stays in a relatively comfortable position while waiting for rescue. Selecting the right full-body harness depends on a combination of job requirements and personal preference. Harness manufacturers normally provide selection guidelines in their product literature.

Personal-positioning systems allow workers to hold themselves in place, keeping their hands free to accomplish a task. A personal-positioning system should not allow a worker to free-fall more than 2 feet. The anchor point should be able to support at least twice the impact load of a worker's fall or 5,000 pounds, whichever is greater.

3.2.2 Lanyards

Lanyards are short, flexible lines with connectors on each end. They are used to connect a body harness or body belt to a lifeline, deceleration device, or anchor point. There are many kinds of lanyards made for different uses and climbing situations. All must have a minimum breaking strength of 5,000 pounds. They come in both fixed and adjustable lengths and are made out of steel, rope, or nylon webbing. Some have a shock absorber (*Figure 11*) that absorbs up to 80 percent of the arresting force when a fall is being stopped. When choosing a lanyard for a particular job, always follow the manufacturer's recommendations. Never connect two lanyards together end to end.

28106-13_F11.EPS

Figure 11 Lanyard with a shock absorber.

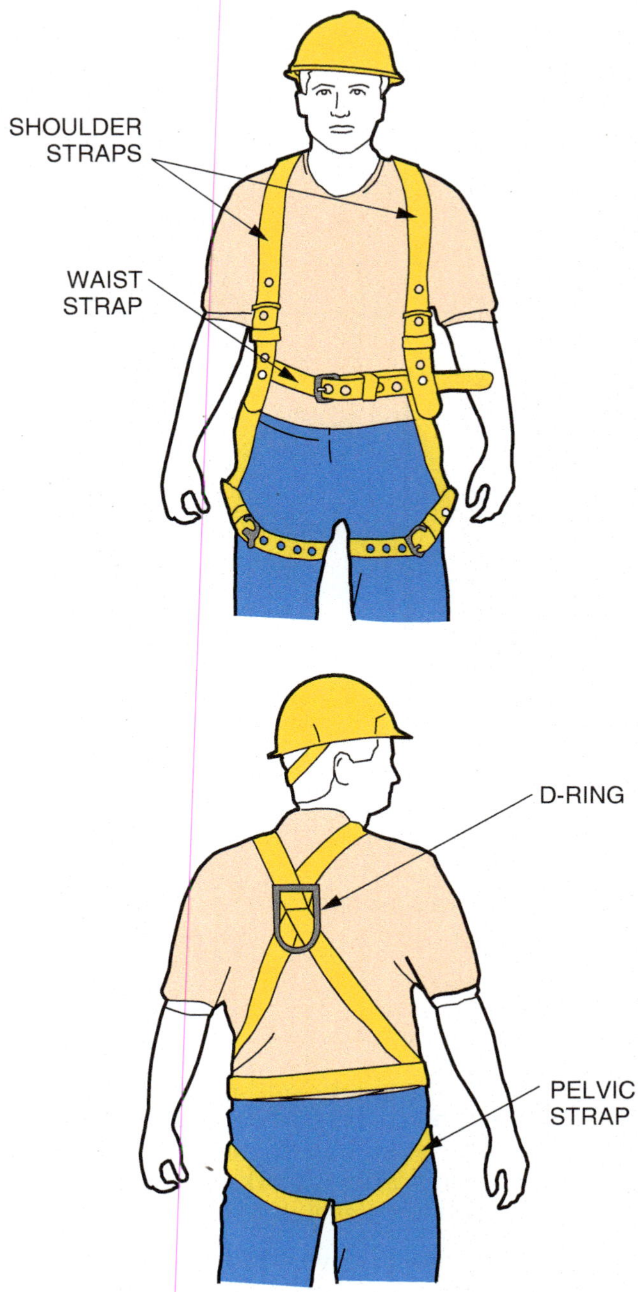

28106-13_F10.EPS

Figure 10 Full-body harnesses with a sliding D-rings in the back.

Fall Protection

Most workers who die from falls were wearing harnesses but had failed to tie off properly. Always follow the manufacturer's instructions when wearing a harness. Know and follow your company's safety procedures when working on roofs, ladders, and other elevated locations.

NOTE

In the past, body belts were often used strictly as a fall arrest system. As of January 1, 1998, however, they have been banned from such use. This is because body belts concentrate all of the arresting force in the abdominal area, which can cause significant injuries. It also causes the worker to hang in an uncomfortable and potentially dangerous position while awaiting rescue.

WARNING!

When activated during the fall arresting process, a shock-absorbing lanyard stretches in order to reduce the arresting force. This potential increase in length must always be taken into consideration when determining the total free-fall distance from an anchor point.

3.2.3 Deceleration Devices

Deceleration devices limit the arresting force to which a worker is subjected when the fall is stopped suddenly. Rope grabs and self-retracting lifelines are two common deceleration devices (*Figure 12*). A rope grab connects to a lanyard and attaches to a lifeline. In the event of a fall, the rope grab is pulled down by the attached lanyard, causing it to grip the lifeline and lock in place. Some rope grabs have a mechanism that allows the worker to unlock the device and slowly descend down the lifeline to the ground or surface below.

Self-retracting lifelines provide unrestricted movement and fall protection while workers are climbing and descending ladders and similar equipment or when working on multiple levels. Typically, they have a 25- to 100-foot galvanized-steel cable that automatically takes up the slack in the attached lanyard, keeping the lanyard out of the worker's way. In the event of a fall, a centrifugal braking mechanism engages to limit the worker's free-fall distance. Self-retracting lifelines and lanyards must be able to hold a tensile load of at least 5,000 pounds.

Figure 12 Rope grab and self-retracting lifeline.

3.2.4 Lifelines

Lifelines are ropes or flexible steel cables that are attached to an anchor point. They provide a means for tying off personal fall protection equipment. Vertical lifelines (*Figure 13*) are suspended vertically from a fixed anchor point. A fall arrest device such as a rope grab is attached to the lifeline. Vertical lifelines must have a minimum breaking strength of 5,000 pounds. Each worker must use his or her own line. This is because if one worker falls, the movement of the lifeline during the fall arrest may also cause the other workers to fall. A vertical lifeline must be connected in a way that will keep the worker from moving past its end, or it must extend to the ground or the next lower working level.

Case History

Icy Scaffold Can Be Deadly

A laborer was working on the third level of a tubular welded-frame scaffold. It was covered with ice and snow. Planking on the scaffold was weak, and a guardrail had not been set up. The worker slipped and fell headfirst approximately 20 feet to the pavement below.

The Bottom Line: Never work on a wet or icy scaffold. Make sure all scaffold is sturdy and includes the proper guardrails.

Source: Occupational Safety and Health Administration (OSHA)

28106-13_F13.EPS

Figure 13 Vertical lifeline.

Horizontal lifelines (*Figure 14*) are connected horizontally between two fixed anchor points. These lifelines must be designed, installed, and used under the supervision of a qualified, competent person. The more workers who are tied off to a single horizontal line, the stronger the line and anchors must be. Horizontal lines must be able to support a minimum tensile load of 5,000 pounds per person attached to the line.

3.2.5 Anchor Points and Equipment Connectors

Anchor points, commonly called tie-off points, support the entire weight of the fall arrest system. The anchor point must be capable of supporting 5,000 pounds for each worker attached. Eye bolts (*Figure 15*) and overhead beams are considered anchor points to which fall arrest systems are attached.

28106-13_F14.EPS

Figure 14 Horizontal lifeline.

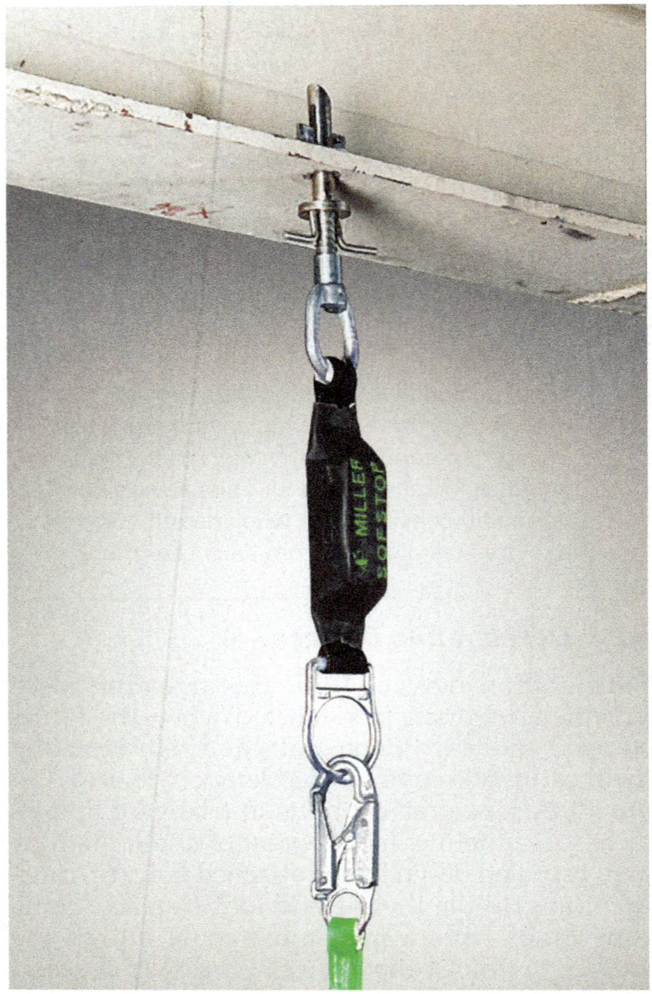

28106-13_F15.EPS

Figure 15 Push-through eye bolt and shock absorber.

The D-rings, buckles, **carabiners**, and snap hooks (*Figure 16*) that fasten and/or connect the parts of a personal fall arrest system are called connectors. There are regulations that specify how they are to be made and that require D-rings and snap hooks to have a minimum tensile strength of 5,000 pounds.

> **NOTE**
> Since January 1, 1998, only locking-type snaphooks are permitted for use in personal fall arrest systems.

3.2.6 Selecting an Anchor Point and Tying Off

Connecting the body harness either directly or indirectly to a secure anchor point is called tying off. Tying off is always done before you get into a

Figure 16 Carabiners and locking snap hook.

position from which you can fall. Follow the manufacturer's instructions on the best tie-off methods for your equipment.

In addition to the manufacturer's instructions, an anchor point should be as follows:

- Directly above the worker
- Easily accessible
- Damage free and capable of supporting 5,000 pounds per worker
- High enough so that no lower level is struck should a fall occur

Be sure to check the manufacturer's equipment labels, and allow for any equipment stretch and deceleration distance.

3.2.7 Using Personal Fall Arrest Equipment

Before using fall protection equipment on the job, you must know the basics and proper usage of fall protection equipment. All equipment supplied by your employer must meet established standards for strength. Before each use, always read the instructions and warnings on any fall protection equipment. Inspect the equipment using the following guidelines:

- Examine harnesses and lanyards for mildew, wear, damage, and deterioration.
- Ensure no straps are cut, broken, torn, or scraped.
- Check for damage from fire, chemicals, or corrosives.
- Make sure all hardware is free of cracks, sharp edges, and burrs.
- Check that snap hooks close and lock tightly and that buckles work properly.
- Check ropes for wear, broken fibers, pulled stitches, and discoloration.
- If the equipment was used in a previous fall, remove it from service until it can be inspected by a qualified person.
- Make sure lifeline anchor points and mountings are not loose or damaged.

WARNING!

Never use fall protection equipment that shows signs of wear or damage.

Safety Net Systems

Safety nets are used for fall protection on bridges and similar projects. They must be installed as close as possible, but not more than 30 feet, beneath the work area. There must be enough clearance under a safety net to prevent a worker who falls into it from hitting the surface below. There must also be no obstruction between the work area and the net.

Depending on the actual vertical distance between the net and the work area, the net must extend 8 to 13 feet beyond the edge of the work area. Mesh openings in the net must be limited to 36 square inches and 6 inches on the side. The border rope must have a 5,000-pound minimum breaking strength, and connections between net panels must be as strong as the nets themselves. Safety nets must be inspected at least once a week and after any event that might have damaged or weakened them. Worn or damaged nets must be removed from service.

Safety nets should be drop-tested at the job site after the initial installation, whenever relocated, after a repair, and at least every six months if left in one place. The drop test consists of a 400-pound bag of sand of 29 to 31 inches in diameter that is dropped into the net from at least 42 inches above the highest walking/working surface at which workers are exposed to fall hazards. If the net is still intact after the bag of sand is dropped, it passes the test.

Do not mix or match equipment from different manufacturers. All substitutions must be approved by your supervisor. All damaged or defective parts must be taken out of service immediately and either tagged as unusable or else destroyed.

3.2.8 Rescue After a Fall

Every elevated job site is required to have an established rescue and retrieval plan. Planning is especially important in remote areas where help is not readily available. Before beginning work, make sure that you know what your employer's rescue plan calls for you to do in the event of a fall. Find out what rescue equipment is available and where it is located. Learn how to use equipment for self-rescue and the rescue of others.

If a fall occurs, any employee hanging from the fall arrest system must be rescued safely and quickly. Your employer should have previously determined the method of rescue for fall victims, which may include equipment that lets the victim rescue himself or herself, a system of rescue by co-workers, or a way to alert a trained rescue squad. If a rescue depends on calling for outside help such as the fire department or rescue squad, all the needed phone numbers must be posted in plain view at the work site. In the event a coworker falls, follow your employer's rescue plan. Call any special rescue service needed. Communicate with the victim, and monitor him or her constantly during the rescue.

3.3.0 Understanding Scaffold Safety

Masons spend a great deal of time working from scaffold (*Figure 17*). The chance of an accident occurring because of poor work practices is greater on scaffold than at any other time or place. Tubular steel scaffold is less likely to give way than the old-fashioned wood scaffold but requires more care in its assembly and use.

Tubular welded-frame scaffold is used in accessible places with fairly level ground conditions. It consists of one or more manufactured platforms supported by welded-end frame sections, horizontal members called bearers, and intermediate members. The members are manufactured in assorted heights and widths that are joined with horizontal and diagonal braces and secured by pins. The braces have a fixed length, which aligns the vertical members so that the erected scaffold always remains plumb, square, and rigid. Additional braces and frames can be used to extend the length of a tubular welded-frame scaffold. Likewise, end frames can be stacked on top of each other to increase the height of the scaffold. To do this, slide the bottoms of the legs of the upper end frames into the tops of the legs of the lower end frames, and secure them with drop locks and coupling pins.

Scaffold provides safe elevated work platforms for people and materials. Scaffold is designed and built to comply with high safety standards, but normal wear and tear or accidentally putting too much weight on it can weaken the scaffold and make it unsafe. That's why it is important to inspect every part of a scaffold system before each use. You will learn more about setting up and using scaffold in the module *Masonry Tools and Equipment*.

CAUTION

Only a competent person has the authority to supervise setting up, moving, and taking down scaffold. Only a competent person can approve the use of scaffold on the job site after inspecting the scaffold.

WARNING!

Never unlock the wheel brakes of a rolling scaffold while anyone is on it. People on a moving scaffold can lose their balance and fall.

The federal safety rules focusing on worker safety when using scaffold are summarized here. These rules are only reminders for the safe assembly and use of scaffold.

- Platforms on all working levels must be fully decked between the front uprights and the guardrail supports.
- Never exceed the weight-bearing capacity of a scaffold or its components, which OSHA has specified as the ability to support, without failure, the scaffold's own weight plus at least four times the maximum intended load.

Case History

Death Due to Unguarded Protruding Steel Bar

A laborer fell approximately 8 feet through a roof opening to a foundation that had about 20 half-inch rebars protruding straight up. The laborer was impaled on one of the bars and died.

The Bottom Line: Even a short-distance fall can be fatal. Use a personal fall protection system, and check the area for potential hazards.

Source: Occupational Safety and Health Administration (OSHA)

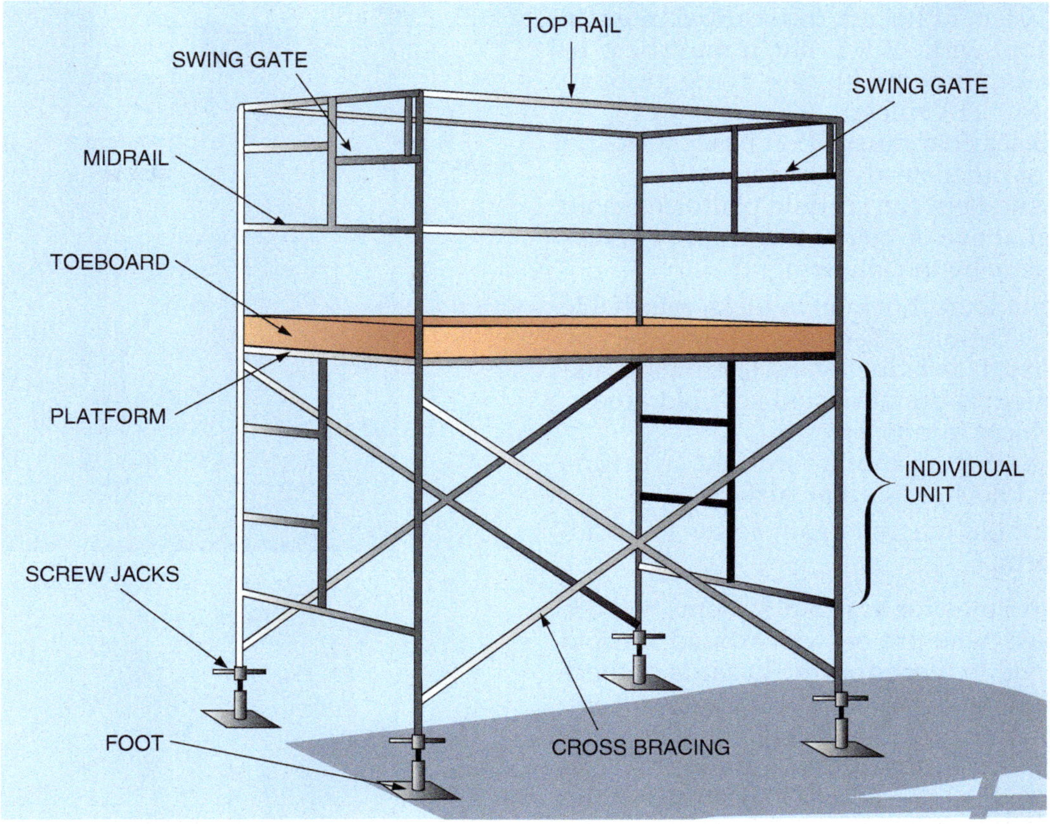

28106-13_F17.EPS

Figure 17 Completed tubular steel scaffold.

- The space between planks and the platform and uprights can be no more than 1 inch wide.
- Ensure that there is sufficient clearance between scaffolds and power lines. *Table 2* lists the proper clearances as defined by OSHA.
- Platforms and walkways must be at least 18 inches wide; the ladder jack, top plate bracket, and pump jack scaffolds must be at least 12 inches wide.

Table 2 Power-Line Clearances for Scaffolds

	Insulated Lines	
Voltage	**Minimum distance**	**Alternatives**
Less than 300 volts	3 feet (0.9 m)	
300 volts to 50 kV	10 feet (3.1 m)	
More than 50 kV	10 feet (3.1 m) plus 0.4 inches (1.0 cm) for each 1 kV over 50 kV.	2 times the length of the line insulator, but never less than 10 feet (3.1 m).
	Uninsulated Lines	
Voltage	**Minimum distance**	**Alternatives**
Less than 50 kV	10 feet (3.1 m)	
More than 50 kV	10 feet (3.1 m) plus 0.4 inches (1.0 cm) for each 1 kV over 50 kV.	2 times the length of the line insulator, but never less than 10 feet (3.1 m).

Exception: Scaffolds and materials may be closer to power lines than specified above where such clearance is necessary for performance of work, and only after the utility company, or electrical system operator, has been notified of the need to work closer and the utility company, or electrical system operator, has de-energized the lines, relocated the lines, or installed protective coverings to prevent accidental contact with the lines.

28106-13_T02.EPS

- For every 4 feet of height, the scaffold must be at least 1 foot wide. If it is not, it must be protected from tipping by guying, tying, or bracing per the OSHA rules.
- Supported scaffold must sit on baseplates and mud sills or other steady foundations.
- Access to and between scaffold platforms more than 2 feet above or below the point of access must be made by the following:
  - Portable ladders, hook-on ladders, attachable ladders, scaffold stairways, stairway-type ladders (such as ladder stands), ramps, walkways, integral prefabricated scaffold access, or equivalent means; or
  - Direct access from another scaffold, structure, personnel hoist, or similar surface
- Never use cross braces to gain access to a scaffold platform.

The OSHA rules for scaffold systems require that all workers who use or work around scaffold must be trained to recognize the hazards of these systems, including proper erection and use. All employees who work on scaffold must be trained by a person qualified in scaffold hazards and uses, so they understand the procedures to control or minimize any hazards. All personnel who erect, take down, move, operate, maintain, repair, or inspect scaffold must be trained by a competent person.

Guardrails (*Figure 18*) protect workers by providing a barrier between the work area on scaffold and the ground or lower work areas. Install guardrails on all scaffold erected 6 feet or higher from the ground. Guardrails may be made of wood, pipe, steel, or wire rope, and must be able to support 200 pounds of force applied to the top rail. They should be at least two inches by four inches thick, and should include both a top rail and a midrail between 36 and 42 inches above the planking. Open scaffold sides also require the installation of four-inch toeboards. Mesh screens running from the toeboards to the guardrails should be installed if people will be working or walking underneath the scaffold.

28106-13_F18.EPS

Figure 18 Guardrails.

Planking used as floorboards on scaffold must be able to support its own weight and at least four times the intended load. Planks can be made from solid sawn wood and fabricated planking or platform materials if the manufacturer, lumber grading association, or the inspection agency has approved them for such use.

As noted in the list above, scaffold must be guyed, tied, or braced to a rigid structure if its height exceeds four times the narrowest dimension of its base. The lowest and highest guys, ties, or braces should be installed at the closest horizontal member to the 4:1 height and repeated vertically in between. To prevent movement, built-up scaffolds must be secured to the structure at vertical intervals not to exceed 20 feet for scaffolds with a base 3 feet wide or less, and 26 feet for scaffolds with a base over 3 feet wide. The horizontal interval is 30 feet regardless of the width of the base.

3.4.0 Protecting against Falling Objects

Falling objects are a real danger on the job site. Follow these guidelines to stay safe when working around overhead hazards:

- Always wear a hard hat.
- Keep the working area clear by removing excess mortar, broken or scattered masonry units, and all other materials and debris on a regular basis.
- Keep openings in floors covered. When guardrail systems are used to prevent materials from falling from one level to another, any openings must be small enough to prevent the passage of potential falling objects.
- Do not store materials other than masonry and mortar within 4 feet of the working edges of a guardrail system.
- Be very careful around operating cranes. Stay clear of the crane's working area.
- Never work or walk under loads that are being hoisted by a crane.
- Erect toeboards or guardrail systems to protect yourself from objects falling from higher levels. Toeboards should be erected along the edges of the overhead walking/working surface for a distance that is sufficient to protect the workers below.
- Erect paneling or screening from the walking/working surface or toeboard to the top of a guardrail system's top rail or midrail if tools, equipment, or materials are piled higher than the top edge of the toeboard.
- Raise or lower tools or materials with a rope and bucket or other lifting device. Never throw tools or materials.
- Never put tools or materials down on ladders or in other places where they can fall and injure people below. Before moving a ladder, make sure there are no tools left on it.

Case History

Inspect All Materials

A crew laying brick on the upper floor of a three-story building built a 6-foot platform to connect two scaffolds. The platform was correctly constructed of two 2-inch by 12-inch planks with standard guardrails. One of the planks however, was not scaffold-grade lumber. It also had extensive dry rot in the center. When a bricklayer stepped on the plank, it disintegrated, and he fell 30 feet to his death.

The Bottom Line: Make sure that all planking is sound and secure. Your life depends on it.

Source: Occupational Safety and Health Administration (OSHA)

Additional Resources

"Online Safety Library: Scaffold Safety." Oklahoma State University. **www.ehs.okstate.edu/links/scaffold.htm**

"Scaffolding." OSHA. **www.osha.gov/SLTC/scaffolding/index.html**

Fall Protection and Scaffolding Safety: An Illustrated Guide. 2000. Grace Drennan Ganget. Government Institutes.

3.0.0 Section Review

1. Fall protection is required when workers are at risk of falling from work areas that are elevated at least _____.
   a. 6 feet
   b. 7 feet
   c. 8 feet
   d. 9 feet

2. The vertical distance a worker moves after a fall before a deceleration device is activated is called the _____.
   a. deceleration distance
   b. free-fall distance
   c. arresting distance
   d. harnessing distance

3. OSHA defines the weight-bearing capacity of a scaffold and its components as the ability to support, without failure, the scaffold's own weight plus at least an increase of the maximum intended load of at least _____.
   a. two times
   b. four times
   c. six times
   d. eight times

4. When working on scaffold, the minimum distance that materials other than masonry and mortar can be stored from the working edges of a guardrail system is _____.
   a. 6 feet
   b. 5 feet
   c. 4 feet
   d. 3 feet

SECTION FOUR

4.0.0 TOOL AND EQUIPMENT SAFETY

Objective

Describe how to use tools and equipment safely.

a. Describe how to use hand tools safely.
b. Describe how to use saws safely.
c. Describe how to use mixers safely.
d. Describe how to use grinders safely.
e. Describe how to work safely around forklifts.
f. List basic electrical safety guidelines.
g. Describe how to use powder-actuated tools safely.

Using your tools and equipment safely will help ensure you have a long, productive career. You will learn how to use hand tools, saws, mixers, grinders, forklifts, and electrical devices in other modules in *Masonry Level One*. But first, it's important for you to understand the basic safety principles for each of these types of tools that you will encounter on the job.

Tool and equipment safety can be divided into the following categories:

- Following safe work practices and procedures
- Inspecting tools and equipment before use
- Using tools and equipment correctly
- Keeping tools and equipment clean and properly maintained

Tool safety means using tools properly, keeping them clean, and being careful not to damage them. Damaged tools and equipment can break, cause injuries, and slow down or stop work. To avoid damaging tools, always use the right tool for the job. The following are some rules for avoiding the misuse of tools:

- Do not use a trowel as a hammer.
- Do not hit your level.
- Do not use a hammer as a chisel.
- Do not use your level or trowel as a pry bar.
- Do not use broken or defective tools.
- Do not cut reinforcement with a hammer, chisel, or trowel.

Accidents also happen when you leave tools and equipment in the way of other workers. Store tools safely, where other people cannot trip over them and where the tools will not get damaged. Clean, well-kept tools work better and are safer.

Do not drop or temporarily store tools or masonry units in pathways or around other workers. Stack masonry units neatly, reversing direction with each layer, to avoid the risk of the masonry units toppling over. A higher stack is more likely to tip over.

4.1.0 Using Hand Tools Safely

Tools and equipment must be in good repair to be safe to use. Repair or replace defective tools and equipment immediately. Clean all tools and equipment that touch mortar or grout immediately after use. Use a bucket of water to soak tools that you will use again in a few minutes. Keep wooden handles out of the water. If mortar dries on a tool, it will harden and make the tool unusable. Be sure that you remove all mortar or grout completely.

If you do not use tools regularly, coat them lightly with oil, grease, or other approved coating to prevent rust and corrosion. After cleaning, check that tool handles are secure and free from cracks or splinters. Oil the pivot joints and wooden parts of the tools. Sharpen blades and cutting edges when they become dull or nicked.

Here are some specific hand-tool care tips:

- Keep your mason's hammer sharp and square.
- Keep folding rules oiled.
- Keep chisel blades sharp.
- Keep chisel heads and blades free of burrs.
- Keep all tool handles tight.

4.2.0 Using Saws Safely

Handheld and table-mounted masonry saws cut more accurately than a mason's hammer or brick set. Unlike hand tools, saws do not weaken or fracture the material. Some larger table saws (see *Figure 19*) can be operated with a foot control, which leaves both hands free for guiding the piece to be cut.

> **WARNING!**
> Always keep your hands away from the cutting area behind the blade when operating a saw. The saw can jump backwards suddenly and could cut your hand.

When using handheld and table-mounted saws of any size, review the saw's control operations before starting to cut. The guidelines for the safe use of saws include the following:

- Wear a hard hat and eye protection to guard against flying chips.

28106-13_F19.EPS

Figure 19 Large masonry saw.

- Wear rubber boots and gloves to reduce the chance of electric shock.
- Wear earplugs and/or other ear protection.
- Wear a respirator when using a dry-cut saw.
- Check the guards to ensure they move and close freely. Never operate a saw with damaged or missing guards.
- Check all adjustment levers before cutting to make sure they are set at the correct bevel and depth.
- Always hold the saw firmly when operating and support the work being cut, to prevent loss of control.
- In between cuts, always release the trigger and hold the saw until the blade comes to a complete stop.
- Check that the blade is properly mounted and tightened before starting to cut.
- Make repeated passes using a light, forward pressure to achieve the desired depth. Do not use excessive pressure.
- Clean masonry dust from the saw's air vents frequently. Always disconnect the saw's plug from the power source and let the blade come to a complete stop before cleaning the air vents.

> **WARNING!**
>
> The lower guard may become sluggish when cutting masonry materials. Clean the guards frequently to allow the guard to operate freely.

> **WARNING!**
>
> When starting a saw, hold it tightly with both hands. Otherwise, the torque from the motor could twist the saw out of your grasp.

Many saws are gasoline powered. When using a gasoline-powered saw, follow these additional safety guidelines:

- Be sure there is proper ventilation before operating the saw indoors.
- Use caution to prevent contact with hot manifolds and hoses.
- Be sure the saw is out of gear before starting it.
- Use the recommended starting fluid.
- Always keep the appropriate fire extinguishers near when filling, starting, and operating gasoline-powered saws. OSHA requires that gasoline-powered equipment be turned off prior to filling.
- Do not pour gasoline into the carburetor or cylinder head when starting the engine.
- Never pour gasoline into the fuel tank when the engine is hot or when the engine is running.
- Do not operate a saw that is leaking gasoline.

4.3.0 Using Mixers Safely

Powered mortar mixers are available in electric- and gasoline-engine varieties (see *Figure 20*). The mixer drum is fitted with a shaft with blades or paddles that revolve through the mix in the drum. Mortar is emptied from the drum onto a pan or board using the mixer's dump handle and drum release. You will learn how to mix mortar using a power mixer later in this curriculum, in the module titled *Mortar*.

Guidelines for the safe use of mortar mixers include the following:

- Always wear appropriate personal protective equipment when operating a mixer.
- Never operate a mixer without proper training first.

28106-13_F20.EPS

Figure 20 Powered mortar mixer.

- Do not use accessories or attachments that were not designed for the particular make and model of mixer that you are using.
- Ensure that all emergency and safety devices are connected and working before operating the mixer.
- Never fill the oil level higher than what is specified in the manufacturer's instructions. Always check the oil level daily or prior to any use.
- Ensure that the belt, blades or paddles, and dumping mechanism are all in working order according to the manufacturer's instructions.
- Place the safety grate into position and ensure that it is fitted properly.
- Ensure that the air filters are clean and properly sized for the mixer.
- If the mixer is gasoline powered, ensure that the engine is properly ventilated to prevent the buildup of toxic exhaust gases. Never touch the manifold, cylinder, or muffler when hot.
- When cleaning the mixer, dispose of all excess mortar properly in according with the manufacturer's instructions and the standards and regulations of your jurisdiction.

WARNING!

Never place your hands or tools inside the drum when the mortar mixer is starting or operating.

WARNING!

Never service a mixer when it is operating. Make sure that the mixer has stopped rotating and is disconnected from its power source and that the engine or motor has been allowed to cool before attempting to service the mixer.

4.4.0 Using Grinders Safely

Masons use handheld, electric grinders called tuckpoint grinders (*Figure 21*) to grind out old joints in masonry walls. Tuckpoint grinders use shatterproof blades and are fitted with a safety guard on the part of the machine that faces the operator.

The following are some general safety guidelines to follow when using a grinder:

- Be sure the grinding head is in good condition and has been properly secured.
- Wear appropriate personal protective equipment (PPE).
- Follow the manufacturer's instructions for installing shrouds and other dust-collection devices to help capture the dust created by the grinding process.

28106-13_F21.EPS

Figure 21 Tuckpoint grinder.

- Use the proper grinding head for the material being ground.
- Ensure that the wheel guard is able to move freely and close during operation.
- Grip the grinder tightly on the main and auxiliary handles when using the grinder. Hold the grinder only on the insulated gripping surfaces.
- Ensure that the grinder is clean before using it, and clean it periodically while using it. Always allow the grinder wheel to come to a complete stop and unplug the grinder before cleaning it.
- Never force the grinder along as you work or apply excessive pressure. Use smooth, straight strokes. Maintain a firm grip on the handles when operating.
- Be sure the grinder head has stopped turning before putting the grinder down.

WARNING!

Before starting a grinder, remove all safety keys and ensure that grinder blades are secured and not in contact with other materials.

WARNING!

When using a tuckpoint grinder as a cutter, use the appropriate blade for the material being cut.

4.5.0 Working Safely around Forklifts

Forklifts are common on many masonry work sites. They are useful for lifting and moving heavy or awkward loads of materials, supplies, and equipment. The most commonly used forklifts on masonry projects are reach-type forklifts (*Figure 22*) and skid-steer forklifts (*Figure 23*). You may also see conventional forklifts on occasion (*Figure 24*). Forklifts present several risks for masons working near them, including hitting other workers, dropping loads, tipping over, and causing fires and explosions.

Forklift accidents can be caused by mechanical and hydraulic problems, but the most common cause of forklift accidents is human error. That means most forklift accidents can be avoided if the operator and other workers in the area stay alert and use caution and common sense. In fact, research by Liberty Mutual Insurance Company shows that drivers with more than a year of experience operating a forklift are more likely to have an accident than someone with little experience. This is because operators tend to become too comfortable and less attentive after they gain experience on the equipment. The same study showed that the most common type of forklift accident is one in which a pedestrian is hit by the truck.

28106-13_F22.EPS

Figure 22 Mason's reach-type forklift.

28106-13_F23.EPS

Figure 23 Skid-steer forklift.

28106-13_F24.EPS

Figure 24 Conventional forklift.

Masons need to take precautions when working around forklifts to ensure their safety. Here are some general guidelines:

- Do not put any part of your body into the mast structure or between the mast and the forklift, or within the reach mechanism.
- Do not ride as a passenger in a forklift unless a safe place to ride has been provided by the manufacturer.
- Do not stand or ride on the forks.
- When walking, always be on the lookout for forklifts that may be moving or operating nearby.
- Do not stand in front of a forklift or a load.
- Stand clear of the rear swing area of a turning forklift.
- Exercise particular care at cross aisles, doorways, and other locations where forklifts may drive across your path.

4.6.0 Observing Electrical Safety

Electrical safety is a concern for masons, not just electricians. Extension cords, power tools, portable lights, and many other pieces of equipment that masons use on the job are powered by electricity. Electric tools and equipment can cause electric shock, electrocution, burns, fires, and explosions.

Not all electrical accidents result in death. There are different types of electrical accidents. Any of the following can happen:

- Burns
- Electric shock
- Explosions
- Falls caused by electric shock
- Fires

> **NOTE**
> Double-insulated electrical tools are becoming more common on the job site because they are safer than those that use a three-wire cord alone.

OSHA and your company have specific policies and procedures to keep the workplace safe from electrical hazards. You can do many things to reduce the chance of an electrical accident. If you ever have any questions about electrical safety on the job site, ask your supervisor. Here are some basic job-site electrical safety guidelines:

- Use three-wire extension cords and protect them from damage. Never fasten them with staples, hang them from nails, or suspend them from wires. Never use damaged extension cords.
- Use three-wire cords for portable power tools and make sure they are properly connected (*Figure 25A*). The three-wire system is one of the most common safety grounding systems used to protect you from accidental electrical shock. The wires are color coded for your safety (black is the hot wire, white is the neutral wire, and green is the ground wire). If the insulation in a tool fails, the current will pass to the electrical ground through the third wire—not through your body.
- Double-insulated cords are also effective in preventing shocks when using power tools. *Figure 25B* shows the double-insulated symbol that can be found on double-insulated tools (*Figure 25C*).
- Use a ground fault circuit interrupter (GFCI) (*Figure 26A*), a portable GFCI (*Figure 26B*), or an ensured grounding program with every tool.
- Make sure that panels, switches, outlets, and plugs are grounded.
- Never use bare electrical wire.
- Never use metal ladders near any source of electricity.
- Never wear a metal hard hat.
- Always inspect electrical power tools before you use them.
- Never operate any piece of electrical equipment that has a danger tag or lockout device attached to it.
- Never use worn or frayed cords (*Figure 27*).
- Make sure all light bulbs have protective guards to prevent accidental contact (*Figure 28*).
- Do not hang temporary lights by their power cords unless they are specifically designed for this use.
- Check the cable and ground prong. Check for cuts in the cords and make sure the cords are clean of grease.

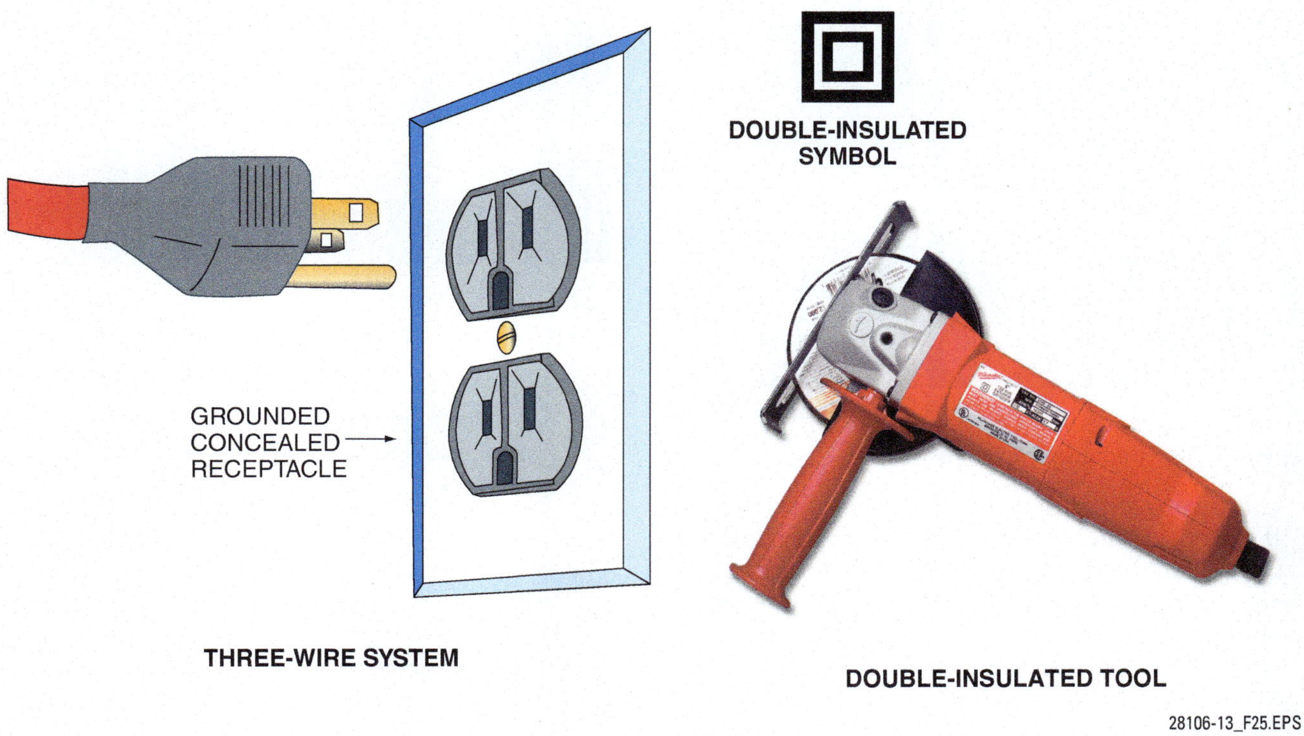

Figure 25 Three-wire system, double-insulated tool, and double-insulated symbol.

28106-13_F26.EPS

Figure 26 GFCI and portable GFCI.

- Use approved concealed receptacles for plugs. If different voltages or types of current are used in the same area, the receptacles should be designed so that the plugs are not interchangeable.
- Any repairs to cords must be performed by a qualified person.
- Always make sure all electrically powered tools are grounded before use.

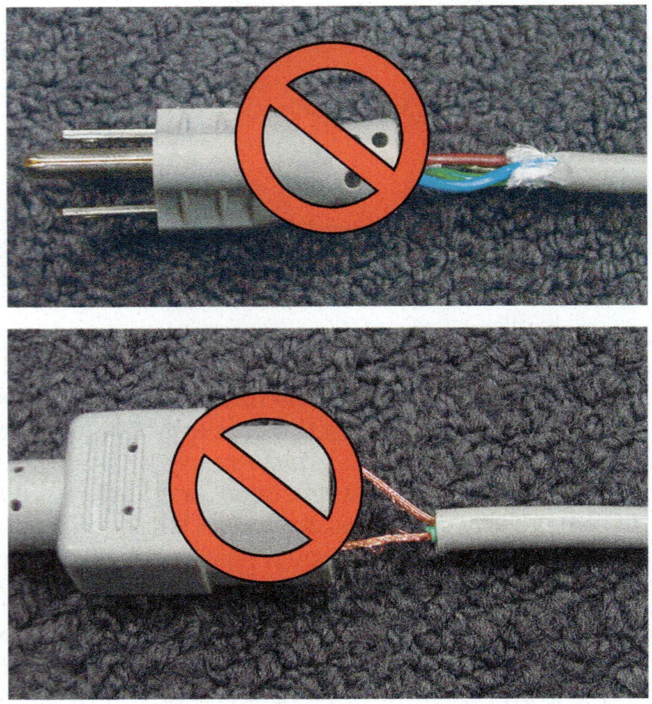

28106-13_F27.EPS

Figure 27 Never used damaged cords.

28106-13_F28.EPS

Figure 28 Work light with protective guard.

CAUTION

All electrical tools must be ground-fault protected. This helps ensure the safety of workers.

WARNING!

Do not perform work on electrical equipment until the circuits have been de-energized, locked out, and confirmed. Assume that all conductors, buses, and connections are energized unless and until proven otherwise.

4.7.0 Using Powder-Actuated Tools Safely

A powder-actuated fastening tool (*Figure 29*) is a low-velocity fastening system powered by gunpowder cartridges, commonly called boosters. Powder-actuated tools are used to drive specially designed fasteners into masonry and steel.

Manufacturers use color-coding schemes to identify the strength of a powder load charge. It is extremely important to select the right charge for the job, so learn the color-coding system that applies to the tool you are using. *Table 3* shows an example of a color-coding system.

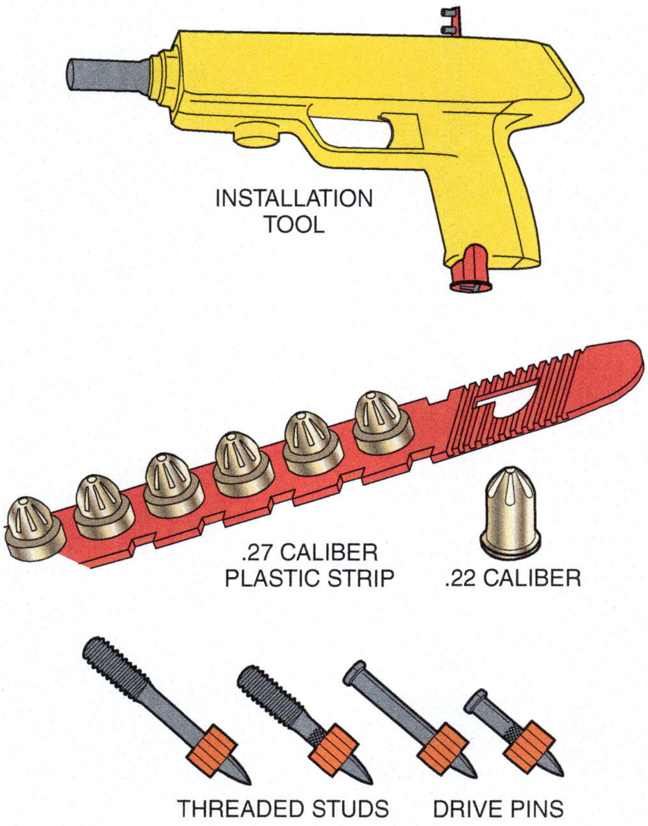

28106-13_F29.EPS

Figure 29 Powder-actuated fastening tool, boosters, and fasteners.

Other rules for safely operating a powder-actuated tool are as follows:

- Do not use a powder-actuated tool unless you are certified.
- Follow all safety precautions in the manufacturer's instruction manual.
- Always wear safety goggles and a hard hat when operating a powder-actuated tool.
- Use the proper-size pin for the job you are doing.
- When loading the tool, put the pin in before the charge.
- Use the correct booster (powder load) according to the manufacturer's instructions.
- Never hold your hand behind or near the material you are fastening.
- Never hold the end of the barrel against any part of your body or cock the tool against your hand.
- Do not shoot close to the edge of concrete.
- Never attempt to pry the booster out of the magazine with a sharp instrument.
- Always wear ear and eye protection.
- Always hold the muzzle perpendicular (90 degrees) to the work.

Table 3 Powder Charge Color-Coding System

Load Color	Powder Level	
Gray	1	Low Power
Brown	2	↓
Green	3	
Yellow	4	
Red	5	
Purple	6	High Power

28106-13_T03.EPS

Additional Resources

"Using Masonry Saws." Norton Construction Products North America. **www.nortonconstructionproducts.com/solutions/masonry-saw.aspx**

"What You Need to Know about Brick Table Saws." Tom Inglesby. *Masonry Magazine*, June 2002. **www.masonrymagazine.com/6-02/cover.html**

Forklift Safety: A Practical Guide to Preventing Powered Industrial Truck Incidents and Injuries, Second Edition. 1999. George Swartz. Government Institutes.

4.0.0 Section Review

1. Mortar can make a tool unusable if it is allowed to _____.
   a. corrode
   b. harden
   c. soften
   d. melt

2. When fueling gasoline-powered equipment such as saws, OSHA requires that the equipment first be _____.
   a. warmed up
   b. inspected
   c. turned off
   d. drained of all fluids

3. To empty mortar from a mortar mixer, use the mixer's dump handle and _____.
   a. safety grate
   b. drum release
   c. a wheelbarrow
   d. wheel guard

4. When operating a tuckpoint grinder, move the grinder along the wall using _____.
   a. smooth, straight strokes
   b. excessive pressure
   c. short, sharp strokes
   d. a circular, sweeping motion

5. The most common cause of forklift accidents is _____.
   a. mechanical and hydraulic problems
   b. operating with insufficient clearance
   c. carrying loads that are too wide or tall
   d. human error

6. In a three-wire system, the green wire is connected to _____.
   a. a nonconducting surface
   b. an electrical ground
   c. the tool
   d. a circuit breaker

7. The gunpowder cartridges that powder-actuated fastening tools use to drive specially designed fasteners into masonry and steel are called _____.
   a. shells
   b. rails
   c. boosters
   d. fasteners

SECTION FIVE

5.0.0 MATERIALS HANDLING

Objective

Explain how to handle materials properly.

a. Describe how to store and stockpile masonry materials.
b. Describe how to stack brick.

Trade Terms

Bundled: To be wrapped or bound in a cube shape.

Working stack: A stack of brick that has been removed from a bundle and set up near where the brick will be used.

Strains, sprains, fractures, and crushing injuries can be minimized with a knowledge of safe lifting and handling procedures and proper ergonomics, which means the optimum design of systems for use by people. General guidelines for you to keep in mind when handling or moving construction materials are as follows:

- Wear appropriate safety boots, at least ankle high, with reinforced toes to prevent crushing.
- Keep floors free of water, grease, and other slippery substances so as to prevent falls.
- Inspect materials for grease, slivers, and rough or sharp edges.
- Determine the weight of the load before applying force to move it.
- Know your own limits for how much you can lift.
- Be sure that your intended pathway is free from obstacles.
- Make sure your hands are free of oil and grease.
- Take a firm grip on the object before you move it, being careful to keep your fingers from being pinched.
- When the load is too large or heavy, get help, or, if possible, simply reduce the load and make more trips.
- Whenever possible, use mechanical means of material handling.
- When stacking materials, be sure to follow OSHA regulations as to the height, shape, and stability of the pile.

Materials in the general working area and on the stockpile should be stacked safely. Masonry units that are not stacked properly and secured in some way are very likely to fall, which could cause injury to you or a fellow worker.

5.1.0 Storing and Stockpiling Masonry Materials

OSHA guidelines for the stockpiling and handling of materials are covered in *19 CFR 1926.250(b)*. Paragraph 6 covers the requirements for brick stacks, while Paragraph 7 discusses the requirements for concrete masonry units (CMUs). You must adhere to these and all OSHA guidelines when storing and stockpiling masonry materials:

- All masonry materials stored in tiers should be stacked, racked, blocked, interlocked, or otherwise secured to prevent sliding, falling, or collapse.
- Maximum safe load limits of floors within buildings and structures should be posted in all storage areas, and maximum safe loads should not be exceeded.
- Aisles and passageways should be kept clear to provide for the free and safe movement of material handling equipment or employees.
- Incompatible materials should be segregated in storage.
- Bagged materials should be stacked by stepping back the layers and cross keying the bags at least every 10 bags high. The industry best practice is to do this every other layer.
- Stockpiles of **bundled** brick, or brick that has been wrapped or bound in a cube shape, should not be higher than 7 feet (see *Figure 30*).
- When a loose brick stockpile reaches a height of 4 feet, it should be tapered back 2 inches for every foot of height above the 4-foot level.

28106-13_F30.EPS

Figure 30 Bundled (cubed) brick.

- When loose masonry block is stockpiled higher than 6 feet, the stack should be tapered back one-half block per tier above the 6-foot level.

5.2.0 Stacking Brick

A stack of brick that masons prepare at the wall, on the ground, or on the scaffold platform is called a working stack. Masons make working stacks by unbundling brick that has been bundled (cubed) and using a cart to move the brick to where it will be needed. When stacking masonry materials in working stacks, keep the stack neat and vertically in line to eliminate the possibility of snagging your clothes. Keep the piles about 3 feet high so that you can easily get to the brick. Brick and other materials stacked too high not only pose a safety hazard, but may reduce productivity.

The most common way to stack brick is to reverse the direction of every other course so that the stack is secure. Such a stack should be no more than 3 feet high and no closer than 2 feet to the wall. Wider stacks can be made by alternating a pattern of brick, as shown in *Figure 31*.

SECOND COURSE

FIRST COURSE

FULL WORKING STACK

28106-13_F31.EPS

Figure 31 Stacking brick.

Additional Resources

"Materials Handling and Storage." OSHA. **www.osha.gov/Publications/OSHA2236/osha2236.html**

5.0.0 Section Review

1. When stockpiling loose brick in piles, the pile should be tapered back _____.
   a. 4 inches for every foot of height above the 6-foot level
   b. 2 inches for every foot of height above the 6-foot level
   c. 4 inches for every foot of height above the 4-foot level
   d. 2 inches for every foot of height above the 4-foot level

2. The height of a working stack of brick should be no more than _____.
   a. 3 feet
   b. 4 feet
   c. 5 feet
   d. 6 feet

SUMMARY

Serious accidents and injuries can occur anywhere on a work site. Whether you work for a large contractor or as a subcontractor, you have a responsibility to work safely and follow the rules and regulations of the site. You also have a responsibility to be aware of what's going on around you and to notice the behavior of your co-workers. Personal injury and damage to equipment is less likely to happen when you do so.

Masons should be able to identify the common causes of accidents and the hazards associated with tools, equipment, mortar, and concrete, and they should also be able to prevent accidents and hazards on the job site. Masons need to know how to use personal protective equipment and how to work safely from elevated surfaces, as well as how to use masonry tools and equipment and how to handle masonry materials safely. The ability to work safely is an essential skill for a mason, and is a key to ensuring a long and successful career.

Review Questions

1. Exit signs typically have _____.
   a. red letters on a white field
   b. white letters on a green field
   c. white letters on a red field
   d. red letters on a yellow field

2. Safety tags are used to _____.
   a. identify no-smoking areas
   b. provide temporary warnings of hazards
   c. warn against high voltage
   d. locate emergency eye wash stations

3. The construction industry annually suffers millions of dollars in accidents, lost time and productivity because of worker _____.
   a. waste of mortar
   b. vehicle accidents
   c. tardiness
   d. substance abuse

4. Accidents caused by lack of skill can be prevented by _____.
   a. proper training
   b. observing warning signs
   c. working more carefully
   d. practice

5. Making excuses for taking risks on the job is known as _____.
   a. cutting corners
   b. rationalizing
   c. hazardous behavior
   d. carelessness

6. Poor housekeeping on the job site is an example of a(n)_____.
   a. unsafe act
   b. management system failure
   c. unsafe condition
   d. communication failure

7. When dumping mortar bags into a mixer, whenever possible you should _____.
   a. stand upwind
   b. warn others to stay away
   c. stand downwind
   d. shake the bag vigorously

8. Flammable liquids should be stored only in _____.
   a. outdoor locations
   b. red plastic gas cans
   c. glass containers with tight-fitting lids
   d. approved safety containers

9. If safety goggles are not available, the best form of eye protection is _____.
   a. regular prescription lenses
   b. safety lenses with side shields
   c. contact lenses
   d. tinted lenses

10. Large, padded covers for the entire ear are called _____.
    a. earplugs
    b. ear shields
    c. earmuffs
    d. ear guards

11. To protect your hands from abrasion, chemical burns, and other masonry hazards, you should wear gloves made from _____.
    a. heavy canvas
    b. nylon and nitrile
    c. leather
    d. reinforced polyester

12. One of the leading causes of fatalities among construction workers is _____.
    a. falls from elevated areas
    b. inhalation of silicon dust
    c. chemical burns from wet concrete
    d. trowel-related injuries

13. The number of groups that falls are classified into is _____.
    a. two
    b. three
    c. four
    d. five

14. When working at an elevation of 6 feet or more, workers are required to wear _____.
    a. safety shoes
    b. suspenders
    c. safety vests
    d. fall protection

15. Lanyards used in fall protection systems must have a minimum breaking strength of _____.
    a. 2,000 pounds
    b. 3,000 pounds
    c. 4,000 pounds
    d. 5,000 pounds

16. When vertical lifelines are used, _____.
    a. two workers may share a line
    b. one line is needed for each four workers
    c. each worker must have his or her own line
    d. a maximum of six may be installed

17. A personal positioning system should limit a worker's fall to _____.
    a. 2 feet
    b. 3 feet
    c. 4 feet
    d. 5 feet

18. The force to which a worker is subjected when the fall is stopped suddenly is limited by the use of a _____.
    a. fall arrester
    b. body harness
    c. deceleration device
    d. shock absorber

19. The entire weight of a fall arrest system is supported by a(n) _____.
    a. counterweight
    b. anchor point
    c. A-frame
    d. support structure

20. An established rescue and retrieval plan is required on every _____.
    a. job site
    b. confined space job site
    c. crane job site
    d. elevated job site

21. If any part of a personal fall arrest system is found to be damaged, the part should be _____.
    a. returned to the manufacturer
    b. temporarily repaired and re-used
    c. either tagged as unusable or else destroyed
    d. reported to a supervisor

22. Walkways and platforms on scaffold must have a width of at least _____.
    a. 18 inches
    b. 24 inches
    c. 36 inches
    d. 48 inches

23. If people will be working or walking beneath scaffold, the space between the guardrail and toeboards should be closed with _____.
    a. solid planking
    b. plastic netting
    c. mesh screen
    d. plywood sheets

24. Tools or equipment used with mortar or grout should be cleaned _____.
    a. at the beginning of the workday
    b. immediately after use
    c. hourly
    d. at the end of the workday

25. For the most accurate results when cutting masonry units, use a _____.
    a. handsaw
    b. hammer and chisel
    c. masonry saw
    d. power shear

26. To avoid a buildup of toxic exhaust gases, a gasoline-powered mortar mixer must be _____.
    a. used outdoors
    b. properly ventilated
    c. provided with a muffler
    d. used only for short periods

27. The material-handling device shown in *Review Question Figure 1* is a _____.

    a. pallet jack
    b. reach-type forklift
    c. skid-steer loader
    d. conventional forklift

28106-13_RQ01.EPS

Figure 1

28. All electrical tools used on the job site must be _____.

    a. OSHA approved
    b. equipped with circuit breakers
    c. ground-fault protected
    d. tuned for altitude and humidity

29. According to industry best practice, when stacking bagged materials, you should step back and cross key the bags every _____.

    a. 2 layers
    b. 5 layers
    c. 10 layers
    d. 12 layers

30. When stacking masonry units, you should reverse the direction of each layer to _____.

    a. use the material most efficiently
    b. keep the work site looking neat
    c. avoid the risk of the stack falling over
    d. allow making a taller stack

Trade Terms Quiz

Fill in the blank with the correct term that you learned from your study of this module.

1. ________ means to be capable of easily igniting and rapidly burning, or a fuel with a flash point below 100°F.
2. Brick that is wrapped or bound in a cube shape is said to be ________.
3. ________ is equipment or clothing designed to prevent or reduce injuries.
4. An elevated platform for workers and materials is called ________.
5. The organization of a company's management, including reporting procedures, supervisory responsibility, and administration is called the ________.
6. ________ is the term used to define something that is capable of causing chemical burns.
7. A synthetic rubberlike material used in masonry gloves to protect hands while permitting a tactile response is called ________.
8. A device that provides clean, filtered air for breathing, no matter what is in the surrounding air is a(n) ________.
9. Having a high initial rate of moisture absorption is called ________.
10. ________ is a respiratory disease caused by the inhalation of silica dust.
11. A stack of brick that has been removed from a bundle and set up near where the brick will be used is called a(n) ________.
12. A(n) ________ is a document that identifies a hazardous substance and gives the exposure limits, the physical and chemical characteristics, the kind of hazard it presents, precautions for safe handling and use, and specific control measures.
13. Water that is safe for cooking and drinking is called ________.
14. The ________ is the vertical distance a worker moves after a fall before a deceleration device is activated.
15. A device such as a shock-absorbing lanyard or self-retracting lifeline that brings a falling person to a stop without injury is called a(n) ________.
16. The ________ is the force needed to stop a person from falling.
17. The distance it takes before a person comes to a stop when falling is called the ________.
18. A coupling link fitted with a safety closure is called a(n) ________.

Trade Terms

Arresting force
Bundled
Carabiner
Caustic
Deceleration device
Deceleration distance
Flammable
Free-fall distance
Hygroscopic
Management system
Nitrile
Personal protective equipment (PPE)
Potable
Respirator
Safety data sheet (SDS)
Scaffold
Silicosis
Working stack

Cornerstone of Craftsmanship

Dennis W. Neal

Regional NCCER Representative
Florida Masonry Apprentice & Educational Foundation, Inc.

Dennis Neal's brother encouraged him to work in the masonry industry because of the range of experience that such a career could offer him. Today, Dennis encourages apprentices to seek careers in masonry to take advantage of the same breadth of opportunities that await them.

How did you get started in the construction industry?
In high school, I worked for my brother's masonry business in Delaware during the summer. After attending college and serving in the Air Force, I returned to Delaware and began working full time at my brother's business.

Who or what inspired you to enter the industry? Why?
My brother encouraged me to work in the masonry industry after I completed my military service. Working in the trade was a good experience for him, and he saw the same possibility for me too.

What do you enjoy most about your career?
I enjoy the daily changes in the tasks to be completed. No two days are ever the same in the construction industry. Weather, job-site conditions, materials, and co-workers all change frequently, as do the tasks that need to be accomplished. That makes every day interesting and challenging. I also enjoy sharing my knowledge and passion for the industry with others.

Why do you think training and education are important in construction?
Now more than ever, training and education are important in the construction trades. There is a severe shortage of skilled workers that needs to be filled. Training and education will increase the knowledge and the skill level of the workforce. The NCCER programs are an excellent tool to use for apprenticeship training and career advancement. The ability to advance your career in construction never ends; there are credentials and certifications to increase your knowledge and standing available from many sources.

Why do you think credentials are important in construction?
The credentials that you earn will increase your knowledge and skills. Credentials that are earned will increase your job security and increase your financial reward.

How has training/construction impacted your life and career?
My training over the years has given me the ability to have a great career in the construction industry and to always have the opportunity to earn a good living, no matter the economic conditions.

Would you recommend construction as a career to others?
Yes! With the proper training, there are great career opportunities in the industry. As I mentioned above, every day is different. At the end of each day, you can see what you have accomplished. The shortage of skilled workers in the construction industry will provide great opportunities for young people willing to work hard and earn their credentials.

What does craftsmanship mean to you?
Craftsmanship is attaining a level of knowledge to be able to perform the required skills on the job. True craftspeople will never stop learning even after the formal training has been completed.

Trade Terms Introduced in This Module

Arresting force: The force needed to stop a person from falling. The greater the free-fall distance, the more force is needed to stop or arrest the fall.

Bundled: To be wrapped or bound in a cube shape.

Carabiner: A coupling link fitted with a safety closure.

Caustic: Capable of causing chemical burns.

Deceleration device: A device such as a shock-absorbing lanyard or self-retracting lifeline that brings a falling person to a stop without injury.

Deceleration distance: The distance it takes before a person comes to a stop when falling. The required deceleration distance for a fall arrest system is a maximum of 3½ feet.

Flammable: Capable of easily igniting and rapidly burning; used to describe a fuel with a flash point below 100°F.

Free-fall distance: The vertical distance a worker moves after a fall before a deceleration device is activated.

Hygroscopic: Having a high initial rate of moisture absorption.

Management system: The organization of a company's management, including reporting procedures, supervisory responsibility, and administration.

Nitrile: A synthetic rubberlike material used in masonry gloves to protect hands while permitting a tactile response.

Personal protective equipment (PPE): Equipment or clothing designed to prevent or reduce injuries.

Potable: Water that is safe for cooking and drinking.

Respirator: A device that provides clean, filtered air for breathing, no matter what is in the surrounding air.

Safety data sheet (SDS): A document that must accompany any hazardous substance. The SDS identifies the substance and gives the exposure limits, the physical and chemical characteristics, the kind of hazard it presents, precautions for safe handling and use, and specific control measures. Formerly known as material safety data sheet (MSDS).

Scaffold: An elevated platform for workers and materials.

Silicosis: A respiratory disease caused by the inhalation of silica dust.

Working stack: A stack of brick that has been removed from a bundle and set up near where the brick will be used.

Additional Resources

This module presents thorough resources for task training. The following resource material is suggested for further study.

Fall Protection and Scaffolding Safety: An Illustrated Guide. 2000. Grace Drennan Ganget. Government Institutes.

Forklift Safety: A Practical Guide to Preventing Powered Industrial Truck Incidents and Injuries, Second Edition. 1999. George Swartz. Government Institutes.

"The Sense of Safety." K. K. Snyder. *Masonry Magazine*, March 2008. **www.masonrymagazine.com/3-08/safety.html**

"Using Masonry Saws." Norton Construction Products North America. **www.nortonconstructionproducts.com/solutions/masonry-saw.aspx**

"Online Safety Library: Scaffold Safety." Oklahoma State University. **www.ehs.okstate.edu/links/scaffold.htm**

OSHA website. **www.osha.gov**

Personal Protective and Life Saving Equipment. OSHA. 29 CFR 1926 Subpart E.

"Materials Handling and Storage." OSHA. **www.osha.gov/Publications/OSHA2236/osha2236.html**

"Standards for Scaffold." OSHA. **www.osha.gov/SLTC/scaffold/index.html**

"What You Need to Know about Brick Table Saws." Tom Inglesby. *Masonry Magazine*, June 2002. **www.masonrymagazine.com/6-02/cover.html**

WorkSAFE masonry safety resources. **www.worksafecenter.com/safety/tutorial/masonry/step-1.page**

Figure Credits

Courtesy of Honeywell Safety Products, Figure 4 (top right, bottom left), Figure 6, Figure 12, Figure 13, Figure 14, Figure 15, Figure 16 (top), E01, E02

West Chester Holdings, Inc., Figure 7

Courtesy of MSA, Figure 8 (top, middle)

US Occupational Safety and Health Administration, Table 2

Construction Safety Association of Ontario, Figure 18 (top)

Fall Protection Systems, Inc., Figure 18 (middle, bottom)

Bon Tool Company, Figure 21, E03

CareLift Equipment Ltd, Figure 22, RQ01

Courtesy of Bobcat Company. Bobcat®, the Bobcat logo and the colors applied to the Bobcat vehicle are registered trademarks of Bobcat Company in the United States and various other countries., Figure 23

Manitou North America, Inc., Figure 24

Courtesy of Dennis Neal, FMA&EF, Figure 26 (bottom), Figure 28

Courtesy of Bryan Light, Figure 31

Section Review Answer Key

Answer	Section Reference	Objective
Section One		
1. a	1.1.0	1a
2. b	1.2.1	1b
3. d	1.3.0	1c
4. b	1.4.0	1d
5. a	1.5.0	1e
6. c	1.6.0	1f
Section Two		
1. b	2.1.0	2a
2. d	2.2.0	2b
3. c	2.3.0	2c
4. b	2.4.0	2d
Section Three		
1. a	3.1.0	3a
2. b	3.2.0	3b
3. b	3.3.0	3c
4. c	3.4.0	3d
Section Four		
1. b	4.1.0	4a
2. c	4.2.0	4b
3. b	4.3.0	4c
4. a	4.4.0	4d
5. d	4.5.0	4e
6. b	4.6.0	4f
7. c	4.7.0	4g
Section Five		
1. d	5.1.0	5a
2. a	5.2.0	5b

NCCER CURRICULA — USER UPDATE

NCCER makes every effort to keep its textbooks up-to-date and free of technical errors. We appreciate your help in this process. If you find an error, a typographical mistake, or an inaccuracy in NCCER's curricula, please fill out this form (or a photocopy), or complete the online form at **www.nccer.org/olf**. Be sure to include the exact module ID number, page number, a detailed description, and your recommended correction. Your input will be brought to the attention of the Authoring Team. Thank you for your assistance.

Instructors – If you have an idea for improving this textbook, or have found that additional materials were necessary to teach this module effectively, please let us know so that we may present your suggestions to the Authoring Team.

NCCER Product Development and Revision

13614 Progress Blvd., Alachua, FL 32615

Email: curriculum@nccer.org

Online: www.nccer.org/olf

❑ Trainee Guide ❑ Lesson Plans ❑ Exam ❑ PowerPoints Other ____________

Craft / Level: Copyright Date:

Module ID Number / Title:

Section Number(s):

Description:

Recommended Correction:

Your Name:

Address:

Email: Phone:

28102-13

Masonry Tools and Equipment

Overview

Masonry tools are the interface between masons and their work. The quality of the tool directly affects the quality of the work. Because masons do many different tasks, they have at their disposal many special-purpose tools. Some hand tools are virtually unchanged since ancient times, while some power tools use the most up-to-date technology. This module introduces the tools and equipment that you will need to lay masonry units. You will learn how to safely use hand tools, measuring tools, power tools, power equipment, lifting equipment, and scaffolds. By the end of this module, you will be able to identify each item and explain what it does.

Module Three

Trainees with successful module completions may be eligible for credentialing through NCCER's National Registry. To learn more, go to **www.nccer.org** or contact us at **1.888.622.3720**. Our website has information on the latest product releases and training, as well as online versions of our *Cornerstone* magazine and Pearson's product catalog.

Your feedback is welcome. You may email your comments to **curriculum@nccer.org**, send general comments and inquiries to **info@nccer.org**, or fill in the User Update form at the back of this module.

This information is general in nature and intended for training purposes only. Actual performance of activities described in this manual requires compliance with all applicable operating, service, maintenance, and safety procedures under the direction of qualified personnel. References in this manual to patented or proprietary devices do not constitute a recommendation of their use.

28102-13
MASONRY TOOLS AND EQUIPMENT

Objectives

When you have completed this module, you will be able to do the following:

1. Identify hand tools used in masonry.
   a. Describe how to use trowels.
   b. Describe how to use hammers and chisels.
   c. Describe how to use jointers and brushes.
   d. Identify other hand tools used in masonry.
2. Identify measures and measuring tools used in masonry.
   a. Describe how to use rules.
   b. Describe how to use levels.
   c. Describe how to use chalk boxes, squares, plumb bobs, and laser levels.
   d. Describe how to use corner poles, lines, and fasteners.
3. Identify mortar equipment used in masonry.
   a. Describe how to use mortar boxes.
   b. Describe how to use mixing accessories.
4. Identify power tools used in masonry.
   a. Describe how to use masonry saws.
   b. Describe how to use splitters.
   c. Describe how to use grinders.
   d. Describe how to use power drills and powder-actuated tools.
5. Identify power equipment used in masonry.
   a. Describe how to use a mortar mixer.
   b. Describe how to use a masonry pump, vibrator, and hydraulic grout placer.
   c. Describe how to use pressurized cleaning equipment.
6. Identify lifting equipment used in masonry.
   a. Describe how to use mounted and portable hoists.
   b. Describe how to use hydraulic-lift materials trucks.
   c. Describe how to use forklifts and pallet jacks.
7. Recognize scaffolds used in masonry.
   a. Identify scaffold systems.
   b. Describe how to assemble and disassemble tubular frame scaffold.

Performance Tasks

Under the supervision of your instructor, you should be able to do the following:

1. Demonstrate the proper use of a rule.
2. Demonstrate the proper use of a level.
3. Use the correct procedures for fueling and starting a mixer.
4. Assemble and disassemble tubular frame scaffold.

Trade Terms

Bed joint
Corner pole
Kickback
Lead
Parge
Pointing
Tempering

Industry-Recognized Credentials

If you're training through an NCCER-accredited sponsor, you may be eligible for credentials from NCCER's Registry. The ID number for this module is 28102-13. Note that this module may have been used in other NCCER curricula and may apply to other level completions. Contact NCCER's Registry at 888.622.3720 or go to **www.nccer.org** for more information.

Code Note

Codes vary among jurisdictions. Because of the variations in code, consult the applicable code whenever regulations are in question. Referring to an incorrect set of codes can cause as much trouble as failing to reference codes altogether. Obtain, review, and familiarize yourself with your local adopted code.

Contents

Topics to be presented in this module include:

Figures

Section One

1.0.0 Hand Tools

Objective

Identify hand tools used in masonry.

a. Describe how to use trowels.
b. Describe how to use hammers and chisels.
c. Describe how to use jointers and brushes.
d. Identify other hand tools used in masonry.

Trade Terms

Bed joint: A horizontal joint between two masonry units.

Parge: A thin coat of mortar or grout on the outside surface of a wall. Parging prepares a masonry surface for attaching veneer or tile, or parging can waterproof the back of a masonry wall.

Pointing: Troweling mortar or a mortar-repairing material, such as epoxy, into a joint after masonry is laid.

The quality and use of masonry hand tools greatly affect the quality of the final work. Hand tools are used to coat, cut, carry, clean, align, and level masonry units. The following sections describe masonry hand tools you will typically use to perform masonry construction.

As you learned in the module *Masonry Safety,* you are responsible for keeping your tools and equipment in good repair. Repair or replace defective tools and equipment immediately. Clean all tools and equipment that touch mortar or grout immediately after use. Use a bucket of water to soak tools that you will use again in a few minutes. Keep wooden handles out of the water. If mortar dries on a tool, it will harden and make the tool unusable.

Wash tools thoroughly with water and a wire brush. Be sure that you remove all mortar or grout completely. If you do not use tools regularly, coat them lightly with oil, grease, or other approved coating to prevent rust and corrosion. After cleaning, check that tool handles are secure and free from cracks or splinters. Oil the pivot joints and wooden parts of the tools. Sharpen blades and cutting edges when they become dull or nicked.

Check Tools Daily

Cleaning and checking your tools at the end of each day is a good habit. This will keep them in good working order and prevent loss.

1.1.0 Using Trowels

The trowel is the mason's most commonly used tool. Trowels are used for placing mortar, an activity commonly called buttering. You were introduced to buttering in the module *Introduction to Masonry*. Masons use trowels to move and shape mortar between masonry units. Trowels are also used to mix, scrape, and shape mortar and clean mortar from masonry units and tools. The trowel's handle is often used to tap units into place.

The mason's trowel comes in many sizes and shapes. The typical trowel (*Figure 1*) consists of a steel blade ground to the proper balance, taper, and shape. The narrow end of the blade is the point, and the wide end is the heel. The blade is connected to the handle by a shank. The handle is made of wood or plastic, and it can be covered with leather or foam for a more comfortable grip. The wood handle has a band, or ferrule, on the shank end to prevent splitting.

The blade can come with a sharply angled heel (London pattern) or a square heel (Philadelphia pattern), as shown in *Figure 2*. Some masons prefer a wider heel, which can make buttering masonry units easier.

Trowels come in different shapes and sizes for different purposes. Trowels can range in width from about 4 to 7 inches and can be up to 13 inches long. *Figure 3* shows different types of trowels. Some trowels have specialized uses and will not

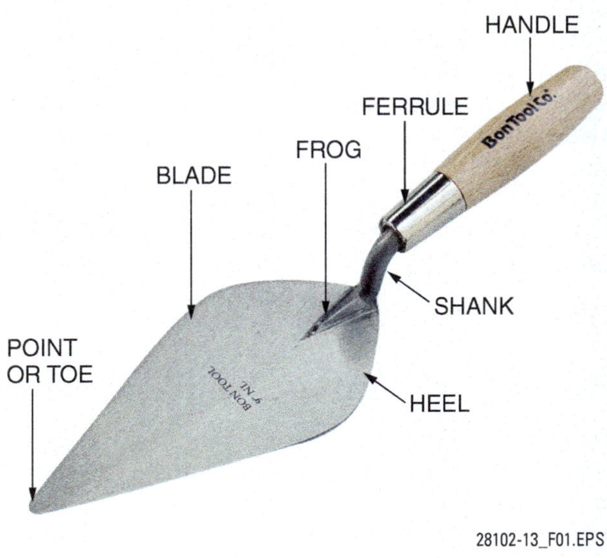

Figure 1 Parts of a trowel.

Figure 2 Brick trowel shapes.

be used as often as a standard brick trowel. Some trowels are particular to certain parts of the country.

The standard trowel, or brick trowel, is the all-purpose model, used for cutting, buttering, and adjusting units. The small **pointing** trowel fits into tight spaces. It is used to point, grout, and tool mortar joints. The margin trowel, with a small square blade, is used to smooth joints and mix grout or other compounds. The tuckpointer trowel is used to shape mortar between joints. It comes in widths ranging from ⅛ to 1 inch, to fit commonly used joint sizes.

The plasterer's, or parging, trowel is a flat piece of steel with the handle mounted on one side. It is used to apply a thin coat, or **parge**, of mortar. The parge coat will hold tile or veneer to the masonry units. The duck bill trowel is narrow, with a long blade up to 13 inches long. It is primarily used for cleanup. The bucket trowel has straight sides and a square toe, which aid the mason in scraping the sides of the bucket. Tile-setting trowels have a wide blade to handle large amounts of mortar or grout.

Hold the Trowel Properly

When holding a trowel, keep your thumb along the top of the handle, not the shank. If your thumb is on the shank, it can get coated with mortar as you work, causing skin irritation.

1.2.0 Using Hammers and Chisels

Mason's hammers fall into two categories: cutting hammers and mauls. Using the right hammer for each job will ensure the most efficient use of your energy. Many types of hammers come with wooden handles. Inspect these handles every day to make sure they are not loose, splintered, or cracked. Chisel edges on hammers need to be inspected and sharpened, just as chisels do.

Masons use chisels to cut masonry units. For everyday work, masons use a hammer to cut brick or block, but for precise, sharp edges, they use a

Stack Brick Safely

A neat brick stack is a safe stack. Stack materials by reversing direction on every other layer, so they will be less prone to tip. Keep the pile neat and vertical to avoid snagging clothes.

28102-13_SA01.EPS

Figure 3 Different types of trowels.

chisel. There are several types of chisels. In this section, you will learn about the various types of hammers and chisels that you will use on the job.

> **WARNING!**
>
> If the wooden handle on a hammer becomes loose, replace it immediately. If the handle breaks, the head could fly off, causing serious damage or injury.

1.2.1 Cutting Hammers

The brick hammer is a double-headed hammer with a chisel head on one side, as shown in *Figure 4*. This is the most commonly used, everyday mason's hammer. The brick hammer drives nails and strikes chisels with one head. It can break, cut, or chip masonry material with the other head. Use it for cutting masonry units, setting line pins, and nailing wall ties.

Figure 4 Brick hammers.

Brick hammers are one-piece, drop-forged steel tools, or they can be steel heads on wooden or fiberglass shafts. Steel hammers need a comfortable grip sleeve over the shaft. The weight of this hammer usually ranges from 12 to 24 ounces. If

> **Choose Wisely**
>
> A trowel should be sturdy, light, and balanced. It should have a comfortable grip and a flexible steel blade. There are many options to consider, so choose a tool that is right for you.

28102-13_F05.EPS

Figure 5 Tile hammer.

you are buying a hammer, pick a weight that is comfortable for your hand.

The tile hammer is smaller than a standard brick hammer with a smaller, thinner cutting head. *Figure 5* shows this hammer, which usually weighs about 9 ounces. Use it for cutting and trimming tile where you need more precision than you can get with a brick hammer. Do not drive nails, strike chisels, or do other heavy work with a tile hammer.

The stonemason's hammer, shown in *Figure 6*, resembles an ax. This heavy hammer has a cutting blade on one end of the head. Use this special-purpose hammer for dressing, cutting, splitting, or trimming stone.

> **WARNING!**
>
> Never strike the heads of two hammers together. A chip could fly off one of the heads.

1.2.2 Mashes and Mauls

Mashes (also called two-pound hammers) and mauls (also called bushhammers) are heavy club-headed hammers used for tooling. They usually have two striking heads.

The mason's mash is a drop-forged, one-piece tool with a grip sleeve over the shaft, as shown in *Figure 7*. It can also have a steel head mounted on a wooden handle. It resembles a short-handled, double-headed sledgehammer. The mash is most commonly used to strike chisels to cut masonry units. This hammer is too heavy to use for setting line pins or nailing wall ties.

28102-13_F06.EPS

Figure 6 Stonemason's hammers.

28102-13_F07.EPS

Figure 7 Mash.

The maul is a square-ended, rectangular, double-headed sledgehammer. It has a heavy head with toothed ends and a long wooden handle, as shown in *Figure 8*. It is a stonemason's tool used to face block, stone, or concrete.

The rubber mallet (*Figure 9*) has a rubber head with two flat ends and a wooden handle. This hammer can tap or drive something without leaving marks. Use the rubber mallet for setting stone, marble, tile, or other finely finished units into place.

28102-13_F08.EPS

Figure 8 Toothed maul, or bushhammer.

28102-13_F09.EPS

Figure 9 Rubber mallets.

When working with wooden-handled hammers, be sure to inspect them daily. Check that the handle is securely seated in the head. If the handle is loose, replace it before using the hammer.

1.2.3 Chisels

Chisels become damaged after prolonged use. Cutting edges get notches or burrs from striking rough surfaces or metal. Chisels are cutting tools, so keep them sharp. The best way to keep chisels sharp is to take them to a blacksmith. Like the mason's hammer, tempered steel, such as that used for chisels, must be sharpened carefully so it does not shatter. Blacksmiths are trained to do this properly.

Heads flatten after long use. The striking head mushrooms out, and metal burrs form at the edges of the head. These deformed edges can fly off and may cause injury. Inspect your chisels every day for dullness or deformation. To prevent injury, grind off the deformed part of the chisel head on a grinding wheel. As you grind, cool the chisel with water to keep it from overheating. If it gets overheated, the steel loses its temper and may unexpectedly shatter. Do not repair or sharpen the blade yourself; take it to a blacksmith.

> **WARNING!**
>
> Do not use a chisel with a deformed blade or a mushroomed head. Repair it or replace it immediately.

Figure 10 shows a mason's steel chisel, used for general cutting. It has a narrow blade of 1½ to 3 inches for a neat, clean cut. Use it to cut out masonry and make repairs. Use it for brick, block, and veined stone.

Figure 11 shows a brick set chisel. This is a wider chisel, up to 7 inches wide, with a thicker, beveled cutting blade. Traditionally, the brick set

28102-13_F10.EPS

Figure 10 Mason's chisel.

Protect Your Hands

Some brick chisels have a protective grip. The grip is made of rubber and has a rubber hood to protect your knuckles from misplaced hammer strikes.

28102-13_F11.EPS

Figure 11 Brick set chisel.

is as wide as the brick it cuts. A wider version of the brick set is called the blocking chisel, or bolster chisel. This is usually 8 inches wide and used for cutting block.

The rubber-grip mason's chisel is shown in *Figure 12*. This chisel has a rubber cushion on the handle to soften the impact of the hammer blow. It also has a wider cutting blade than the standard mason's chisel and is used for cutting brick, block, or stone.

The tooth chisel (*Figure 13*) has a toothed edge. This chisel is designed to cut soft stone and shape it to fit. It should not be used for hard stone. The pitching tool (also shown in *Figure 13*) is used for sizing, trimming, and facing hard stone.

The plugging chisel, or joint chisel, (*Figure 14*) has a sharply tapered blade. It is also called a tuckpointer's chisel. Use this chisel for cleaning mortar joints, for chipping mortar out of a joint, or for removing a brick or block from a wall. Each chisel is made for a specific job. Using the right tool for the job will make your work easier and more professional looking.

1.3.0 Using Jointers and Brushes

Before the mortar is set, it can be troweled or tooled. Special tools called jointers are used to finish, or point, the surface of the mortar joint. They are called jointers because they are designed to fit into the joint between brick or block. Brushes are used to remove any burrs or excess mortar from masonry work. Brushing is the finishing process for the wall or floor. In this section, you will learn about commonly used jointers and brushes.

28102-13_F12.EPS

Figure 12 Rubber-grip mason's chisel.

1.3.1 Jointers

Jointers, slickers, rakers, beaders, and sled runners are used for finishing or pointing the surface of mortar joints. Also called joint tools or finishing tools, jointers are available to fit a range of joint sizes. The standard-size jointer will fit into the standard joint between brick or block. Jointers come in various shapes to give different effects to the finished joint. *Figure 15* shows how the final surface appearance is determined by the shape of the jointer used.

Jointing compresses the mortar and decreases moisture absorption at the surface, so it adds water protection. Struck and raked joints are not recommended for exterior walls as they do not offer good water protection.

The jointers are usually cast or forged, shaped steel rods with or without wooden handles. *Figure 16* shows some simple jointers for compressing and waterproofing joints. These convex, round, flat, and V-jointers are used on the short vertical or head joints. You can tell from the profile of the jointer what the tooled mortar joint will look like.

28102-13_F13.EPS

Figure 13 Tooth chisel and pitching tool.

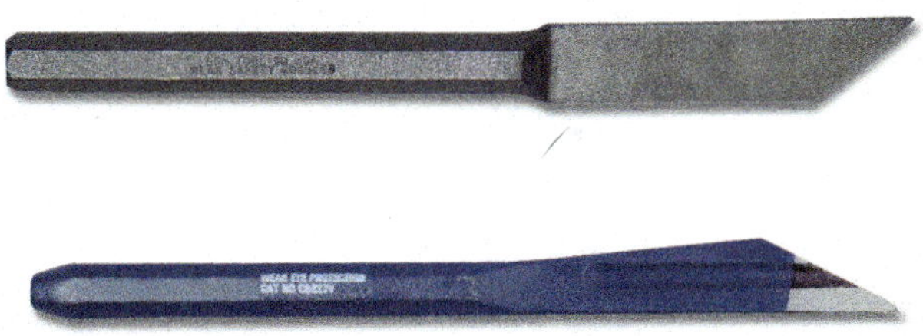

28102-13_F14.EPS

Figure 14 Plugging, or joint, chisel.

The **bed joints**, or the long horizontal joints, are tooled after the head joints. Longer jointers with wooden handles are used for the horizontal joints. As shown in *Figure 17*, these sled runner jointers come in a variety of shapes also, to match the head joints.

Another type of jointer is the raker or rake-out jointer; these also come mounted on skate wheels. *Figure 18* shows a skatewheel and a hand raker. The skatewheel raker is used for bed joints. The hand raker is used for joints the skatewheel raker will not reach. Note that the raking mechanism has an adjustable setscrew to rake different depths. The advantage of the skatewheel raker is the speed with which it forms a neat, hole-free joint.

All jointers need to be cleaned after they are used. If mortar hardens on them, they are not usable for smoothing.

1.3.2 Brushes

Brushing masonry work removes any burrs or excess mortar. This is the finishing process for the wall or floor. The brush should have stiff plastic or wire bristles. *Figure 19* shows several brushes commonly used by masons.

A stove brush has a longer handle to keep fingers clear of the work area. Brushes are also useful for brushing off footings before laying masonry units and for cleaning the work area. For cleaning stains, brushes are used to apply proprietary cleaner to the face of masonry units. A cleaning brush may also have a scraper on one edge. Some brushes are designed to fit a long extension handle. Brushes for the application of proprietary cleaner are made of stiff plastic or fiber and have longer handles.

> **WARNING!** When applying proprietary cleaner, use a long-handled brush, and wear gloves and eye protection.

1.4.0 Using Other Masonry Tools

Brick tongs are designed to carry brick without chipping or breaking it, as shown in *Figure 20*. Most are made of an adjustable steel clamp with a locking nut and a handle. They can be adjusted for different sizes and will hold 6 to 11 masonry units. Some brick carriers are steel rods that fit into the holes in the brick. These carriers do not have a clamp and must be held upright. Smaller carriers are designed to carry larger individual block. With practice, the mason can carry loaded brick tongs in one hand, freeing the other hand for other activities.

In addition to their other tools, masons use bulk guns, grout bags, pinch bars, and bolt cutters (*Figure 21*). The bulk gun, also called a caulking gun, is used to add caulking to expansion joints.

Multipurpose Jointers

Some jointers have different heads on each end; for example, a round head on one end and a V-head on the other end. Using double-headed jointers allows you to do more joints with fewer tools.

The flat jointer is sometimes called a slicker. Either end of a slicker can be used to shape mortar; usually the ends are different sizes. The slicker is handy for working on inside corners or other tight places where a neat joint is needed. A bull horn is a common round jointer that is larger on one end and tapered on the other.

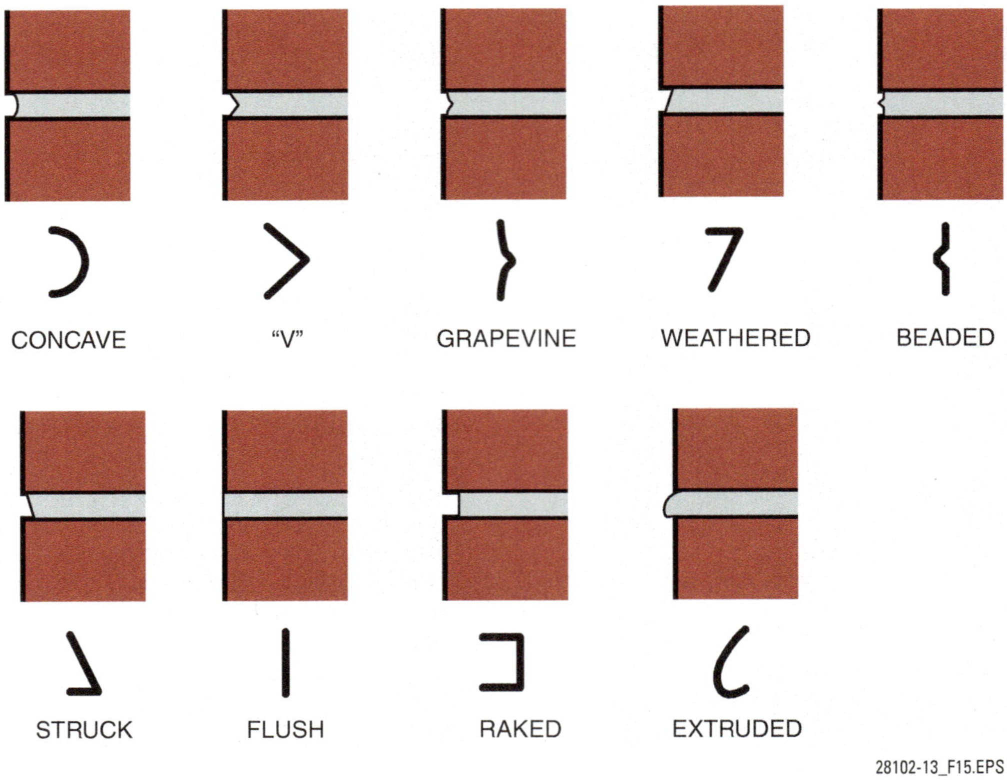

Figure 15 Tooled mortar joints.

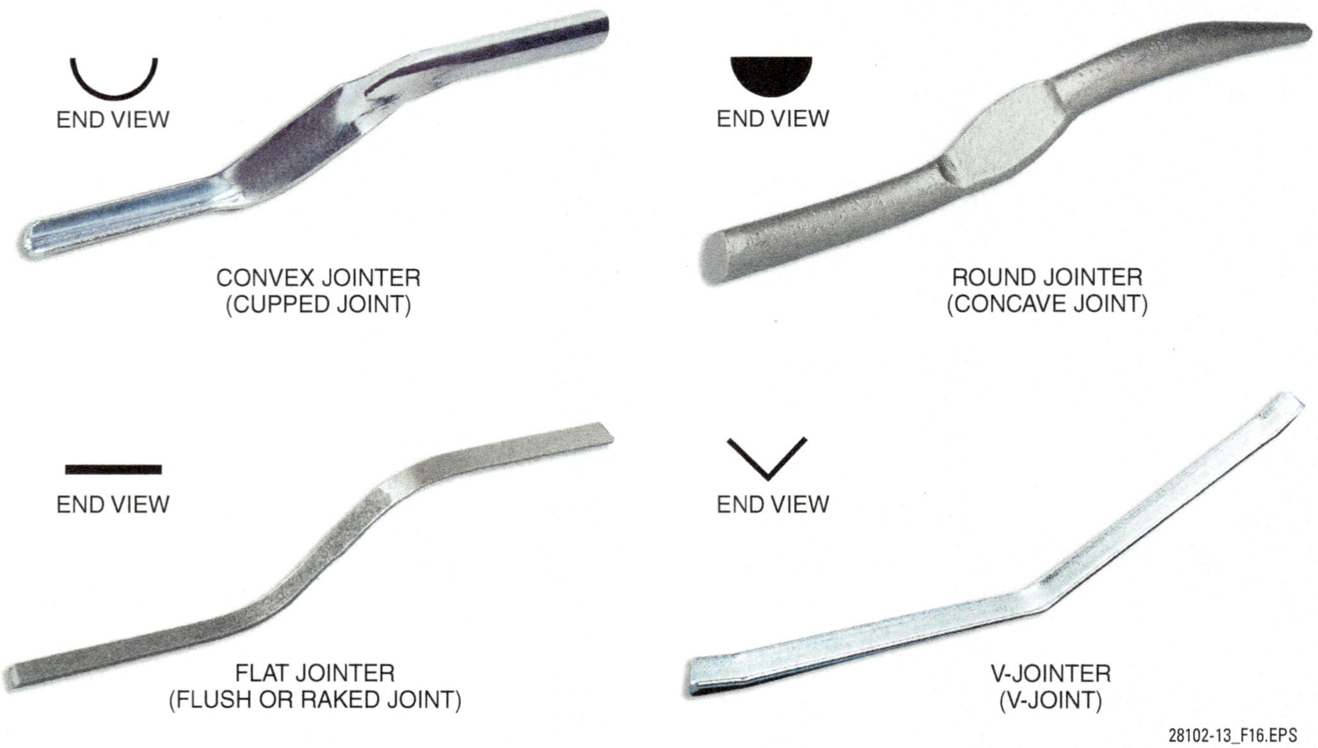

Figure 16 Jointers.

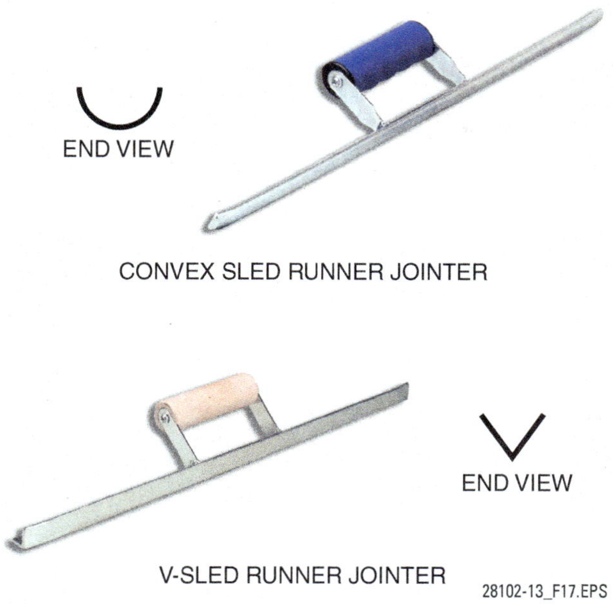

Figure 17 Runner jointers.

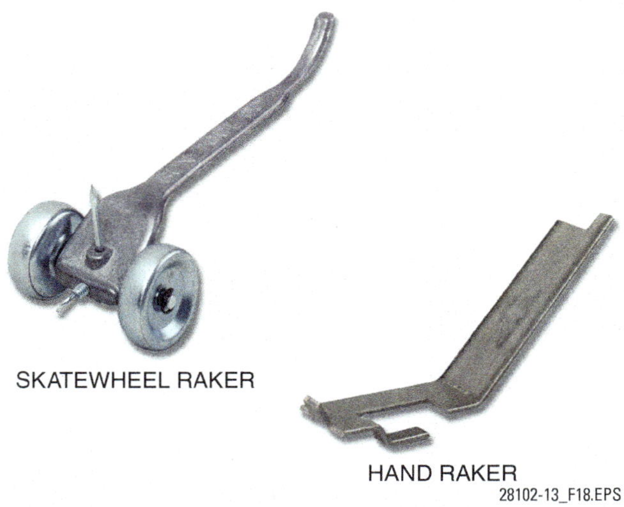

Figure 18 Rakers.

Figure 19 Mason's brushes.

Figure 20 Brick tongs.

The suction gun is used to propel abrasive material at masonry surfaces. The grout bag, with a metal or plastic tip, is squeezed to apply grout between masonry units. The pinch bar is used to pry out masonry units. The bolt cutter is used to cut reinforcing wire or ties. These and other hand tools will fit into the mason's tool bag.

Figure 22 shows a mason's canvas tool bag with leather straps, handles, and reinforced bottom. The tool bag keeps tools together and within reach as the mason moves around the job site. Buckets or other open containers will not keep tools dry. It is important to keep your tools dry and protect them from damage. Levels, rules, and

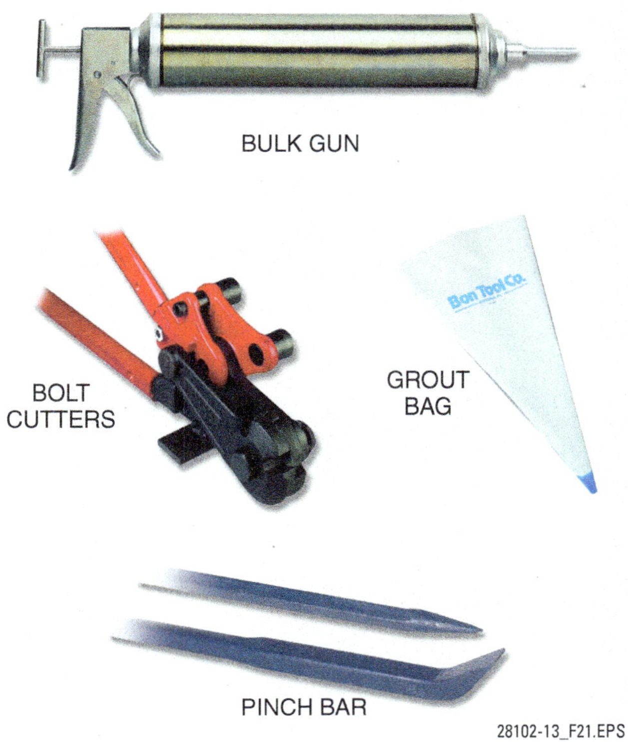

Figure 21 Common mason's hand tools.

Figure 22 Mason's tool bag.

squares must be properly maintained if they are to remain accurate.

Typically, a tool bag measures 14 to 18 inches across and has inside pockets for small items. The steel square and long level will fit under the leather straps on the outside of the bag. Some tool bag handles convert to shoulder straps, so masons will have hands free for climbing ladders or scaffold. A good tool belt can also keep tools handy and your hands free.

Additional Resources

Bricklaying: Brick and Block Masonry. Reston, VA: Brick Industry Association.

Concrete Masonry Handbook. Skokie, IL: Portland Cement Association.

Masonry Construction. David L. Hunter, Sr. Upper Saddle River, NJ: Prentice-Hall.

1.0.0 Section Review

1. The wide end of a mason's trowel is called the _____.
   a. point
   b. heel
   c. shank
   d. ferrule

2. The most commonly used mason's hammer is the _____.
   a. tile hammer
   b. stonemason's hammer
   c. two-pound hammer
   d. brick hammer

3. A joint that is *not* recommended for exterior walls is the _____.
   a. concave joint
   b. weathered joint
   c. raked joint
   d. extruded joint

4. To pry out masonry units, use a(n) _____.
   a. tong
   b. extractor bar
   c. lever
   d. pinch bar

SECTION TWO

2.0.0 MEASURES AND MEASURING TOOLS

Objective

Identify measures and measuring tools used in masonry.

a. Describe how to use rules.
b. Describe how to use levels.
c. Describe how to use chalk boxes, squares, plumb bobs, and laser levels.
d. Describe how to use corner poles, lines, and fasteners.

Performance Tasks 1 and 2

Demonstrate the proper use of a rule.
Demonstrate the proper use of a level.

Trade Terms

Corner pole: Any type of post braced into a plumb position so that a line can be fastened to it. Also called a *deadman*.

Lead: The two corners of a structural unit or wall, built first and used as a position marker and measuring guide for the entire wall.

Measuring is an important part of the work that masons do. Accuracy is important when making measurements, for reasons of safety as well as for aesthetics. Masons use a variety of tools for making measurements on the job, including rules and levels, chalk boxes, squares, plumb bobs, laser levels, **corner poles**, lines, and fasteners. Take the time to learn how to use these measuring tools well, because they are just as important as any other tool you will use.

2.1.0 Using Rules

Masons use two kinds of rules, a 6-foot folding rule, and a 10-foot retractable tape. Mason's rules have special marking scales on them.

The brick spacing rule, as shown in *Figure 23*, has markings for different sizes of joints. It shows the course spacing for each joint size. This measure is also called the course counter rule. Use it to lay out and space standard brick courses to dimensions that are not modular. It is useful for spacing mortar joints around door tops, window tops, or other uneven spaces.

Figure 23 Spacing rules.

The modular spacing rule, as shown in *Figure 23*, is based on a module of 4 inches. It has six scales ranging from 8 to 2, with scale 6 representing the size of a standard brick. The modular rule can be used for block as well as brick. There is a slight disparity between the brick spacing rule and the modular spacing rule.

The mason's steel tape and folding rule are both marked with spacing measures. They are available with either the modular or the course markings. *Figure 24* shows both these rules, which also have inch measurements. Folding rules are usually made of wood, with brass joints and piv-

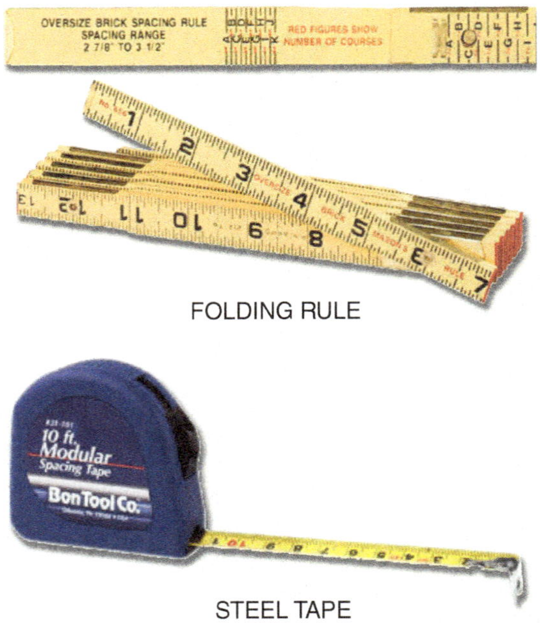

Figure 24 Folding rule and steel tape.

ots between the sections. An oversized brick ruler has 11 scales for queen- or king-size brick spacing, ranging from 2⅞ to 3½ inches on one side, and inches, eighths, and sixteenths on the other side. These rules are usually marked with letters and/or numerals to show the number of courses.

Masons prefer a folding rule because one person can easily and accurately extend it to the needed length. Tapes must be stretched without bowing, which may need an extra hand over longer distances. Both rules and tapes should be inspected and cleaned daily. Use a very light touch of linseed oil on wood rules, and be sure to clean dust and grit from the pivots. A belt pouch or clip will keep your rule handy and clean.

2.2.0 Using Levels

Levels were introduced in *Core Curriculum, Introduction to Hand Tools*. The level, sometimes called a plumb rule, establishes two measures:

- A plumb line that is vertical to the surface of Earth
- A level line that is horizontal to the surface of Earth

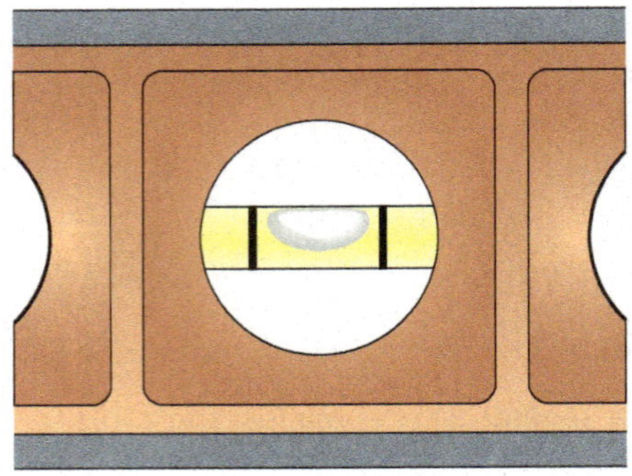

Figure 25 An air bubble shows level or plumb.

The level contains air bubbles in sealed vials filled with oil or alcohol, as shown in *Figure 25*. When the bubble in the vial is centered between the vial's two markers, the object being checked is level or plumb.

Levels can come with single or double sets of bubble vials. Two common levels are shown in *Figure 26*. They are made of hardwood, metal, or plastic to make them as light as possible without sacrificing strength. Levels come in a wide range of sizes. A smaller 8-inch size, often called a torpedo level, is used for individual block or smaller spaces. Longer levels can be up to 72 inches for leveling wall sections or spanning distances. Masons use levels continually to check the plumb and level of individual masonry units and the entire length of a wall.

Figure 26 Torpedo level and standard level.

Take Care of Your Tape

Steel tapes are wound with a spring, so they will retract into the housing. Never let a retractable steel tape snap back into the housing. The end tab can easily break off, making the tape useless. Always guide the tape back into the housing.

String Line Level

A line level is only a few inches long and usually made of plastic. They are designed to hang on a string line to check the accuracy of guidelines.

The level is usually the most expensive of a mason's tools. Always check the action of the bubble vials against another level before purchasing a level. Check the accuracy of your level daily, especially if it is dropped or jarred. Do not tap a level with a hammer to level a brick. Some levels can be adjusted if they are no longer accurate.

Clean your level carefully at the end of the day to keep any mortar from hardening on it. Wipe off wooden levels with a rag dampened with linseed oil to preserve the wood. Clean a metal level with a dry cloth, so dust and grit do not stick to it. You may want to oil it occasionally as well.

2.3.0 Using Chalk Boxes, Squares, Plumb Bobs, and Laser Levels

A chalk box (*Figure 27*) is a metal or plastic case with finely ground chalk and about 50 feet of twisted cotton line wound on a spool inside. As the line is drawn out of the box, it picks up powdered chalk. The line can be stretched between two points and snapped. This will leave a chalk mark exactly under the line. Masons use chalklines to establish straight lines for the first course of brick in a wall. The chalkline can be snapped against the foundation slab or footing. Some chalk boxes can also be used as a plumb bob.

Masons use several types of squares. The framing, or steel, square (*Figure 28*), looks like a carpenter's square. Use this for laying out **leads** and other corners and for checking that corners are square.

As shown in *Figure 29A*, the combination square, or T-square, has a movable crosspiece and a built-in 45-degree angle. Use this for marking right-angle and 45-degree angled cuts and

28102-13_F27.EPS

Figure 27 Chalk box.

28102-13_F28.EPS

Figure 28 Framing, or steel, square.

for checking these angles on cuts. The sliding T-bevel, as shown in *Figure 29B*, has a setscrew in the crosspiece. The screw fixes the angle of the crosspiece so that the same angle can be used for marking many cuts. The bevel is useful when the job calls for an odd angle other than 90 degrees or 45 degrees. Use the bevel for marking these kinds of cuts, for marking skewbacks, and for checking the angles on cuts.

Check Your Squares

Squares may be damaged accidentally from time to time. Make sure that you periodically check your squares with a known straightedge. This will ensure accuracy and square walls. A small error is compounded over the length of a wall. This can result in serious error in the finished structure.

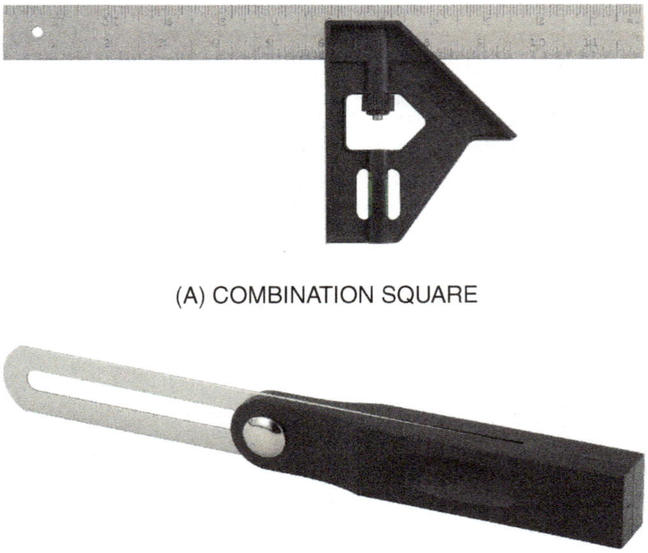

(A) COMBINATION SQUARE

(B) SLIDING T-BEVEL

28102-13_F29.EPS

Figure 29 Combination square and sliding T-bevel.

Cold beam laser levels are used to mark plumb (*Figure 30*). This leveling device projects a thin beam of low-wattage light. Laser leveling instruments are more expensive than other leveling tools (although they are becoming more affordable), but they offer several advantages. Lasers are more accurate over longer distances. Lasers are very easy to use when establishing vertical or horizontal lines and are used for measuring or laying out openings.

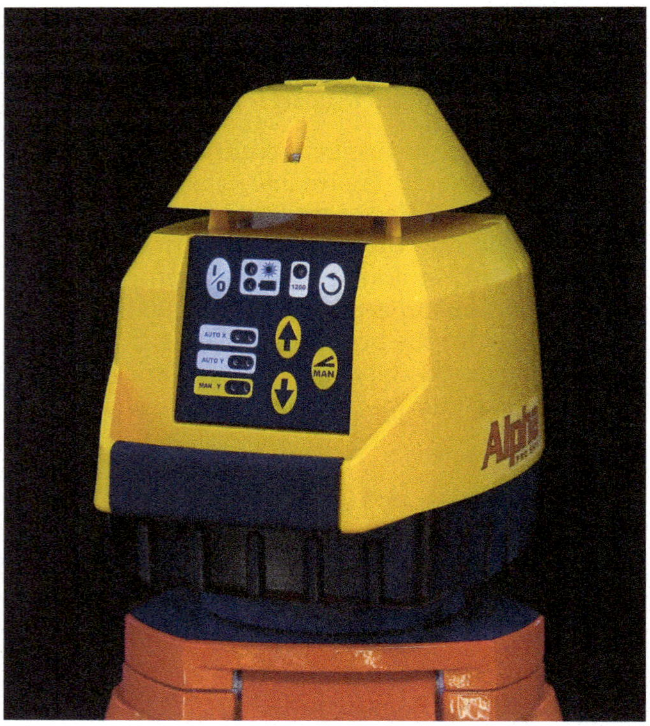

28102-13_F30.EPS

Figure 30 Cold beam laser level.

Chalk Box Maintenance

Before using a chalkline, check that there is enough chalk in the reservoir and that the string is not frayed. Both of these parts should be replaced or refreshed periodically.

CAUTION

Use cold beam lasers with extreme care. Although low-power lasers can be used, arcs can still be unpredictable. Always wear appropriate personal protective equipment and refer to the manufacturer's specifications.

Some laser level tools can project beams up to 1,500 feet to hit a specific point. They can have a variety of components designed to get the most accurate readings possible:

- Large vertical and horizontal indicator bubbles
- Lens refractors that show line or dot images
- Dual diodes for increased beam visibility
- Wall, floor, or tripod mounts, brackets, or clamps
- Built-in bump sensors that will automatically notify the user, or shut off the device, if the instrument has been bumped hard enough to affect the accuracy of its readings
- Electronic level vials that dampen vibrations from conditions at the job site, to ensure the stability of the laser beam over distance
- Remote control capability
- Optional receivers that increase the laser's operating range still more

Using this tool, you can tell immediately if something has been aligned. If you are installing gravity-flow pipelines, for instance, you can use this tool as an alignment guide and lay the pipe along the beam of light emitted by the laser. Other applications of cold beam lasers include installing and aligning such building elements as walls, partitions, and access floors.

Portable, handheld lasers that feature plumb (both up and down), level, and square capabilities can greatly enhance productivity, which saves time and money. Lasers can be set up more quickly than other leveling devices, and because of their accuracy, they require fewer repeat measures. The time saved when using a laser leads to cost savings. The initial expense of a laser is quickly made up in increased productivity.

These devices require relatively little maintenance. Like other levels, lasers should be cleaned off after use and stored in a safe, dry place. Once a year, the unit should be cleaned, checked, and adjusted by a professional. Heat, cold, and moisture—conditions common to plumbing sites—can affect a laser's operating performance. The degree to which these conditions affect the laser's stability and accuracy will depend on the quality of the instrument itself. Higher-quality lasers are generally more resilient to adverse environmental conditions. Some lasers are specifically designed and tested to endure shock, moisture, and temperature changes. The performance of all lasers, however, begins to deteriorate rapidly at temperatures of 110°F or higher.

The light emanating from a cold beam laser will not instantly injure humans, but long-term exposure could injure the eyes. Therefore, never point a laser level at another person. Only qualified and trained masons should use this equipment. Cold beam lasers must meet all of the Electronic Product Radiation Control provisions of the federal Food, Drug, and Cosmetic Act.

WARNING!

Never look directly into the laser beam or point the beam at a co-worker. Over time, exposure to the beam can injure eyes.

Sometimes plumb is marked with a plumb bob (*Figure 31*). The plumb bob is a pointed weight attached to a length of mason's line. The length of the line is easily changed. A plumb bob can establish vertical plumb points. Use it to mark a point directly under another measured point. A plumb bob is also useful for checking the plumb of a story pole or corner pole. Today, laser levels have largely replaced plumb bobs in the field, but because of their ease of use they are still sometimes used for simple tasks.

28102-13_F31.EPS

Figure 31 Plumb bob.

2.4.0 Using Corner Poles, Lines, and Fasteners

The next in the series of course alignment measures is the corner pole, or deadman. The corner pole is any type of post braced into a plumb position so that a line can be fastened to it. This allows a wall to be built without building the corners first. This tool is useful when building a veneer wall. The corner pole can be braced against an existing part of a structure or the block foundation.

Corner poles can be made from dimensional lumber and braces. Commercial corner poles (*Figure 32*) are metal with proprietary line blocks and supporting hardware. Commercial corner poles have masonry units marked on them. A handmade corner pole can be marked off in masonry units with a grease pencil, marker, or carpenter's pencil. If you make a deadman from scrap wood, make sure that the wood is straight and not badly warped. A corner pole with markings for different stories of a building is called a story pole.

While masons use levels to lay out structural masonry elements 4 feet in length or shorter, they use a mason's line to lay out masonry structures longer than 4 feet. The line itself is twisted or braided nylon cord. The typical line is about 150 pounds test and should be used at 20 percent of its test. This means it is good for a pull of up to 30 pounds. Braided line is preferred because it will not sag as much when pulled tight and will last longer.

The line is stretched tight as it is strung between two fixed points. For most jobs, the fixed points are on the wall corners, or leads, which have already been laid. The line becomes the guide for laying the course of masonry units between the leads, or fixed points. Using the line properly will result in a wall without bulges or hollows. The

Blowing in the Wind

When using a plumb bob outside, you must be aware that the wind may blow it out of true vertical. The longer the drop line, the more likely that wind will affect your measurement.

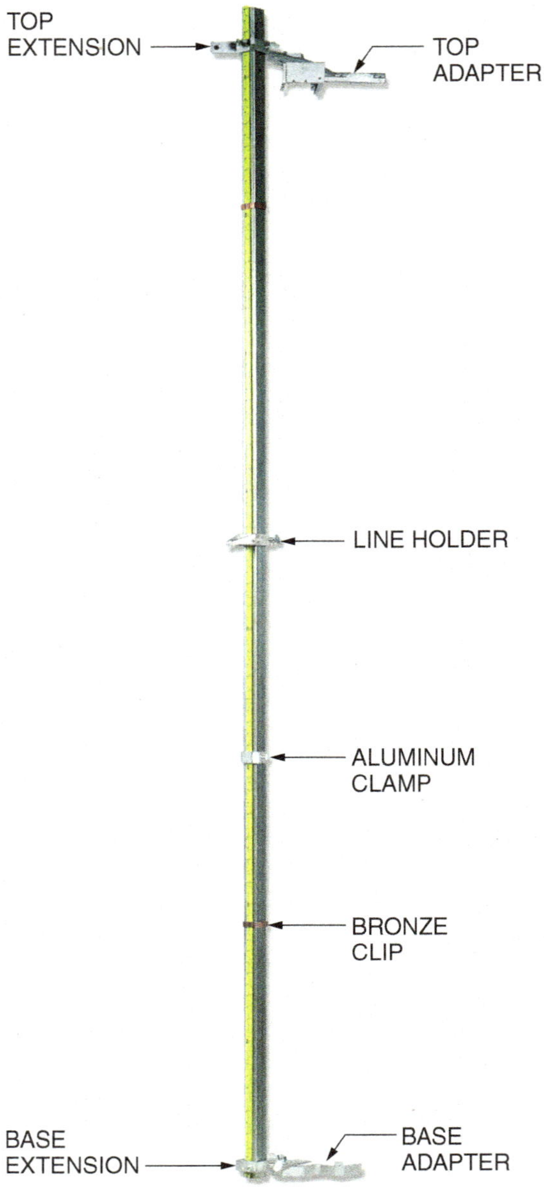

28102-13_F32.EPS

Figure 32 Corner pole or deadman.

masonry units are laid under the line, and the line is moved up for each course.

There are three methods of fastening the line in place: pins, blocks, and stretchers. Line pins (*Figure 33*) are about 4 inches long and made of steel. Drive them into the wall or structure at the marked point, and string the line tightly between them. Line pins leave holes, and the string must be remeasured or retightened for each course. However, line pins are harder to accidentally knock off than line blocks.

> **WARNING!** When using line pins, remember that the line is under tension. Should the line snap or the line pin come out while the line is under tension, the line or the pin can whip around and cause injury.

Line blocks are also called corner blocks (*Figure 34*). They are made of wood, plastic, or metal; however, wood grips the corner better (*Figure 35*). The knotted line passes through the slit in the back of the block. The line tension between the two blocks holds the blocks in place.

28102-13_F33.EPS

Figure 33 Line pin.

Don't Tangle Your Line

Usually, the line is bought wound around a core. If it is rewound on a shuttle or holder, it will not tangle as it is used and re-used. Lines are made in several bright colors, so they can be easily seen.

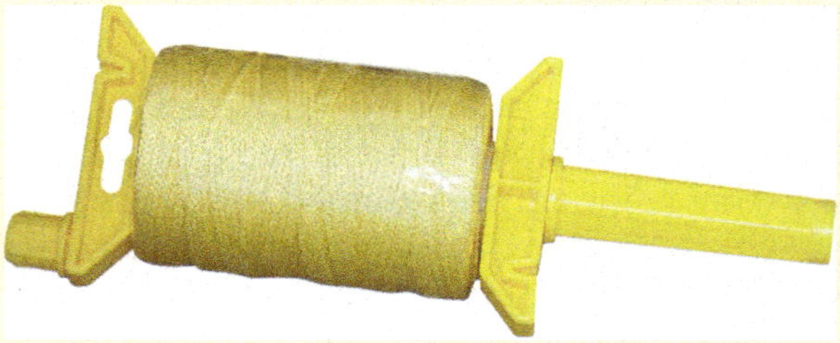

28102-13_SA02.EPS

28102-13_F34.EPS

Figure 34 Corner block.

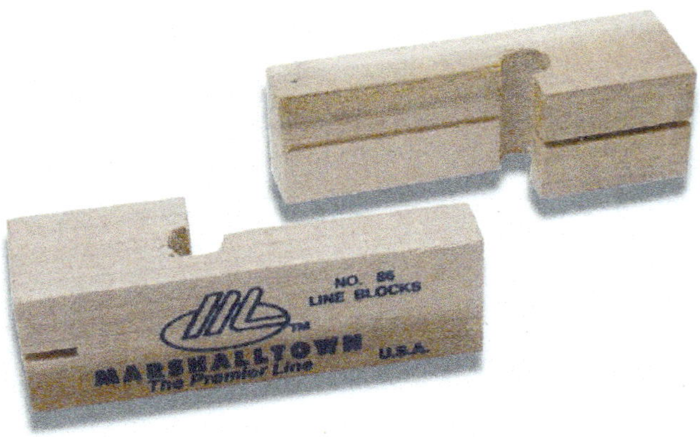

28102-13_F35.EPS

Figure 35 Wood line block.

Line stretchers are shown in *Figure 36*. They have a flatter profile and also use the line tension to hold them in place. They come in standard and adjustable sizes and fit more snugly to the lead than corner blocks.

Unlike pins, blocks and stretchers must have a corner to be held snugly against. They do not leave holes in brick or walls, but they do project slightly from the corners of the wall. They can be accidentally knocked off. Because of the tension on the line, a flying block or stretcher can severely injury someone in its path.

WARNING!

Use extra care working with corner blocks or stretchers to avoid knocking them off.

Once the line is in place, it may need more support. Line trigs (*Figure 37*), sometimes called twigs, are steel fasteners that hold the line in position and keep it from sagging. A very long line

28102-13_F36.EPS

Figure 36 Line stretchers.

28102-13_F37.EPS

Figure 37 Line trig, or twig.

may need several trigs, even if it is stretched tight. The trig slips over the line and rests on a masonry unit that has been put in place to support it. A half brick or piece of block can anchor the trig to ensure the line is not disturbed.

Check with Your Supplier

Masonry materials suppliers sometimes give away trigs and wood line blocks with larger orders.

Additional Resources

Complete Masonry: Building Techniques, Decorative Concrete, Tools and Materials. Des Moines, IA: Oxmoor House.

Building with Masonry: Brick, Block, and Concrete. Dick Kreh. Newtown, CT: Taunton Press.

2.0.0 Section Review

1. The number of scales on an oversized brick ruler for queen- or king-size brick spacing is _____.
   a. 7
   b. 9
   c. 11
   d. 13

2. Eight-inch levels are often called _____.
   a. torpedo levels
   b. standard levels
   c. plumb levels
   d. line levels

3. To lay out leads and check that corners are square, use a _____.
   a. combination square
   b. framing square
   c. laser level
   d. plumb bob

4. Steel fasteners that hold a line in position and keep it from sagging are called _____.
   a. line pins
   b. line blocks
   c. line stretchers
   d. line trigs

SECTION THREE

3.0.0 MORTAR EQUIPMENT

Objective

Identify mortar equipment used in masonry.

a. Describe how to use mortar boxes.
b. Describe how to use mixing accessories.

Trade Term

Tempering: Adding water to mortar to replace evaporated moisture and restore proper consistency. Any tempering must be done within the first 2 hours after mixing.

Hand-powered mortar equipment is used on both small and large jobs. It complements the hand tools and completes the tool set needed to build a structural masonry unit.

3.1.0 Using Mortar Boxes

Mortar is mixed in a wheelbarrow or mortar box. A mortar box can be steel or plastic and is about 32 inches × 60 inches. Some mortar boxes have wheels for easy movement (*Figure 38*). Make sure the box is set level so that water does not collect in one end. After it is mixed, the mortar may be moved to a mortar pan (also shown in *Figure 38*) or smaller mortarboard.

The mortar pan and mortarboard can be located next to the working mason. They can be placed on a stand, so they are convenient for working higher courses. The metal or plastic mortar pan fits more securely on the stand, but mortar tends to stick in the corners. Mortarboards are available in plastic or plywood and average about 2 feet on a side. A plywood board needs to be wetted thoroughly before mortar is put on it to keep the board from absorbing moisture and drying the mortar.

A cubic-foot measuring box (*Figure 39*) is sometimes used to prepare mortar. This box measures 1 cubic foot of sand or cement. It is used to measure ingredients by volume in order to proportion the mortar correctly. As prepackaged mortar has become more popular, use of the cubic-foot box is uncommon, but sometimes it is specified by architects for measuring sand.

Mortar containers and carriers should be cleaned out with water between loads of mortar. The mixing pan should be cleaned out after each batch and cleaned thoroughly at the end of each day.

MORTARBOARD

MORTAR PAN

MORTAR BOX WITH WHEELS

28102-13_F38.EPS

Figure 38 Mortarboard, mortar pan, and mortar box with wheels.

3.2.0 Using Mixing Accessories

Several of the tools used in mixing mortar are shown in *Figure 40*. A long-handled pointed-end shovel is used to measure, shovel, and mix dry ingredients in the mortar box. A square-end, short-handled shovel is used to move mortar from the mixing box to a pan, stand, or hod. The square end helps you to scrape the bottom of the mixing pan.

A large hoe is used for mixing wet mortar and **tempering** mortar. The mortar hoe has a 10-inch blade with two holes in the blade to make it easier

28102-13_F39.EPS

Figure 39 Cubic-foot measuring box being emptied into a mortar container.

to pull the hoe through the mix. Some hoes have shorter handles for use in smaller spaces. Remember to clean mixing tools with a stiff brush and water immediately after use.

3.2.1 Water Bucket and Barrel

The water bucket and barrel (*Figure 41*) are essential in masonry. Made of steel, plastic, or galva-

28102-13_F40.EPS

Figure 40 Mixing aids.

nized metal, both are necessary equipment. Half filled with water, the bucket provides a ready bath to clean mortar off hand tools. Masons frequently use 5-gallon buckets to measure water for mixing mortar. The bucket can also serve as a carrier for small amounts of mortar or grout. The barrel pro-

Hawks and Hods

As a mason, you won't see hawks and hods on the job very often, but when you do, it's important to recognize how to use them.

The hawk is a smaller version of the mortarboard. This small board has a pole-grip handle underneath and is used to carry small amounts of mortar. A close relative of the plasterer's hawk, it is useful for tasks such as pointing. A wooden hawk needs to be wetted thoroughly before mortar is put on it.

The hod is used to carry mortar from the mixer to the mason. It is practical for moving material in tight spaces or on scaffold. It is an aluminum trough with a long pole handle. The mortar can be lifted up to a mason on a scaffold by using a hod.

HAWK

HOD

28102-13_SA03.EPS

28102-13_F41.EPS

Figure 41 Water bucket and barrel.

vides a bath for larger tools and a ready supply of water for tempering mortar.

Be sure to wet the inside surface of the bucket before filling it with mortar or grout. Always wash empty buckets out immediately to prevent mortar from hardening inside. Wash the barrel out at the end of the day.

3.2.2 Barrows

Masons use two types of barrows (*Figure 42*). The standard contractor's wheelbarrow is made of steel with wooden handles. It can carry about 5 cubic feet of masonry or mortar. The second type of barrow is a pallet on wheels. Its wooden body has no sides, so it is convenient to unload masonry units from either side. However, it must be unloaded so that it does not become unbalanced and tip over. This brick barrow works well for moving bags of cement, piles of block, and other bulky materials.

CONTRACTOR'S WHEELBARROW

BRICK AND TILE BARROW

28102-13_F42.EPS

Figure 42 Barrows for masonry units.

Both types of barrows need large, air-filled tires to move over bumpy ground without tipping. Check the wheels before using any barrow.

Brick Carts

A commercial cart is also very useful for moving packaged brick or block. Most carts have air-filled tires and can carry up to 100 standard brick.

28102-13_SA04.EPS

Additional Resources

Masonry Basics: The Tools You Need and How to Use Them. Masonry Guild of Arizona. **www.masonryforlife.com/howtobasics.htm**

3.0.0 Section Review

1. When using a plywood mortarboard, be sure to keep the board from absorbing moisture and drying the mortar by _____.
   a. placing mortar only in the center of the board
   b. lining the board with a nonabsorbent plastic
   c. ensuring the board remains level while using it
   d. wetting the board thoroughly before putting mortar on it

2. Before filling a bucket with mortar or grout, be sure to _____.
   a. wet the inside surface of the bucket
   b. fill the bucket at least one-quarter of the way with water
   c. ensure the inside of the bucket is completely dry
   d. put a plastic liner inside the bucket to protect it from contamination

Section Four

4.0.0 Power Tools

Objective

Identify power tools used in masonry.

a. Describe how to use masonry saws.
b. Describe how to use splitters.
c. Describe how to use grinders.
d. Describe how to use power drills and powder-actuated tools.

Trade Term

Kickback: A reaction caused by a pinched, misaligned, or snagged tuckpoint grinder wheel that causes the wheel to stop momentarily, propelling the grinder away from the surface and toward the operator.

Power tools bring time and labor savings to masonry work. Mechanical and power tools, most of which were developed after the 1920s, have become an integral part of masonry work.

When using power tools and equipment, always follow power-tool safety rules. Inspect items before using them to make sure they are clean and functional. Disconnect power cords, and turn off engines before inspecting or repairing power equipment. As you learned in the module *Masonry Safety*, there are certain safety guidelines that you should follow when using gasoline-powered tools:

- Be sure there is proper ventilation before operating gasoline-powered equipment indoors.
- Use caution to prevent contact with hot manifolds and hoses.
- Be sure the equipment is out of gear before starting it.
- Use the recommended starting fluid.
- Always keep the appropriate fire extinguishers near when filling, starting, and operating gasoline-powered equipment. OSHA requires that gasoline-powered equipment be turned off prior to filling.
- Do not pour gasoline into the carburetor or cylinder head when starting the engine.
- Never pour gasoline into the fuel tank when the engine is hot or when the engine is running.
- Do not operate equipment that is leaking gasoline.

CAUTION

Always be sure the power is off before the final cleaning at the end of the day.

4.1.0 Using Masonry Saws

Masonry saws can make more accurate cuts than a mason's hammer or brick set. The saw does not weaken or fracture the material as hand tools do. Masonry saws are available in handheld or table-mounted models. Some larger table saws can be operated with a foot control, which leaves both hands free for guiding the piece to be cut. Masonry saws are available in gas-, electric-, and water-powered models.

Figure 43 shows a medium and a large masonry saw. Note that the large saw is used with a carrier or conveyor tray on tracks. The masonry unit to be cut is placed on the carrier tray, aligned, and carried against the blade.

WARNING!

Always keep your hands away from the cutting area behind the blade when operating a saw. The saw can jump backwards suddenly and could cut your hand.

An electrical masonry saw (*Figure 44*) is smaller and more portable than a large masonry saw. It is easily moved about and quickly set up at different areas of the job site. It is operated by hand controls. The blade on some smaller saws is mounted on a rotating arm. They are called chop saws because the blade is lowered (chopped) onto the brick.

Masonry table saws use diamond, silicon carbide (also called Carborundum™) or other abrasive blades. Diamond blades can be irrigated to prevent them from overheating and burning up. The irrigating water wets the masonry unit, cools the blade, and controls dust. The wet units must be allowed to dry before they can be laid. The dry abrasive blades cut slowly and cannot cut as thinly, but the masonry can go directly to the mortar bed. Dry cutting produces dust that must be vented away from workers and the work site.

Know Your Cutting Depth

A 14-inch blade has approximately a 5-inch cutting depth. To cut all the way through an 8-inch block, use a 20-inch blade.

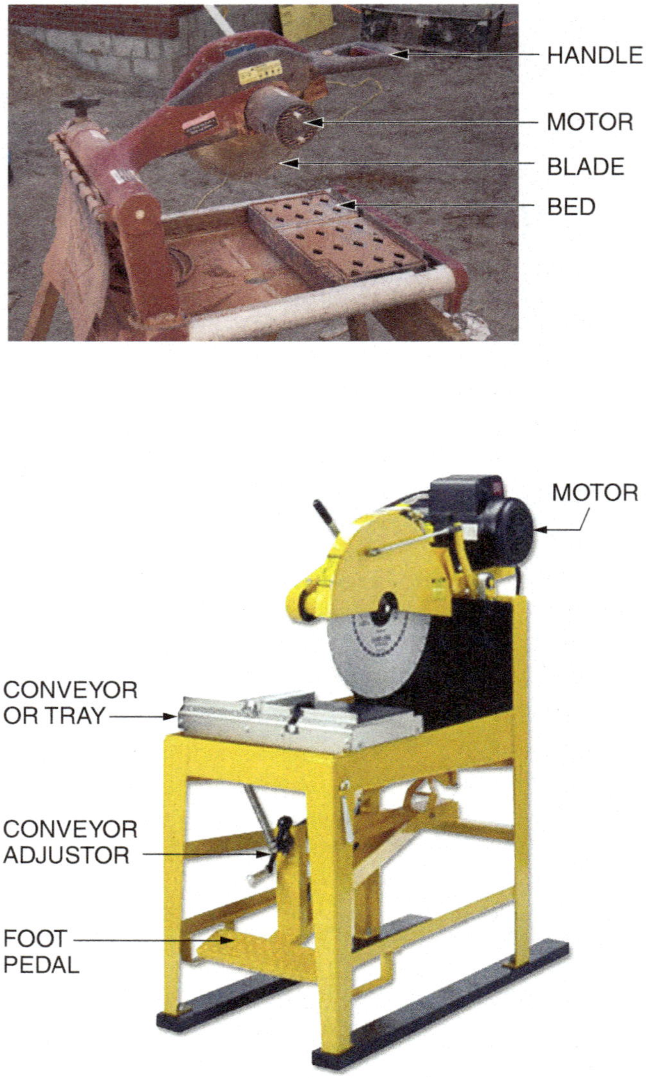

28102-13_F43.EPS

Figure 43 Medium and large masonry saws.

> **WARNING!**
>
> Inhaling dust or sand that contains silica, a natural crystal commonly found in sand, can cause silicosis, which is scarring and inflammation (swelling) of the lungs. Symptoms of silicosis include chronic coughing and shortness of breath, and low levels of oxygen in the blood. Silicosis can be caused by short-term exposure to large amounts of silica dust, or by long-term exposure to small amounts of silica dust.

When using handheld and table-mounted saws of any size, review the saw's control operations before starting to cut. The guidelines for the safe use of saws include the following:

- Wear a hard hat and eye protection to guard against flying chips.

28102-13_F44.EPS

Figure 44 Small masonry saw and integral dust collector.

- Wear rubber boots and gloves to reduce the chance of electric shock.
- Wear earplugs and/or other ear protection.
- Wear a respirator when using a dry-cut saw.
- Check the guards to ensure they move and close freely. Never operate a saw with damaged or missing guards.
- Check all adjustment levers before cutting to make sure they are set at the correct bevel and depth.
- Always hold the saw firmly when operating and support the work being cut to prevent loss of control.
- In between cuts, always release the trigger and hold the saw until the blade comes to a complete stop.
- Check that the blade is properly mounted and tightened before starting to cut.
- Do not cut masonry with excessive pressure per pass. Make repeated passes using a light, forward pressure to achieve the desired depth.
- Clean masonry dust from the saw's air vents frequently. Always disconnect the saw's plug from the power source and let the blade come to a complete stop before cleaning the air vents.

Handheld masonry saws with 14-inch blades (*Figure 45*) are also known as rapid-cut saws.

28102-13_F45.EPS

Figure 45 Handheld masonry saw.

Handheld saws with 12-inch blades are sometimes called cutoff saws or target saws. These smaller saws can be powered by gasoline, electricity, or hydraulics. The diameter of a blade on a handheld saw can be up to 14 inches. Using a handheld saw calls for extra caution.

> **WARNING!**
> Make sure electrical saws are grounded, especially if you are using water to cool the blade and control dust.

4.2.0 Using Splitters

Masonry splitters are mechanical cutters for all types of masonry units. They do not have engines, but some use hydraulic power. Splitters range from small hand-operated units, such as the one shown in *Figure 46*, to massive foot-operated hydraulic splitters. These mechanical units offer more precision than a brick hammer or brick set. They are also faster, especially when many brick need to be cut.

Unlike saws, splitters do not create as much dust and do not have high-speed blades. Splitters do make cuts as neatly as saws. Larger units (*Figure 47*) can deliver as much as 20 tons of cutting pressure but are precise enough to shave ¼ inch off a brick or stone. Large hydraulic splitters can accommodate larger masonry units than most large saws.

28102-13_F46.EPS

Figure 46 Hand-operated masonry splitter.

4.3.0 Using Grinders

The tuckpoint grinder (*Figure 48*) is a handheld, electric grinding tool designed to grind out old bed and head joints in a masonry wall. It has shatterproof blades and a safety guard on the part of the machine facing the mason. Its abrasive wheels can be used to cut and score concrete, block, brick, and stone.

When using a grinder, use both hands to grip the handles and position your body and arms so that you can absorb **kickback** from the grinder. Kickback is a sudden jump that happens when the wheel is pinched, misaligned, or snagged. This causes the wheel to stop momentarily, propelling the grinder away from the surface and toward the operator. Kickback can also cause damage to the surface that is being ground, and in some cases can cause abrasive wheels to shatter with explosive force.

Power Saw Safety

When using power saws of any size, always follow these rules:

- Wear a hard hat and eye protection to guard against flying chips.
- Wear rubber boots and gloves to reduce the chance of electric shock.
- Wear earplugs and/or other ear protection.
- Wear a respirator when using a dry-cut saw.
- Check that the blade is properly mounted and tightened before starting to cut.

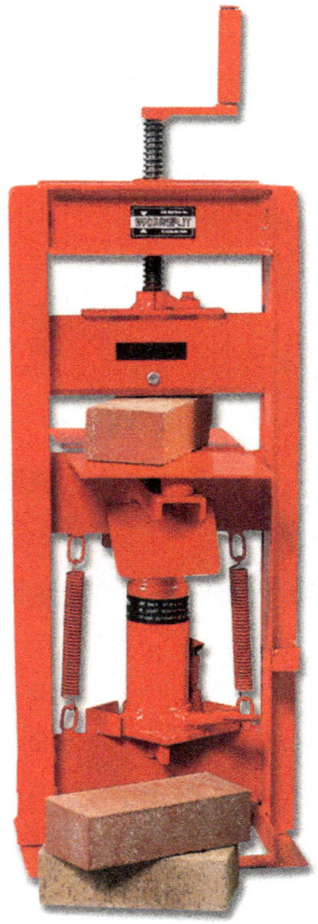

28102-13_F47.EPS

Figure 47 Large foot-operated hydraulic splitter.

WARNING!

Dry cement and wet concrete are harmful. Dry cement dust can enter open wounds and cause blood poisoning. Cement dust, when it comes in contact with body fluids, can cause chemical burns to the membranes of the eyes, nose, mouth, throat, or lungs. It can also cause a fatal lung disease known as silicosis.

Wet cement or concrete can also cause chemical burns to the eyes and skin. Always wear appropriate personal protective equipment when working with dry cement or wet concrete. If wet concrete enters waterproof boots from the top, remove the boots and rinse your legs, feet, boots, and clothing with potable water as soon as possible. Repeated contact with cement or wet concrete can also cause an allergic skin reaction known as cement dermatitis.

Before restarting a grinder, ensure that the wheel is centered in the kerf, or groove, but not bound in the material. Also, ensure that the wheel depth lever is secured, to prevent the grinder from digging too deep. Before adjusting the grinder, always ensure that the power cord has been disconnected from the power source and that the wheel has come to a complete stop.

28102-13_F48.EPS

Figure 48 Tuckpoint grinder.

To change a wheel once it has come to a complete stop and the power has been disconnected, follow the manufacturer's instructions to remove the guard cover and push the spindle lock that keeps the wheel from rotating. Then loosen and remove the wheel locknut and the wheel. Clean the inside of the guard housing before installing the new wheel. Place the wheel on the spindle according to the manufacturer's directions, install the locknut, and replace the guard cover. Always allow the grinder to reach full speed before touching the wheel to the surface being ground. Always use appropriate personal protective equipment when operating a grinder, and follow the manufacturer's instructions when servicing it.

WARNING!

Wear a hard hat and eye protection when using a tuckpoint grinder.

You already learned the following rules for the safe use of a grinder in the module *Masonry Safety*, but they are worth repeating here:

- Be sure the grinding head is in good condition and has been properly secured.
- Wear gloves and other personal protective equipment when appropriate.
- Follow the manufacturer's instructions for installing shrouds and other dust-collection devices to help capture the dust created by the grinding process.
- Make sure the power switch is off before plugging in the grinder.
- Remove all safety keys before starting the grinder.
- Use the proper grinding head for the material being ground.

- Ensure that the wheel guard is able to move freely and close during operation.
- Grip the grinder tightly on the main and auxiliary handles when using the grinder. Hold the grinder only on the insulated gripping surfaces.
- Ensure that the grinder is clean before using it, and clean it periodically while using it. Always allow the grinder wheel to come to a complete stop and unplug the grinder before cleaning it.
- Never force the grinder along as you work or apply excessive pressure. Use smooth, straight strokes. Maintain a firm grip on the handles when operating.
- Be sure the grinder head has stopped turning before putting the grinder down.

WARNING!

When using a tuckpoint grinder as a cutter, use the appropriate blade for the material being cut. For grinding, use 1/4-inch dry-diamond segmented wheels.

4.4.0 Using Power Drills and Powder-Actuated Tools

Usually, masons set bolts to anchor metalwork or wood when the mortar is still pliable. When the mortar is hardened, a power drill or powder-actuated tool is needed. Drilling in masonry takes a combination of power, speed, and hammering, so a ¼-inch drill is not adequate. A ⅜-inch or ½-inch hammer drill, such as the one shown in *Figure 49*, delivers enough power. Carbide-tipped drill bits are available in several standard sizes. Drills are used to attach wall ties or other anchors to a concrete wall.

Powder-actuated tools (*Figure 50*) are usually designed to drive a specific line of bolts, anchors, and other fasteners. These tools use a small explosive charge to drive a pin or stud into masonry. Specially hardened pins or studs are used. Different charge sizes are available. Manufacturers color-code the powder load charges to identify the strength of the charge. Be sure to learn the color code specific to your tool's manufacturer.

WARNING!

Most hammer drills have enough torque to break your wrist. Make sure that you have a firm grip on the side handle when using a hammer drill. You should never hold on to just the main handle. Use both hands to equalize the rotation of the drill.

Some models of powder-actuated tools use compressed air. Do not operate a powder tool without the proper training and credentialing. Always wear appropriate personal protective equipment when working with or around powder-actuated tools.

NOTE

OSHA requires that all operators of powder-actuated tools must be qualified or certified by the tool manufacturer. Operators must carry certification cards whenever using the tool.

28102-13_F49.EPS

Figure 49 Hammer drill.

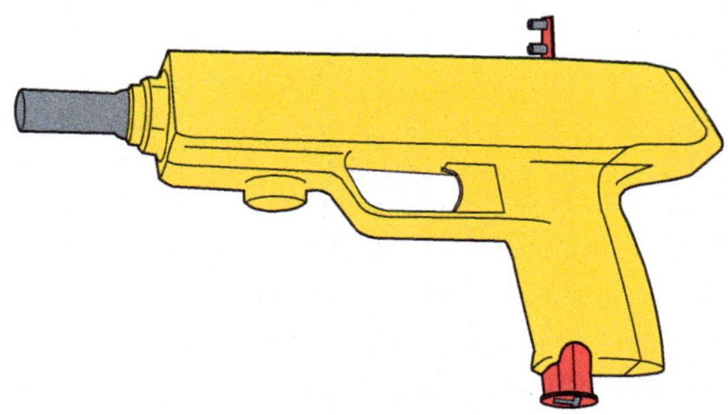

28102-13_F50.EPS

Figure 50 Powder-actuated tool.

Powder-Actuated Tools

A powder-actuated fastening tool is a low-velocity fastening system powered by gunpowder cartridges, commonly called boosters. Boosters often come in a strip called a magazine. This cut-away diagram of a powder-actuated fastening tool shows how the components work together.

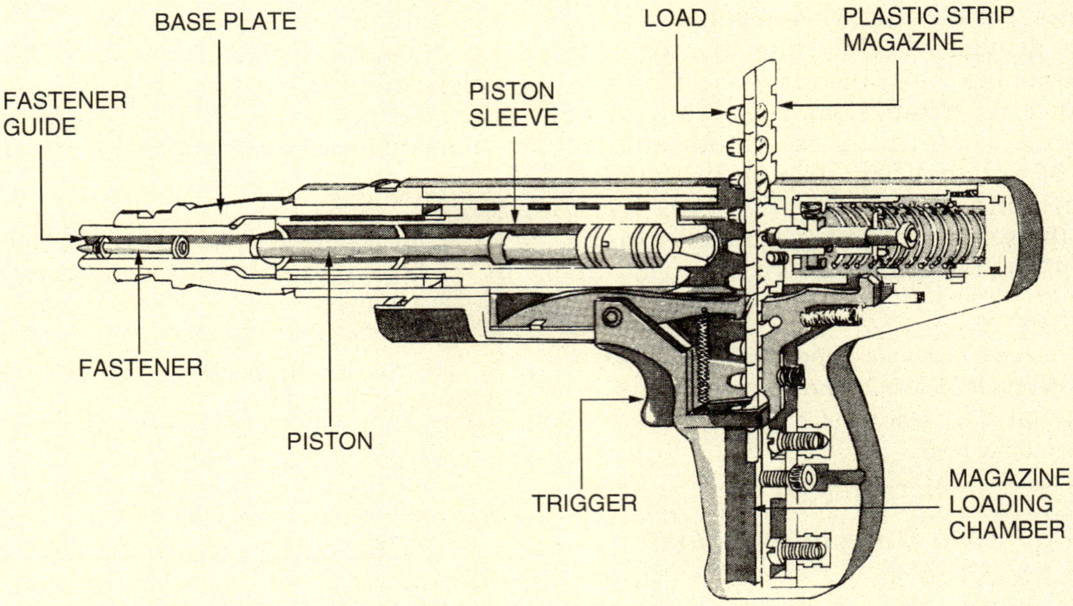

28102-13_SA05.EPS

Powder-Actuated Tool Safety

When using powder-actuated tools, always follow these rules:

- Do not use a powder-actuated tool unless you are certified.
- Follow all safety precautions in the manufacturer's instruction manual.
- Always wear safety goggles and a hard hat when operating a powder-actuated tool.
- Use the proper-size pin for the job you are doing.
- When loading the tool, put the pin in before the charge.
- Use the correct booster (powder load) according to the manufacturer's instructions.
- Never hold your hand behind or near the material you are fastening.
- Never hold the end of the barrel against any part of your body or cock the tool against your hand.
- Do not shoot close to the edge of concrete.
- Never attempt to pry the booster out of the magazine with a sharp instrument.
- Always wear ear protection.
- Always hold the muzzle perpendicular (90 degrees) to the work.

Additional Resources

"Tuckpointing and Repointing Tools: Specialized Tools for Removing and Replacing Mortar." Brett Martin. *Masonry Magazine*, September 2008. **www.masonrymagazine.com/9-08/tools.html**

"Using Masonry Saws." Norton Construction Products North America. **www.nortonconstructionproducts.com/solutions/masonry-saw.aspx**

4.0.0 Section Review

1. To prevent them from overheating and burning up, diamond blades can be _____.
   a. sanded
   b. lubricated
   c. balanced
   d. irrigated

2. To cut brick with more precision than a brick hammer or brick set, use a _____.
   a. chop saw
   b. tuckpoint grinder
   c. masonry splitter
   d. powder-actuated tool

3. The groove cut by a grinder is called a ____.
   a. kerf
   b. tuck
   c. point
   d. maul

4. Usually, masons set bolts to anchor metalwork or wood when _____.
   a. the mortar is dry
   b. the metalwork or wood is installed
   c. the mortar is still pliable
   d. holes have been drilled first

SECTION FIVE

5.0.0 POWER EQUIPMENT

Objective

Identify power equipment used in masonry.

a. Describe how to use a mortar mixer.
b. Describe how to use a masonry pump, vibrator, and hydraulic grout placer.
c. Describe how to use pressurized cleaning equipment.

Performance Task 3

Use the correct procedures for fueling and starting a mixer.

Power equipment, like power tools, bring speed and economy to the masonry building process. When using power equipment, follow the general rules for power-tool safety. Fuels such as gasoline, liquid propane (LP) gas, and diesel fuel are capable of causing a fire or explosion if not handled properly. In addition, LP gas is stored in cylinders under pressure, creating an explosion hazard if the cylinder is exposed to extreme heat or fire. It is very important to keep these fuels away from any source of fire and to keep the areas in which the forklift is used free of any flammable materials. There are specific precautions that must be taken to avoid the possibility of a fire or explosion.

The best way to prevent a fire is to make sure that the three elements needed for fire (fuel, heat, and oxygen) are never present in the same place at the same time. Here are some basic safety guidelines for fire prevention:

- Always work in a well-ventilated area, especially when you are using flammable materials.
- Never smoke or light matches when you are working with flammable materials.
- Keep oily rags in approved, self-closing metal containers.
- Store combustible materials only in approved containers.
- Know where to find fire extinguishers, what kind of extinguisher to use for different kinds of fires, and how to use the extinguishers.
- Keep open fuel containers away from any sources of sparks, fire, or extreme heat.
- Make sure all extinguishers are fully charged. Never remove the tag from an extinguisher; it shows the date the extinguisher was last serviced and inspected.
- Don't fill a gasoline or diesel fuel container while it is resting on a truck bed liner or other ungrounded surface. The flow of fuel creates static electricity that can ignite the fuel if the container is not grounded.
- Always use approved containers, such as safety cans, for flammable liquids.

5.1.0 Using a Mortar Mixer

On most commercial jobs, mortar is mixed in a powered mortar mixer. The mixer has an electric or gasoline engine and is usually on a set of wheels, as shown in *Figure 51*. The mixer portion consists of a drum with a turning horizontal shaft inside. Blades are attached to the shaft and revolve through the mix. The dump handle and drum release are used to empty the mortar onto a pan or board. Mixing mortar using a power mixer is explained later in the module entitled *Mortar*.

CAUTION

Check the oil, gas, and other fluid levels every time you use a mixer with a gasoline engine. Running out of gas with a batch of mortar in progress will spoil the batch and could damage the machine.

WARNING!

Always wear eye protection and other appropriate personal protective equipment when using a power mixer. Never place any part of your body in the mixer.

28102-13_F51.EPS

Figure 51 Mortar mixer.

Mixers have capacities ranging from 1 to 12 cubic feet; the typical mixer holds about 4 cubic feet. The mixer drum needs to be washed out immediately after each use to keep mortar from hardening inside it.

WARNING!

Never place your hands inside the drum when the mortar mixer is starting or operating.

WARNING!

Never service a mixer when it is operating. Make sure that the mixer has stopped rotating and is disconnected from its power source and that the engine or motor has been allowed to cool before attempting to service the mixer.

5.2.0 Using a Masonry Pump, Vibrator, and Hydraulic Grout Placer

The masonry pump (*Figure 52*) is used to deliver mortar or grout to a high location. Grout is usually pumped when it is used to fill the cores in a block wall. Grout is pumped from the mixer to the intake hopper of the grout pump to the delivery hose.

At the deposit site, the mason guides the grout into the cores. After the grout is delivered, it is sometimes vibrated to eliminate air holes. A typical handheld vibrator is shown in *Figure 53*. The snake-like end of the vibrator is inserted into the core. The mason inserts the vibrator into each core to make sure that air pockets are removed and the grout is consolidated. Steel reinforcement may also be placed in the cores before the grout is added.

28102-13_F52.EPS

Figure 52 Masonry pump.

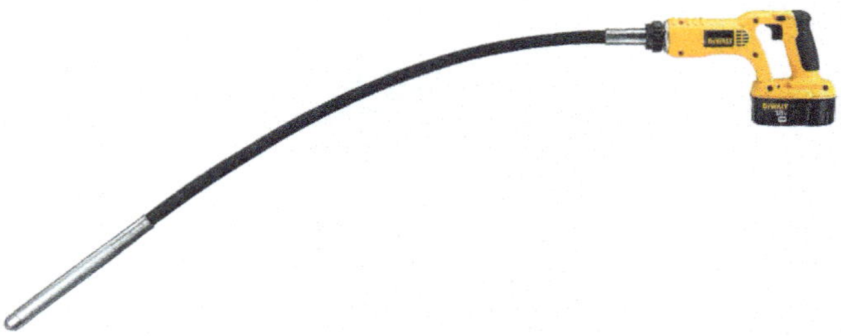

28102-13_F53.EPS

Figure 53 Handheld vibrator.

A hydraulic grout placer uses a piston pump operated by a gasoline engine to pump grout, concrete, and other aggregate materials through a hose, which the mason directs (*Figure 54*). Hydraulic grout placers are used to pump grout to high elevations, over long distances, and around obstacles, which saves time and labor.

Mortar Mixer Safety

When using a mortar mixer, always follow these rules:

- Always wear appropriate personal protective equipment when operating a mixer.
- Never operate a mixer without proper training first.
- Do not use accessories or attachments that were not designed for the particular make and model of mixer that you are using.
- Ensure that all emergency and safety devices are connected and working before operating the mixer.
- Never fill the oil level higher than what is specified in the manufacturer's instructions.
- Ensure that the belt, blades or paddles, and dumping mechanism are all in working order according to the manufacturer's instructions.
- Place the safety grate into position and ensure that it is fitted properly.
- Ensure that the air filters are clean and properly sized for the mixer.
- If the mixer is gasoline powered, ensure that the engine is properly ventilated to prevent the buildup of toxic exhaust gases. Never touch the manifold, cylinder, or muffler when hot.
- When cleaning the mixer, dispose of all excess mortar properly in according with the manufacturer's instructions and the standards and regulations of your jurisdiction.

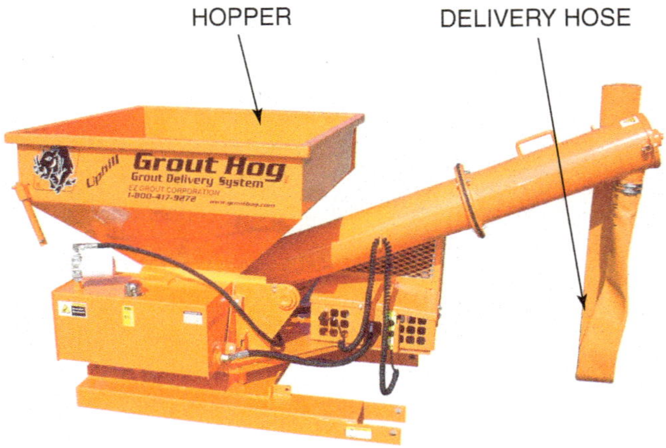

28102-13_F54.EPS

Figure 54 Hydraulic grout placer.

5.3.0 Using Pressurized Cleaning Equipment

Pressurized cleaning equipment has effectively replaced bucket-and-brush cleaning for masonry structures. Pressurized cleaning uses abrasive material under pressure to scour the face of the masonry. The generic equipment for pressurized cleaning includes an air compressor, a tank or reservoir for pressurizing, a delivery hose, and a nozzle or tip. There are two types of pressurized cleaning systems: pressure washing and sandblasting. The systems differ primarily in the type of abrading material they deliver. The following sections describe these two pressurized systems in detail.

> **WARNING!**
> Pressurized cleaning equipment creates dangerous conditions. Read the manufacturer's operating manual before using. Make sure that you know how to use all pressure-release valves and safety switches. Wear safety glasses and other appropriate personal protective equipment.

5.3.1 Pressure Washing

Pressure washing (*Figure 55*) can be the gentlest method for cleaning masonry structures. Also called high-pressure water cleaning, this is a newer cleaning technique. The pressure washer uses a compressed air pump to pressurize water and to deliver it in a focused, tightly controlled area. Sometimes pressure washing is done after manual cleaning.

28102-13_F55.EPS

Figure 55 Typical pressure washing operation.

Pressure washing has the best results when the operator uses a fan-type tip, dispersing the water through 25 to 50 degrees of arc. The amount or volume of water has more effect than the amount of pressure. The minimum flow should be 4 to 6 gallons per minute (gpm). Usually, the compressor should develop from 400 to 800 pounds per square inch (psi) water pressure for the most effective washing. It is important to keep the water stream moving to avoid damaging the wall.

Pressure washing can be used in combination with various cleaning compounds. Training and practice are necessary to properly control the mix of proprietary cleaners, pressure, and spray pattern. *Figure 56* shows what can happen when pressure washing is done incorrectly. Improper pressure washing technique can remove finish and even score brick, resulting in costly repairs.

5.3.2 Sandblasting

Sandblasting is the oldest method of pressurized cleaning. It has the most capability of damaging or scarring the brick face, and so it is rarely used. Because sandblasting can deface brick, it is best done by a trained operator.

Sandblasting employs abrasives, such as wet or dry grit, round or sharp-grained sand, crushed nut shells, rice hulls, egg shells, silica flour, ground corncobs, and other softer abrasives.

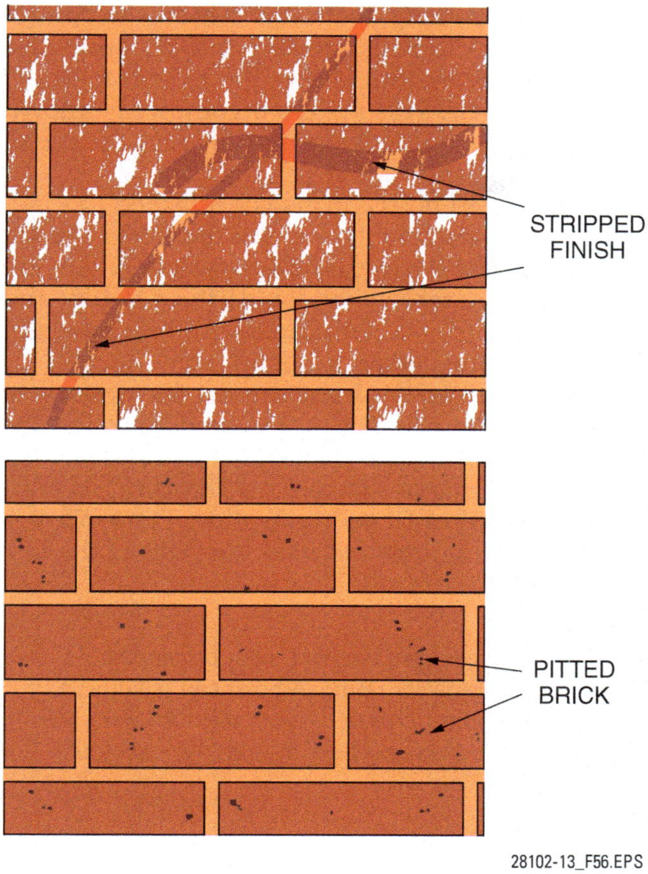

Figure 56 Results of improper pressure washing.

Additional Resources

"Gilson Mortar Mixers by Cleform Tool." 2005. Construction Complete. **www.youtube.com/watch?v=ppMp8Upibz4**

"Pressure Washing Tips, Tricks And Techniques." 2011. PowerWash.com. **www.powerwash.com/articles/pressure-washing-tips-tricks-and-techniques-2.html**

5.0.0 Section Review

1. Wash mixer drums _____.
   a. immediately after each use
   b. at the end of the workday
   c. after every third or fourth use
   d. within 2 hours of each use

2. To remove air bubbles from grout after it has been placed in a core, use a _____.
   a. mixer
   b. vibrator
   c. grout pump
   d. compressor

3. When pressure washing brick, the amount of pressure has less effect than _____.
   a. the height of the wall being washed
   b. the dispersal pattern of the water spray
   c. the rating of the compressor
   d. the amount or volume of water

SECTION SIX

6.0.0 LIFTING EQUIPMENT

Objective

Identify lifting equipment used in masonry.

a. Describe how to use mounted and portable hoists.
b. Describe how to use hydraulic-lift materials trucks.
c. Describe how to use forklifts and pallet jacks.

Once the mason's work reaches higher than 4 feet, it is not efficient to work standing on the ground. Lifting equipment provides the mason with an elevated work area, and in some cases can lift the mason, tools, and materials to the work location. An above-grade masonry workstation usually has a mortar pan or mortarboard, the mason's tools, and a stack of masonry units.

Typically, masonry units arrive at the job site bound, bundled, or palletized. Each bundled cube may contain 500 standard brick, or 90 standard block, depending on the manufacturer. Depending on the job, different types of equipment will move the cubes to the workstation. The following sections describe lifting equipment for material handling.

WARNING!

Material-handling procedures are some of the most hazardous activities on the job site. Stay clear of moving equipment and material in transport.

The following are some general rules for moving materials safely:

- Establish clear pathways for materials movement.
- Use a consistent set of signals to alert workers to materials movement.
- Do not ride on materials as they are moved.
- Stay out of the area between the moving materials and any wall or heavy equipment.

6.1.0 Using Mounted and Portable Hoists

A pulley system, or block and tackle, is the oldest aid to moving materials. However, simple pulley systems do not have safety features to control heavy loads. Adding a motor and supports to a simple pulley system allows the mason to safely move materials. This system is commonly known as a hoist. The hoist can be mounted on a scaffold, as shown in *Figure 57.*

A motorized hoist can also be attached to a ladder to lift materials. This small hoist has a gasoline or electric engine, a pulley system, a take-up reel for lifting cable, a ladder, a lifting trolley, and hand controls. The combination is lightweight enough for one person to set up.

The typical ladder hoist can lift 400 pounds to a height of 16 to 40 feet. A plywood board can be put over the steel trolley for raising nonpalletized materials. Some hoists come with gravel or mortar hoppers that fit on the trolley. The trolley is raised by the pulley and uses the sides of the ladder as rails.

WARNING!

Every hoist and scaffold must be marked with a capacity rating. A hoist can fail due to overloading. This will cause serious injury and damage. Always check the rated capacity of a hoist before use.

A portable materials hoist, also called a buck hoist, is used to lift materials up to a mason on a scaffold (*Figure 58*). The materials hoist includes a lift platform, lift cabling, and a gasoline, diesel, or electric motor. There is usually a pulley system

28102-13_F57.EPS

Figure 57 Hoist.

28102-13_F58.EPS

Figure 58 Materials hoist.

as well. The lift platform may have a cage around it. Materials hoists can lift from 1,000 to 5,000 pounds over a vertical distance of up to 300 feet.

Hoists can be mounted on wheels and towed to the job site. Portable hoists can also be mounted on a truck bed. The larger hoists are usually not portable but are attached to the side of the structure being built.

WARNING!

Never ride on a materials hoist. Do not use the materials hoist as a work platform. A materials hoist has no safety features or brakes.

Materials hoists are not for lifting people. Personnel hoists have guardrails, doors, safety brakes, and hand controls in addition to the features of the materials hoist. Personnel hoists can be used for materials, but material hoists cannot be used for personnel. A personnel hoist is also known as a man-lift.

6.2.0 Using Hydraulic-Lift Materials Trucks

The hydraulic-lift materials truck has a hydraulic boom arm for unloading masonry and other materials. The truck operator lowers stabilizing arms to prevent the truck from overturning. The hydraulic lift arm attaches to the load and swings out to unload each cube of masonry. If the masonry is not palletized, the cubes have openings in the bottom to accommodate the lift arm attachment prongs.

The hydraulic-lift materials truck generally is not used for lifting materials above ground level. It is used to stage materials close to the work site or into a stockpile.

6.3.0 Using Forklifts and Pallet Jacks

The most common method for lifting heavy materials on a job site is a forklift. Forklifts have hydraulic lifting arms that move up and down. They are called forklifts because of the fork shape of the prongs on the lifting arm. The prongs fit into the openings in a pallet, a large mortar pan, or the bottom of a cube of masonry. These mobile lifters are used to move masonry from a stockpile to a workstation. The most commonly used forklifts on masonry projects are reach-type forklifts (*Figure 59*) and skid-steer forklifts (*Figure 60*). You may also see straight-mast forklifts on occasion (*Figure 61*). Scissor lifts, boom lifts (also called cherry pickers), and mast lifts are other common types of forklift-style equipment for lifting materials and people.

28102-13_F59.EPS

Figure 59 Mason's reach-type forklift.

28102-13_F60.EPS

Figure 60 Skid-steer forklift.

28102-13_F61.EPS

Figure 61 Straight-mast forklift.

Pallet jacks are specialized for masonry handling. They carry material from a stockpile to a workstation. Some pallet jacks have gasoline, diesel engines, or electric motors, with a hydraulic forklift. There is no seat for the operator, who stands. The load capacity of a typical pallet jack varies from one-half to one cube of masonry. If less than a full cube of masonry is needed, the materials must be stacked on a pallet. Make sure that the materials are secure before they are raised.

Motorized buggies carry mortar and masonry units around the job site. As shown in *Figure 62*, they are large, motorized wheelbarrows with two or four wheels in front and one wheel in back. The bin dumps its load mechanically. There is a shelf in the back for the operator to stand on as the buggy moves.

28102-13_F62.EPS

Figure 62 Motorized buggy.

Additional Resources

"Forklift Safety." 2009. EDG Safety Series. **www.youtube.com/watch?v=yx67JVlc2Gw**

Forklift Safety Guide, Publication F417-031-000 (04/07). 2007. State of Washington Department of Labor and Industries. **www.lni.wa.gov/IPUB/417-031-000.pdf**

6.0.0 Section Review

1. The typical ladder hoist can lift _____.
   a. 300 pounds
   b. 400 pounds
   c. 500 pounds
   d. 600 pounds

2. Hydraulic-lift materials trucks are generally used for lifting materials _____.
   a. above ground level
   b. on inclined surfaces
   c. on roofs
   d. at ground level

3. Pallet jacks do *not* have _____.
   a. engines
   b. lifting arms
   c. seats
   d. wheels

Section Seven

7.0.0 Scaffolds

Objective

Recognize scaffolds used in masonry.

a. Identify scaffold systems.
b. Describe how to assemble and disassemble tubular frame scaffold.

Performance Task 4

Assemble and disassemble tubular frame scaffold.

Scaffold is any elevated, temporary work platform. Most masonry jobs require some type of scaffold. As the working level rises above 4 feet, masons cannot work efficiently on the ground. It is cost- and time-effective to raise the mason and the materials close to the work. Masons sometimes erect the scaffold they work on, to make sure it is safe and stable.

All scaffold must be erected, moved, and disassembled under the supervision of a competent person. All scaffold must be assembled according to federal safety regulations. Safety regulations prohibit anyone from working on scaffold without first attending a safety class and obtaining a certification.

7.1.0 Identifying Scaffold Systems

Scaffold can be classified into three broad categories:

- Supported scaffold, which has one or more platforms supported by rigid loadbearing members, such as frames, built on the ground
- Suspended scaffold, which is composed of one or more platforms suspended by ropes anchored to the roof
- Personnel hoists, and other machinery

This section looks at some of the types of scaffold that are most commonly used by masons. These include steel tower, swing-stage, and powered scaffold. When using scaffold of any kind, always follow the manufacturer's instructions and wear appropriate personal protective equipment to ensure that you and your fellow workers stay safe.

7.1.1 Steel Tower Scaffold

Adjustable steel tower, or self-climbing, scaffold (*Figure 63*) is a type of supported scaffold built of vertical members braced or tied to the structure itself. The ties must be positioned 30 feet apart or less. The scaffold framing is attached to the vertical members. Metal planking is used for a working platform, which has guardrail and toeboard safety features, and a winch mechanism. The working platform can be raised as the work progresses. Because it requires attachment to the building, tower scaffold must be put up by specialized workers.

7.1.2 Swing-Stage Scaffold

Swing-stage scaffold is a type of suspended scaffold used on multistory buildings. Steel beams are fastened to the roof, and steel cables are dropped to the ground. A steel cage is suspended from the cables with hangers, and a planking floor is added over the frame. Guardrails, toeboards, and an overhead canopy of plywood or metal mesh

28102-13_F63.EPS

Figure 63 Adjustable steel tower scaffold.

Scaffold Safety

When using scaffold, always follow these rules:

- Platforms on all working levels must be fully decked between the front uprights and the guardrail supports.
- Never exceed the weight-bearing capacity of a scaffold or its components, which OSHA has specified as the ability to support, without failure, the scaffold's own weight plus at least 4 times the maximum intended load.
- The space between planks and the platform and uprights can be no more than 1 inch wide.
- Ensure that there is sufficient clearance between scaffolds and power lines.
- Platforms and walkways must be at least 18 inches wide; the ladder jack, top plate bracket, and pump jack scaffolds must be at least 12 inches wide.
- For every 4 feet of height, the scaffold must be at least 1 foot wide. If it is not, it must be protected from tipping by guying, tying, or bracing per the OSHA rules.
- Supported scaffold must sit on baseplates and mud sills or other steady foundations.
- Access to and between scaffold platforms more than 2 feet above or below the point of access must be made by the following:
  - Portable ladders, hook-on ladders, attachable ladders, scaffold stairways, stairway-type ladders (such as ladder stands), ramps, walkways, integral prefabricated scaffold access, or equivalent means; or
  - Direct access from another scaffold, structure, personnel hoist, or similar surface
- Never use cross braces to gain access to a scaffold platform.

complete the cage. A winch moves the cage up and down, so it is always at the level of the work.

Unlike the other types of scaffold, the swing stage is usually erected by specialists. This type of scaffold is also the safest because the mason is completely enclosed by steel, and there are redundant backup systems on the cabling and brakes.

7.1.3 Powered Scaffold

Powered scaffold allows the worker to position a platform in the proper location quickly and easily, perform the necessary work, and then move the platform to the next work location without having to disassemble and reassemble the supporting scaffold each time (*Figure 64*). Powered scaffold uses electrical and gasoline engines to power them.

Powered scaffold is designed for use on paved/slab surfaces, or off-slab. Off-slab lifts are fitted with truck-type tires, or even deep-tread mud-type tires. Some units have oscillating axles and four-wheel drive to maneuver more easily on rough terrain. These lifts have a wide variety of uses in construction and outdoor maintenance. It must be attached to the building at intervals of no more than 30 feet.

Some lifts designed for use on paved/slab surfaces are designed for indoor use and are small enough to fit through interior doorways. Lifts designed for indoor use are battery powered to avoid problems caused by exhaust fumes.

28102-13_F64.EPS

Figure 64 Example of powered scaffold.

Powered scaffolds have two sets of controls, one on the work platform and a second on the base of the unit. The work platform controls are the primary control station. The base-mounted controls are primarily used in emergency situations. If the worker is injured, the base controls can be used to lower the worker to the ground to provide the necessary assistance. As an additional safety feature, all boom lifts can be lowered even when there is no power to the unit. Be sure to read and understand the manufacturer's operating manual before attempting to use any type of powered scaffold.

7.2.0 Assembling and Disassembling Tubular Frame Scaffold

Tubular steel sectional scaffold is the most common type of scaffold. It is strong, light-weight, durable, and easy to erect. The steel frame sections come in several heights, with a typical 4-foot or 5-foot width. As shown in *Figure 65*, two commonly used scaffold frame types are the ladder-braced and the walk-through frames. The walk-through allows the mason an easy passage along the scaffold planks. Tubular steel frame scaffold is used in accessible places with fairly level ground conditions.

Tubular frame scaffold components include frames, locking devices, braces, baseplates, putlogs, casters, platforms and planks, guardrails and gates, and ladders and stairs, and other components. These components may be assembled in a variety of arrangements to meet almost any design configuration. Each manufacturer offers a complete line of components. Never mix components from different manufacturers, because they may not be compatible, or they may be made from dissimilar metals. When dissimilar metals come in contact with each other, they may corrode. When using tube-and-clamp scaffolds, always use dual-purpose clamps.

> **WARNING!**
> Exceeding the weight limits of scaffold can cause the scaffold to fail, which can result in injury or death.

As tubular steel sectional scaffold is set up, each frame is connected to the next frame at ground level with a horizontal diagonal brace. In addition, frames are connected at upper levels by overlapping diagonal braces. These overlapping braces must be secured to the frames by a wing nut or bolt or by a lock device that fits over the frame coupling pin.

To connect each section of scaffold framing vertically, a steel pin, or nipple, is inserted into the hollow tubing at the top of the lower section. The bottom of the upper section fits over this pin and is secured with a slip bolt. Each level must be braced as it is installed. All connections must be secured as they are made. Coupling pins (*Figure 66*) must be secured to each of the scaffold frames. A rivet and locking pin is then run through the aligned holes in the scaffold legs and coupling pins. The frames also have attachment points for the diagonal braces. These braces must also be locked in place.

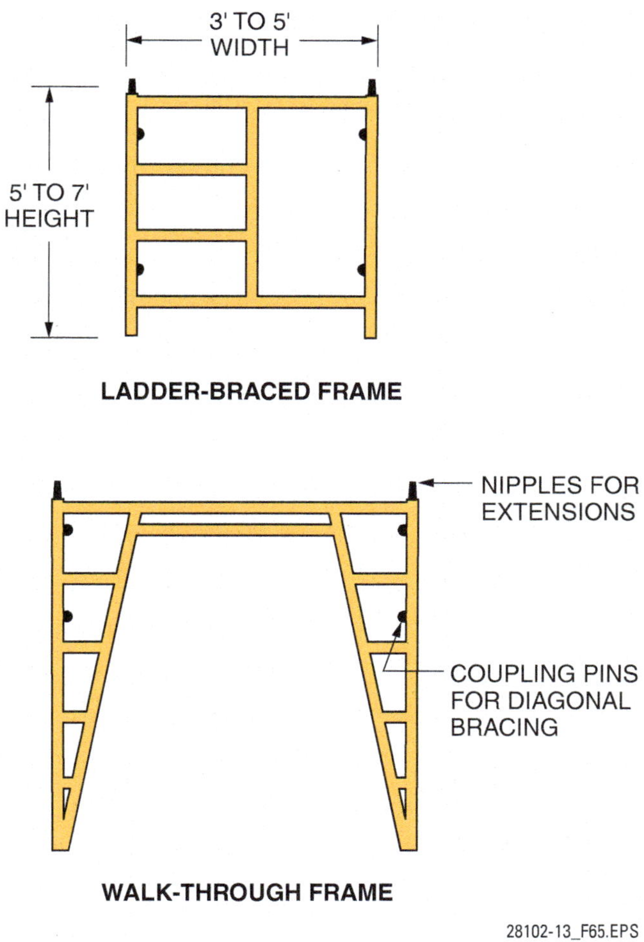

Figure 65 Scaffold frame types.

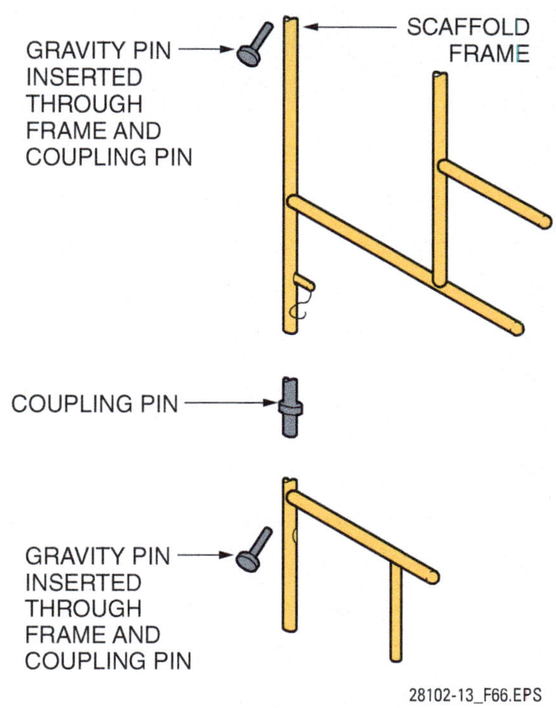

Figure 66 Connecting scaffold sections.

After bracing, the platform flooring or decking is put in place. Regular dimensional lumber planking can be used for the platforms or hook-mounted scaffold planks can be used (*Figure 67*).

Toeboards keep materials from sliding off the platform. Guardrails, to protect masons, must be placed and secured before work from the scaffold can start. Under some conditions, roofboards can also be added. The finished product looks like *Figure 68*.

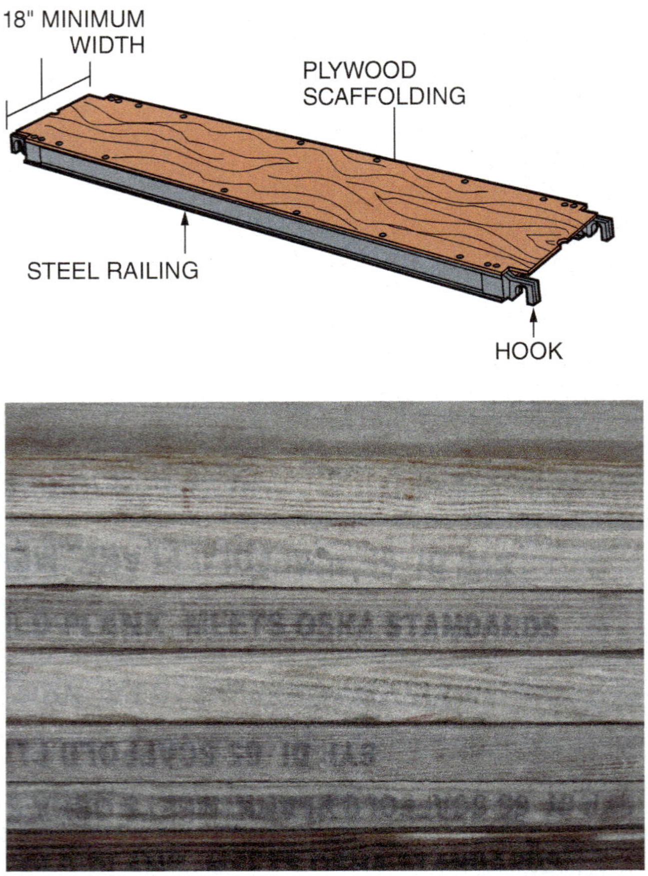

28102-13_F67.EPS

Figure 67 Hook-mounted scaffold plank.

Note that screw jacks have been put under the baseplates. This is a good practice to level the framework after it is assembled. A sill board can be added for ground that is soft, damp, or liable to shift.

Tubular steel framing is very versatile and can be configured many ways with different accessories. Mason's side brackets, or extenders, extend the working area. Wheels are available and can be inserted into the bottom of framing members so that the structure becomes a rolling scaffold. Rolling scaffolds should only be used for pointing and striking joints, not for laying units. Even with wheel locks, they tend to move when heavily loaded with masonry units.

7.2.1 Putlog

A putlog, sometimes called a bridge, is a wooden beam used to support scaffold. Putlogs are used where there is no solid base on which to set the scaffold frame. Because of grade conditions at the job site, one or more putlogs may be needed to stabilize scaffold. Putlogs are set with their greater thickness vertical and must extend at least 3 inches past the edge of the scaffold frame.

If there is not enough space to rest the scaffold, holes are left in the first few courses of a wall. The putlog is inserted in the hole with the other end resting on solid ground (*Figure 69*) or on a constructed base.

The scaffold frame is erected on the putlogs, which provide a solid footing. The scaffold is braced as usual, and the base plates may be fastened to the putlogs. After the scaffold is disassembled, the putlogs are withdrawn from the holes, and the holes are filled with masonry units.

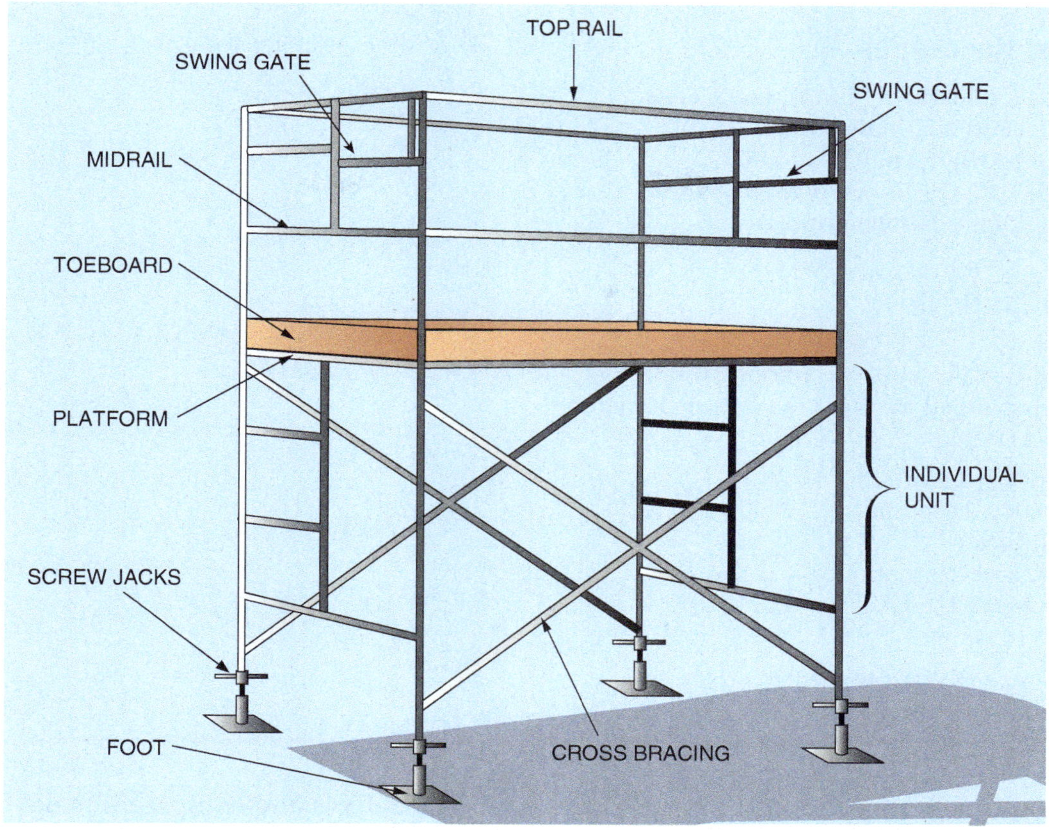

28102-13_F68.EPS

Figure 68 Completed tubular steel scaffold.

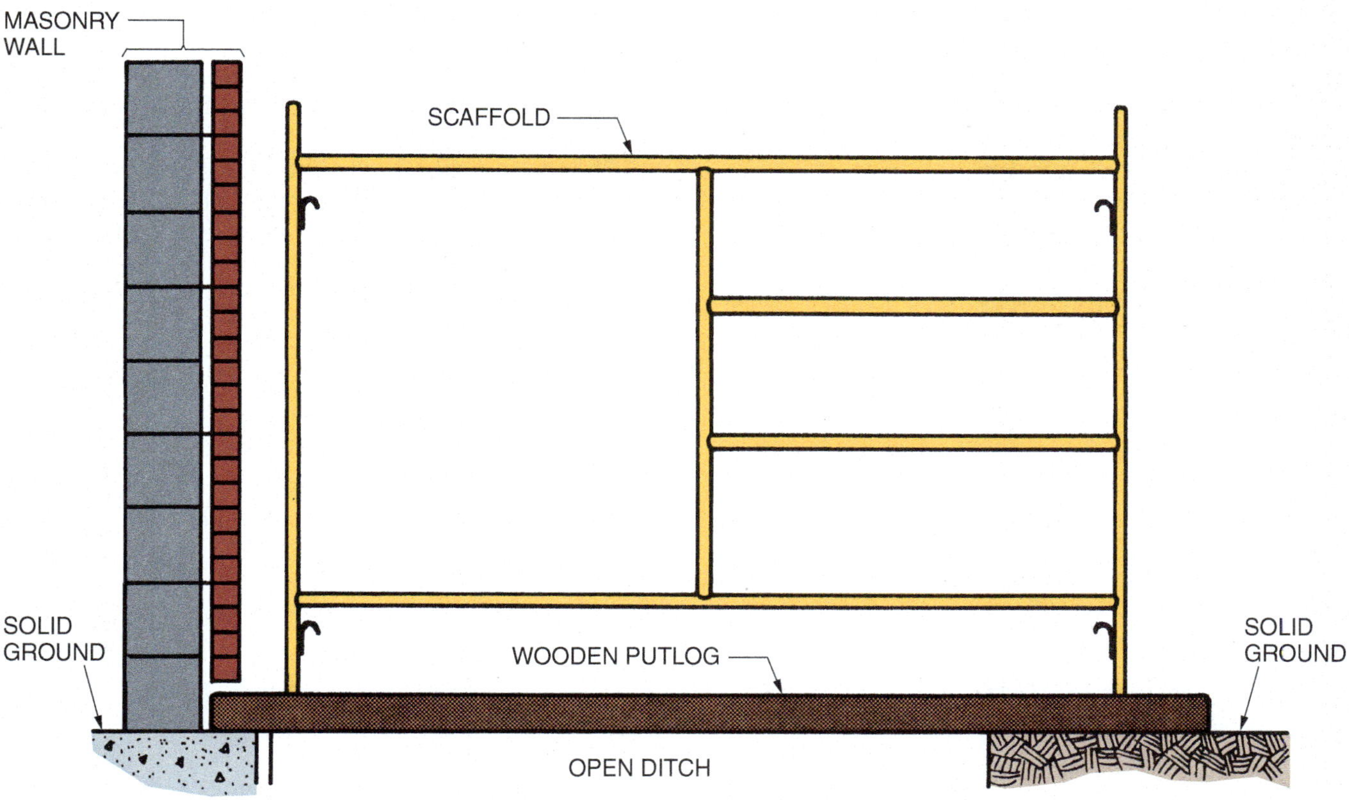

28102-13_F69.EPS

Figure 69 Putlog at base of scaffold.

Additional Resources

"Standards for Scaffold." OSHA. **www.osha.gov/SLTC/scaffolding/index.html**

"Scaffold PowerPoint Presentation." Washington, DC: OSHA. **www.osha.gov/dte/library/scaffolds/scaffolding/notes.pdf**

7.0.0 Section Review

1. Scaffold that is composed of one or more platforms suspended by ropes that are anchored to the roof is called _____.
   a. suspended scaffold
   b. supported scaffold
   c. man-lifts
   d. self-climbing scaffold

2. Another term for *putlog* is _____.
   a. plank
   b. frame
   c. brace
   d. bridge

SUMMARY

Masons use a variety of hand and power tools on the job, as well as power equipment, lifting equipment, and scaffold. It is up to you to learn to use and maintain these tools and pieces of equipment properly. Masons use course layout, alignment, and measuring tools constantly during a workday to check that a masonry structure is plumb. Measures and measuring tools include levels, squares, mason's line and fasteners, corner poles, chalkline, and two types of mason's spacing rules. Mortar mixing equipment includes water buckets and barrels, hoes, shovels, and mortarboards, mortar boxes, and mortar pans. Boxes, boards, and pans used for mortar must be cleaned after use, as should hand and power tools and power equipment.

Power tools and equipment make the mason's job easier. Power mixers quickly and thoroughly mix mortar. Power saws and hydraulic splitters are used to cut masonry units. Material needs to be lifted to the work location by hoists, cranes, forklifts, and pallet jacks. OSHA has strict rules for scaffold safety. All scaffold must be erected, moved, and disassembled under the supervision of a competent person.

The mason's job is one that takes great skill and practice. Learning to properly use and maintain your tools and equipment is the first step toward becoming a successful mason.

Review Questions

1. To tap masonry units into place, masons often use their _____.
   a. brick hammer
   b. trowel handle
   c. gloved fist
   d. brick set chisel

2. A London-pattern trowel blade has a _____.
   a. rounded heel
   b. square heel
   c. flared heel
   d. sharply angled heel

3. A duck bill trowel, used for cleanup, may have a blade length of up to _____.
   a. 8 inches
   b. 11 inches
   c. 13 inches
   d. 15 inches

4. An example of a cutting hammer is the _____.
   a. mash hammer
   b. brick hammer
   c. bushhammer
   d. mallet

5. A maul is used to _____.
   a. face block, stone, or concrete
   b. trim tile
   c. strike chisels for cutting
   d. drive line pins

6. The wooden handle of a hammer should be inspected _____.
   a. before each use
   b. daily
   c. twice each week
   d. weekly

7. Chisels used in masonry work are made of _____.
   a. stainless steel
   b. hardened aluminum
   c. alloyed brass
   d. tempered steel

8. The thick, beveled blade of a brick set may be up to _____.
   a. 5 inches wide
   b. 6 inches wide
   c. 7 inches wide
   d. 8 inches wide

9. The tooth chisel should be used only when working on _____.
   a. soft stone
   b. brick
   c. hard stone
   d. block

10. *Review Question Figure 1* shows a tooled mortar joint called a _____.
    a. weathered joint
    b. struck joint
    c. V-joint
    d. raked joint

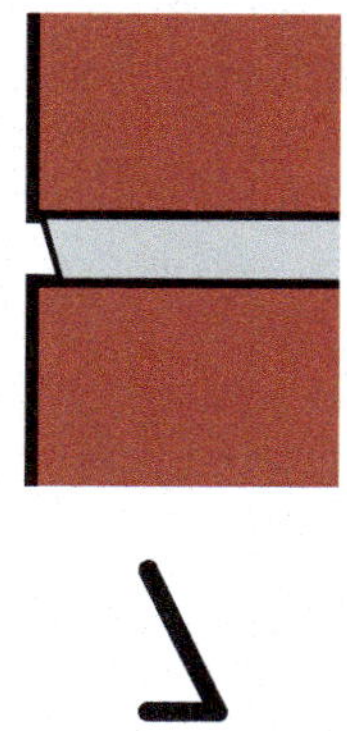

28102-13_RQ01.EPS

Figure 1

11. The jointers used to tool long horizontal mortar joints are known as _____.
    a. bedding jointers
    b. ski-type jointers
    c. extension jointers
    d. sled runner jointers

12. Masons' tool bags are typically made from _____.
    a. nylon
    b. canvas
    c. reinforced polyester
    d. fiberglass mesh

13. A brick spacing rule is also known as a(n) _____.
   a. course counter rule
   b. modular rule
   c. horizontal joint rule
   d. interval rule

14. Typically, the most expensive of a mason's hand tools is the _____.
   a. jointer
   b. brick spacing rule
   c. trowel
   d. level

15. The wooden parts of a folding wood rule should be coated periodically with _____.
   a. spar varnish
   b. mineral oil
   c. polyurethane
   d. linseed oil

16. In addition to checking for level, a mason's level is also used to _____.
   a. measure joint spacing
   b. check for plumb
   c. check brick height
   d. check squareness

17. The tool used for laying out a 45-degree angle cut is the _____.
   a. sliding T-bevel
   b. combination square
   c. chalk box
   d. framing square

18. Compared to a conventional level, the cold beam laser is_____.
   a. more difficult to use
   b. less accurate for plumb measurements
   c. less expensive
   d. more accurate over longer distances

19. Some laser levels can project an accurate beam up to _____.
   a. 750 feet
   b. 1,200 feet
   c. 1,500 feet
   d. 3,000 feet

20. In the field, the laser level has almost completely replaced the _____.
   a. framing square
   b. plumb bob
   c. torpedo level
   d. chalk box

21. Braided mason's line is preferable to twisted line because _____.
   a. it will sag less when pulled tight
   b. it abrades less easily
   c. it is less likely to tangle
   d. it costs less per foot

22. Corner block are held in place by _____.
   a. dabs of mortar
   b. tension of the mason's line
   c. anchor pins
   d. suction cups

23. A mortarboard is approximately _____.
   a. 1' × 2'
   b. 1.5' × 1.5'
   c. 2' × 2'
   d. 3' × 3'

24. A cubic-foot measuring box is sometimes used to measure _____.
   a. water
   b. prepackaged mortar mix
   c. sand
   d. aggregate

25. Mortar-mixing tools should be cleaned with _____.
   a. dilute muriatic acid
   b. a stiff brush and water
   c. water and coarse steel wool
   d. a strong hose stream

26. An advantage of using a foot control with a masonry saw is _____.
   a. use of blade guards can be eliminated
   b. higher productivity
   c. both hands are free to guide the piece being cut
   d. blade speed can be controlled

27. Handheld cutoff saws use blades with a diameter of _____.
   a. 7¼ inches
   b. 9 inches
   c. 12 inches
   d. 15 inches

28. A tuckpoint grinder scores or cuts stone and masonry materials using _____.
   a. abrasive wheels
   b. nitrile blades
   c. tempered steel blades
   d. fused silicon wheels

29. Large hydraulic splitters can exert a pressure of up to _____.
    a. 10,000 pounds
    b. 10 tons
    c. 20 tons
    d. 50,000 pounds

30. Drilling in masonry or hardened mortar requires use of a powder-actuated tool or a(n) _____.
    a. water-cooled drill
    b. hammer drill
    c. high-speed drill bit
    d. anchor setter

31. The tag must never be removed from a fire extinguisher because _____.
    a. the extinguisher can't be activated without it
    b. it shows the expiration date for the contents
    c. it contains the operating instructions
    d. it shows when the extinguisher was last serviced and inspected

32. The minimum flow rate for a pressure washer should be _____.
    a. 1–3 gpm
    b. 4–6 gpm
    c. 7–9 gpm
    d. 10–12 gpm

33. Materials hoists can raise a load of 1,000–5,000 pounds to a height of as much as _____.
    a. 2 floors
    b. 160 feet
    c. 10 floors
    d. 300 feet

34. Powered lifts for indoor use run on battery power to _____.
    a. keep them small enough to fit through doorways
    b. avoid the need for long extension cords
    c. eliminate exhaust fumes
    d. lower noise levels inside the building

Trade Terms Quiz

Fill in the blank with the correct term that you learned from your study of this module.

1. Adding water to mortar to replace evaporated moisture and restore proper consistency is called ________.
2. A(n) ________ is any type of post braced into a plumb position so that a line can be fastened to it. Also called a *deadman*.
3. The act of troweling mortar or a mortar-repairing material, such as epoxy, into a joint after masonry is laid is called ________.
4. A(n) ________ is a horizontal joint between two masonry units.
5. A(n) ________ is a thin coat of mortar or grout on the outside surface of a wall that prepares a masonry surface for attaching veneer or tile.
6. The two corners of a structural unit or wall, built first and used as a position marker and measuring guide for the entire wall, are called the ________.
7. ________ is the reaction caused by a pinched, misaligned, or snagged tuckpoint grinder wheel that causes the wheel to stop momentarily, propelling the grinder away from the surface and toward the operator.

Trade Terms

Bed joint
Corner pole
Kickback
Lead
Parge
Pointing
Tempering

Cornerstone of Craftsmanship

Kenneth Cook

Vice President

Pyramid Masonry

Ken Cook went to his first masonry project site when he was six years old and has never looked back. Today he is vice president of operations for a major masonry contracting company.

Describe your job.
As vice president of operations, I handle the day-to-day running of the business. I do sales and estimating work as well. I keep my hand in by spending as much time as I can on job sites. I make sure the work is being done to the client's specifications, and I do whatever is necessary to make sure the job stays on schedule and within budget.

How did you get started in the construction industry?
My father was a mason, and he began teaching me the trade when I was just a child. I learned on the job, but I had a long time to learn and a patient teacher to bring me along. Not everyone can count on having the same opportunity, so it's best to learn the proper techniques through a training program before you go on a job.

What do you think it takes to become a success?
In a word, teamwork. There are many tasks to do on the site: mixing mortar, laying block and brick, erecting and moving scaffolds, and keeping masons supplied with materials. Everyone has to do their part, and everyone has to pitch in and do what is needed to keep the job flowing smoothly. If there's one prima donna on a crew, it will drag down the entire crew.

What do you enjoy most about your career?
I enjoy working directly with the crews on the job, especially teaching them proper technique and helping them improve their skills. A company is only as good as its employees, so it's important to make sure every employee is as effective and productive as possible.

What would you say to someone just entering the trade?
Get training. Do it through an apprentice program if you can, but get training in any way possible. Also, listen to experienced people. They can pass on knowledge and tricks of the trade they learned from experience. After you've had a couple of years of experience, you may think you know everything there is to know about the trade. In reality, just about everyone you encounter on the job will have a different perspective. No matter how long you've been in the trade, you will meet people who have ideas and techniques you've never thought of.

What does craftsmanship mean to you?
Craftsmanship is when a person practices a craft with great skill and feels pride upon completion. A craftsman finishes each step correctly before proceeding to the next step. Remember: planning, patience, and accuracy. Finish each step, and then finish the finish. "Close enough" is not an option for a true craftsman.

Trade Terms Introduced in This Module

Bed joint: A horizontal joint between two masonry units.

Corner pole: Any type of post braced into a plumb position so that a line can be fastened to it. Also called a *deadman*.

Kickback: A reaction caused by a pinched, misaligned, or snagged tuckpoint grinder wheel that causes the wheel to stop momentarily, propelling the grinder away from the surface and toward the operator.

Lead: The two corners of a structural unit or wall, built first and used as a position marker and measuring guide for the entire wall.

Parge: A thin coat of mortar or grout on the outside surface of a wall. Parging prepares a masonry surface for attaching veneer or tile, or parging can waterproof the back of a masonry wall.

Pointing: Troweling mortar or a mortar-repairing material, such as epoxy, into a joint after masonry is laid.

Tempering: Adding water to mortar to replace evaporated moisture and restore proper consistency. Any tempering must be done within the first 2 hours after mixing, as mortar begins to harden after 2½ hours.

Additional Resources

This module presents thorough resources for task training. The following reference material is suggested for further study.

A Guide to Scaffold Use in the Construction Industry. 2002. OSHA. **http://www.osha.gov/Publications/osha3150.pdf**

Brick Industry Association YouTube Channel. **www.youtube.com/user/BrickIndustry**

Bricklaying: Brick and Block Masonry. Reston, VA: Brick Industry Association.

Building with Masonry: Brick, Block, and Concrete. Dick Kreh. Newtown, CT: Taunton Press.

Complete Masonry: Building Techniques, Decorative Concrete, Tools and Materials. Des Moines, IA: Oxmoor House.

Concrete Masonry Handbook. Skokie, IL: Portland Cement Association.

"Forklift Safety." 2009. EDG Safety Series. **www.youtube.com/watch?v=yx67JVIc2Gw**

Forklift Safety Guide, Publication F417-031-000 (04/07). 2007. State of Washington Department of Labor and Industries. **www.lni.wa.gov/IPUB/417-031-000.pdf**

"Gilson Mortar Mixers by Cleform Tool." 2005. Construction Complete. **www.youtube.com/watch?v=ppMp8Upibz4**.

"Masonry Basics: The Tools You Need and How to Use Them." Masonry Guild of Arizona. **www.masonryforlife.com/howtobasics.htm**

Masonry Construction. David L. Hunter, Sr. Upper Saddle River, NJ: Prentice-Hall.

"Standards for Scaffold." OSHA. **www.osha.gov/SLTC/scaffolding/index.html**

"Pressure Washing Tips, Tricks And Techniques." 2011. **PowerWash.com**. **www.powerwash.com/articles/pressure-washing-tips-tricks-and-techniques-2.html**

"Scaffold PowerPoint Presentation." OSHA. **www.osha.gov/dte/library/scaffolds/scaffolding/notes.pdf**

"Tuckpointing and Repointing Tools: Specialized Tools for Removing and Replacing Mortar." Brett Martin. *Masonry Magazine*, September 2008. **www.masonrymagazine.com/9-08/tools.html**

"Using Masonry Saws." Norton Construction Products North America. **www.nortonconstructionproducts.com/solutions/masonry-saw.aspx**

Figure Credits

Courtesy of Bryan Light, SA01

Bon Tool Co., Figure 1, Figure 38 (middle)

Courtesy of Marshalltown Company, Figure 2

Bon Tool Company, Figures 3 – 11, Figure 13, Figure 14, Figures 16 – 22, Figure 24, Figure 26, Figure 28, Figure 32, SA02, SA03 (hod), Figure 40, Figure 42, SA04, Figure 43 (bottom), Figure 47, Figure 48

The Stanley Works, Figure 27, Figure 31

Courtesy of Irwin Tools, Figure 29

Laser Reference, Inc., Figure 30

Marshalltown Company, Figure 35

Courtesy Kraft Tool Co., Figure 36 (top), Figure 38 (bottom)

Courtesy of iQ Power Tools, Figure 44

Courtesy of Dennis Neal, FMA&EF, Figure 45, Figure 54 (bottom), Figure 61, Figure 64, Figure 67 (bottom)

Granite City Tools, Figure 46

Courtesy of DEWALT Industrial Tool Co., Figure 49, Figure 53

MULTIQUIP INC., Figure 51, Figure 52, Figure 54 (top)

Courtesy of Northern Tool + Equipment, Figure 55

Beta Max Hoists, Inc., Figure 57

Courtesy of Alimak Hek Group, Figure 58

CareLift Equipment Ltd, Figure 59

Courtesy of Bobcat Company. Bobcat®, the Bobcat logo and the colors applied to the Bobcat vehicle are registered trademarks of Bobcat Company in the United States and various other countries, Figure 60

Courtesy of Multiquip Inc., Figure 62

Photo Courtesy of Safway Group Holding LLC, Figure 63

Unidex, Inc, Figure 69

Section Review Answer Key

Answer	Section Reference	Objective
Section One		
1. b	1.1.0	1a
2. d	1.2.1	1b
3. c	1.3.1	1c
4. d	1.4.0	1d
Section Two		
1. c	2.1.0	2a
2. a	2.2.0	2b
3. b	2.3.0	2c
4. d	2.4.0	2d
Section Three		
1. d	3.1.0	3a
2. a	3.2.1	3b
Section Four		
1. d	4.1.0	4a
2. c	4.2.0	4b
3. a	4.3.0	4c
4. c	4.4.0	4d
Section Five		
1. a	5.1.0	5a
2. b	5.2.0	5b
3. d	5.3.1	5c
Section Six		
1. b	6.1.0	6a
2. d	6.2.0	6b
3. c	6.3.0	6c
Section Seven		
1. a	7.1.0	7a
2. d	7.2.1	7b

NCCER CURRICULA — USER UPDATE

NCCER makes every effort to keep its textbooks up-to-date and free of technical errors. We appreciate your help in this process. If you find an error, a typographical mistake, or an inaccuracy in NCCER's curricula, please fill out this form (or a photocopy), or complete the online form at **www.nccer.org/olf**. Be sure to include the exact module ID number, page number, a detailed description, and your recommended correction. Your input will be brought to the attention of the Authoring Team. Thank you for your assistance.

Instructors – If you have an idea for improving this textbook, or have found that additional materials were necessary to teach this module effectively, please let us know so that we may present your suggestions to the Authoring Team.

NCCER Product Development and Revision
13614 Progress Blvd., Alachua, FL 32615

Email: curriculum@nccer.org
Online: www.nccer.org/olf

❑ Trainee Guide ❑ Lesson Plans ❑ Exam ❑ PowerPoints Other ____________

Craft / Level: ____________ Copyright Date: ____________

Module ID Number / Title: ____________

Section Number(s): ____________

Description: ____________

____

____

____

____

Recommended Correction: ____________

____

____

____

____

Your Name: ____________

Address: ____________

____

Email: ____________ Phone: ____________

28103-13

Measurements, Drawings, and Specifications

Overview

This module covers math tools commonly used by masons. It includes concepts used for calculations, reading plans and drawings, and reading and meeting specifications. From understanding a set of drawings, to measuring and mixing mortar, math skills are needed at every step. A mason who has mastered math skills will save time, energy, materials, and money on the job.

Module Four

Trainees with successful module completions may be eligible for credentialing through NCCER's National Registry. To learn more, go to **www.nccer.org** or contact us at **1.888.622.3720**. Our website has information on the latest product releases and training, as well as online versions of our *Cornerstone* magazine and Pearson's product catalog.

Your feedback is welcome. You may email your comments to **curriculum@nccer.org**, send general comments and inquiries to **info@nccer.org**, or fill in the User Update form at the back of this module.

This information is general in nature and intended for training purposes only. Actual performance of activities described in this manual requires compliance with all applicable operating, service, maintenance, and safety procedures under the direction of qualified personnel. References in this manual to patented or proprietary devices do not constitute a recommendation of their use.

28103-13
Measurements, Drawings, and Specifications

Objectives

When you have completed this module, you will be able to do the following:

1. Recognize the mathematical concepts used in masonry.
   a. Explain how to read a six-foot rule.
   b. Explain how to read other measuring devices.
   c. Explain how to read mason's rules.
   d. Recognize modular increments.
   e. Describe how to determine areas and circumferences.
   f. Explain how to use the 3-4-5 ratio to square a corner.
2. Identify the basic parts of a set of drawings and list the information found on each type.
   a. Identify lines, symbols, and abbreviations used on drawings.
   b. Identify scales and dimensions used on drawings.
   c. Identify types of construction drawings.
3. Identify the purpose of specifications, standards, and codes used in the building industry and the sections that pertain to masonry.
   a. Explain the purpose of specifications, standards, and codes.
   b. Describe the purpose of inspections and testing.

Performance Tasks

Under the supervision of your instructor, you should be able to do the following:

1. Use a mason's rule to measure a space and verify its squareness.
2. Use a rule to measure fractional dimensions.
3. Locate information on construction drawings.

Trade Terms

Converting
Legend
MasterFormat™
Nominal dimension

Industry-Recognized Credentials

If you're training through an NCCER-accredited sponsor, you may be eligible for credentials from NCCER's Registry. The ID number for this module is 28103-13. Note that this module may have been used in other NCCER curricula and may apply to other level completions. Contact NCCER's Registry at 888.622.3720 or go to **www.nccer.org** for more information.

Code Note

Codes vary among jurisdictions. Because of the variations in code, consult the applicable code whenever regulations are in question. Referring to an incorrect set of codes can cause as much trouble as failing to reference codes altogether. Obtain, review, and familiarize yourself with your local adopted code.

Contents

Topics to be presented in this module include:

Figures and Tables

SECTION ONE

1.0.0 MASONRY MATH

Objective

Recognize the mathematical concepts used in masonry.

a. Explain how to read a six-foot rule.
b. Explain how to read other measuring devices.
c. Explain how to read mason's rules.
d. Recognize modular increments.
e. Describe how to determine areas and circumferences.
f. Explain how to use the 3-4-5 ratio to square a corner.

Performance Tasks 1 and 2

Use a mason's rule to measure a space and verify its squareness.

Use a rule to measure fractional dimensions.

Trade Terms

Converting: The process of changing from one form of measure to another; for example, from feet to inches or from inches to feet.

Nominal dimension: The size of the masonry unit plus the thickness of one standard (⅜ to ½ inch) mortar joint, used in laying out courses.

In the United States, masonry measurements are made in inches, feet, pounds, and gallons. You can only add and subtract numbers with the same units. When you are finished doing the arithmetic, you may need to convert smaller units into larger ones. The process of changing from one form of measure to another, say from feet to inches or from inches to feet, is called **converting**. The process of converting feet to inches is similar to the process of carrying in addition, as shown in *Figure 1*.

Masons use two numbering systems of their own: the course system and the modular system. As noted in the *Masonry Tools and Equipment* module, masons have two kinds of rules for these measures. When working with these rules (or any measuring tools), it is important to do the following:

- Familiarize yourself with the scale.
- Take readings carefully, and take them again to avoid making costly mistakes.

The old carpentry rule "measure twice and cut once" applies here.

1.1.0 Reading a Six-Foot Rule

Masons use rules to make accurate measurements. One of the most commonly used rules is the six-foot folding rule (*Figure 2*). The markings on the front of a standard six-foot rule are in inches and feet. Some rules have metric markings on the other side. A standard ruler is divided into whole inches and then halves, fourths, eighths, and sixteenths (*Figure 3*). Some rules also include thirty-seconds and sixty-fourths. The most common fractions that masons will encounter on the job are ⅜ and ⅝. You need to pay close attention when measuring. Your projects are only as accurate as your measurements.

In this section, you will learn how to add, subtract, multiply, and divide using the inch and foot measurements on a six-foot rule. Take the time to learn how to measure accurately. The quality of your work depends on accurate measurements.

1.1.1 Addition

This section shows the steps for solving an addition problem using a six-foot rule.

Add these measurements:

```
    2 feet  9 inches
+   2 feet  5 inches
--------------------
```

Step 1 Add the inches.

```
    2 feet  9 inches
+   2 feet  5 inches
--------------------
           14 inches
```

Step 2 There are 12 inches in a foot. When the inch column is 12 or greater, carry 1 foot to the foot column. The rule is to convert inches to feet when the number of inches is greater than 12.

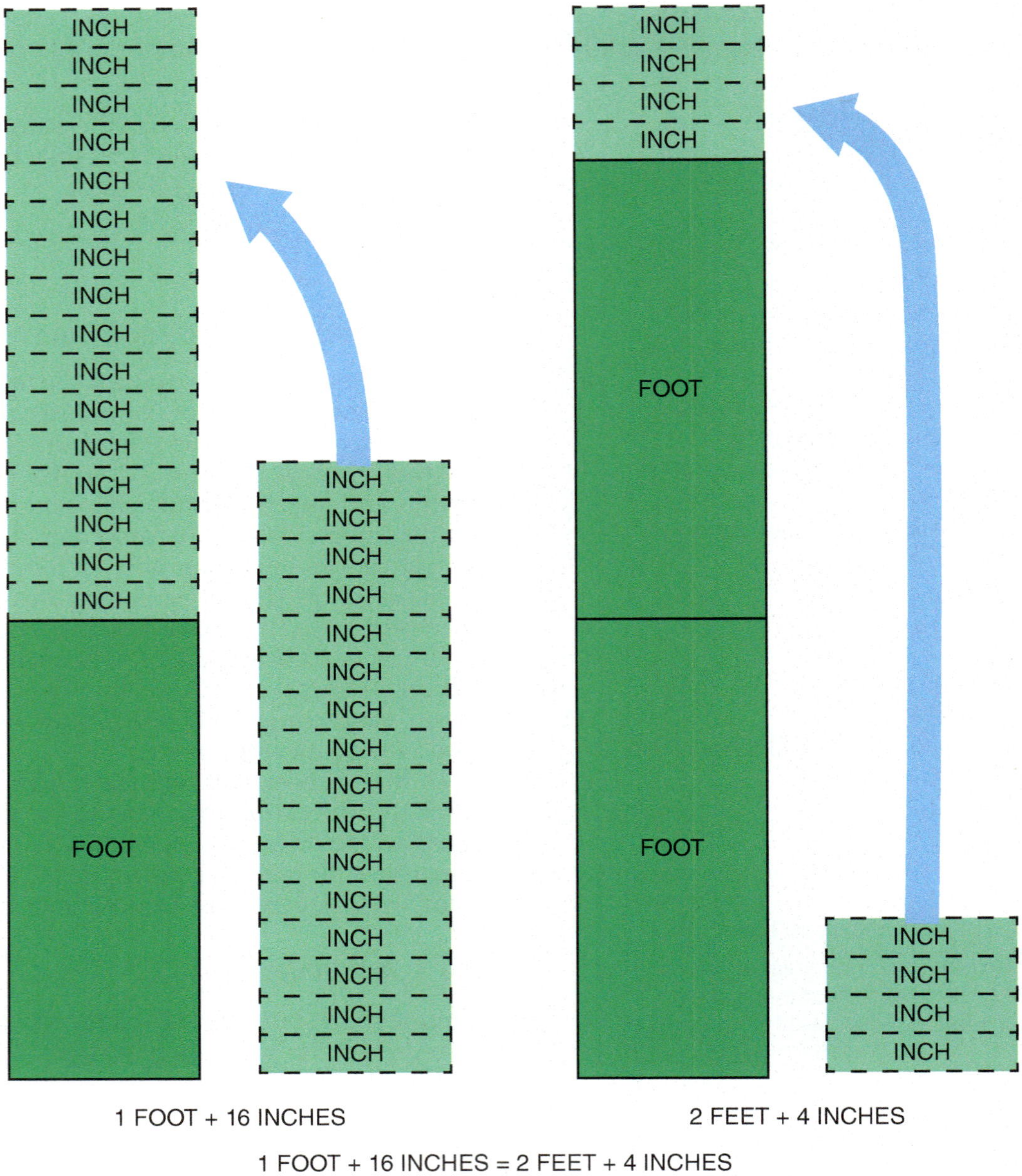

28103-13_F01.EPS

Figure 1 Converting units.

28103-13_F02.EPS

Figure 2 Six-foot folding rule.

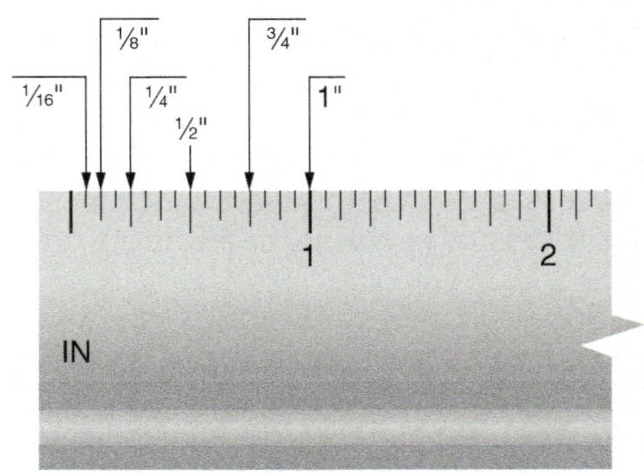

28103-13_F03.EPS

Figure 3 Reading rules accurately.

Step 3 Add the foot column, and subtract 12 from the inches column to account for the carry to the foot column.

	1 foot	
	2 feet	9 inches
+	2 feet	5 inches
	5 feet	14 inches
–		12 inches
	5 feet	**2 inches**

The sum of the two measurements is 5 feet 2 inches.

> **NOTE**
> Inches must be converted to feet when they exceed 12.

1.1.2 Addition Practice Exercises

This exercise will give you practice in adding measurements using a six-foot rule. You can check your work by looking up the answers in *Appendix A*.

1.

	1 foot	10 inches
+	2 feet	5 inches

2.

	2 feet	9 inches
+	2 feet	4 inches

3.

	2 feet	7 inches
+	1 foot	6 inches

4.

	1 foot	9 inches
+	1 foot	7 inches

1.1.3 Subtraction

Subtraction problems call for similar conversions. In subtraction, remember that borrowing does not give the 10 units that you get working with ordinary numbers. The borrowed amount depends on the measure you are converting (*Figure 4*).

The following section shows the steps for solving a subtraction problem with measurements.

Subtract these measurements:

	4 feet	7 inches
–	2 feet	8 inches

Step 1 Subtract the inches. Since 7 is less than 8, borrow 12 inches from the foot column, leaving 3 feet. Add 12 inches to 7 inches to get 19 inches. Now subtract 8 inches from 19 inches.

		(12 + 7) =
	3 feet	19 inches
	~~4 feet~~	~~7 inches~~
–	2 feet	8 inches
		11 inches

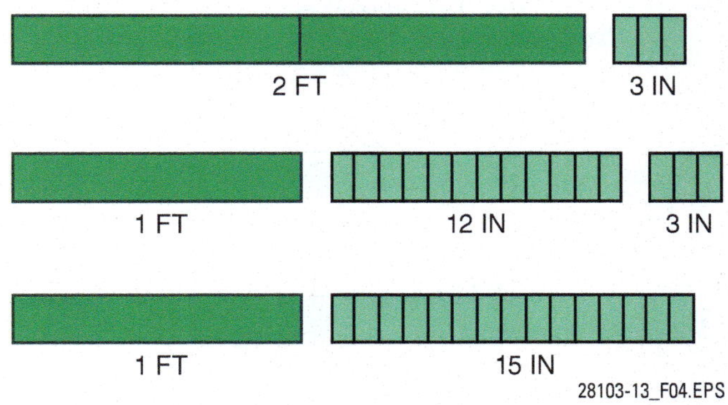

Figure 4 Borrowing units.

Rule of Thumb

Before tape measures, people used their bodies as a gauge. A small length was measured using the thumb. Larger distances were measured in feet or paces. You can still use your body to estimate distances:

- 1 inch = the width of your thumb
- 10 inches = the length of your foot
- 5 feet = a pace (two steps)
- 1 yard = the distance from your nose to your fingertips

You might wish to see how closely these "rules of thumb" apply to you.

Step 2 Subtract the foot column.

	3 feet	19 inches
–	2 feet	8 inches
	1 foot	11 inches

The difference between the two measurements is 1 foot 11 inches.

1.1.4 Subtraction Practice Exercises

This exercise will give you practice in subtracting measurements. You can check your work by looking up the answers in *Appendix A*.

1.

	2 feet	6 inches
–	1 foot	8 inches

2.

	2 feet	5 inches
–	2 feet	7 inches

Sometimes, plans will not give all of the dimensions used. Find the missing dimensions in *Figure 5* using Questions 3 and 4.

3. Subtract the known width from the total width to find A.

	35 feet	4 inches
–	21 feet	6 inches

4. Subtract the known length from the total length to find B.

	21 feet	3 inches
–	11 feet	9 inches

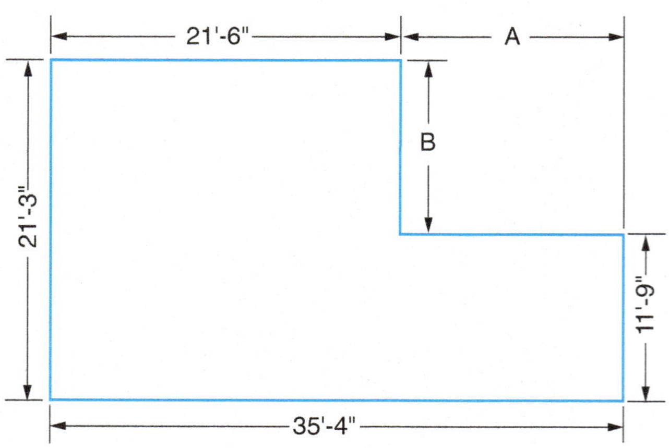

28103-13_F05.EPS

Figure 5 Find missing dimensions.

1.1.5 Other Measures

Measurements commonly used in the United States are listed in *Table 1*.

1.1.6 Fractions

Masons often use fractions in measuring and mixing. A fraction divides whole units into parts. They are usually written as two numbers, such as $\frac{1}{4}$, $\frac{1}{2}$, or $\frac{5}{8}$. The bottom number is the denominator. The upper number is the numerator. This section will review how to add, subtract, divide, and multiply fractions. As noted elsewhere in this module, the most common fractions that masons will encounter on the job are $\frac{3}{8}$ and $\frac{5}{8}$.

As with measurements: you must have the same units in order to perform mathematical operations on fractions. You cannot simply add $\frac{5}{8}$ and $\frac{3}{4}$. The denominators (the bottom numbers) must be the same. The fractions must be converted before they are added together. The conversion process is known as finding a common denominator. The lowest common denominator is the smallest number that the denominators can be evenly divided into.

To find the lowest common denominator, follow these steps:

Step 1 Reduce each fraction to its lowest terms.

Step 2 Find the lowest common multiple of the denominators. Sometimes it is simple; one number is a multiple of the other and the larger is the lowest common denomi-

Table 1 Common Measures

WEIGHT UNITS
1 ton = 2,000 pounds
1 pound = 16 dry ounces
LENGTH UNITS
1 yard = 3 feet
1 foot = 12 inches
VOLUME UNITS
1 cubic yard = 27 cubic feet
1 cubic foot = 1,728 cubic inches
1 gallon = 4 quarts
1 quart = 2 pints
1 pint = 2 cups
1 cup = 8 fluid ounces
AREA UNITS
1 square yard = 9 square feet
1 square foot = 144 square inches

28103-13_T01.EPS

nator. That means you can multiply by a whole number to get the larger number. If this is the case, all you have to do is find the equivalent fraction for the fraction with the smaller denominator.

Step 3 If neither of the denominators is a multiple of the other, you must multiply the two together to get a common denominator.

The following example will walk you through the steps to find the lowest common denominator between $\frac{1}{4}$ and $\frac{1}{3}$:

Step 1 Looking at the denominators, you see that 12 is a multiple of both 4 and 3. You need to find the equivalent fractions for $\frac{1}{4}$ and $\frac{1}{3}$ that have a denominator of 12. Because 4 goes into 12 three times, multiply the numerator in the fraction $\frac{1}{4}$ by 3. Because 3 goes into 12 four times, multiply the numerator in the fraction $\frac{1}{3}$ by 4:

$$\frac{1}{4} = \frac{3}{12}$$

$$\frac{1}{3} = \frac{4}{12}$$

Step 2 Add the two fractions together:

$$\frac{3}{12} + \frac{4}{12} = \frac{7}{12}$$

In this example, since $\frac{7}{12}$ can't be reduced, 12 is the lowest common denominator of the fractions $\frac{1}{4}$ and $\frac{1}{3}$.

1.1.7 Adding Fractions

How many inches will you have if you add $\frac{3}{4}$ of an inch and $\frac{7}{16}$ of an inch? To answer this question, you will have to add the fractions using the following steps:

Step 1 Find the lowest common denominator for the two fractions. Since 4 is a multiple of 16, the lowest common denominator is 16.

Step 2 Convert the fractions to equivalent fractions with the same denominator.

$$\frac{3}{4} \times \frac{4}{4} = \frac{12}{16}$$

Step 3 Add the numerators of the fractions.

$$\frac{12}{16} + \frac{7}{16} = \frac{19}{16}$$

Step 4 Reduce the fraction to its lowest terms. If the numerator is larger than the denominator, the answer is greater than 1.

$$\frac{19}{16} = \frac{16}{16} + \frac{3}{16} = 1\frac{3}{16}$$

1.1.8 Subtracting Fractions

You follow the same steps to subtract fractions. Say you have $\frac{3}{4}$ of a bag of mortar mix. You need $\frac{1}{2}$ a bag for a small batch of mortar. How much mortar mix will you have left?

Step 1 Find the common denominator. In this case, it is 4.

$$\frac{3}{4} - \frac{1}{2}$$

Step 2 Multiply to convert the fractions to equivalent fractions.

$$\frac{1}{2} \times \frac{2}{2} = \frac{2}{4}$$

Step 3 Subtract the numerators.

$$\frac{3}{4} - \frac{2}{4} = \frac{1}{4}$$

How Many Gallons?

The traditional volume units are the names of standard containers. Until the 18th century, it was very difficult to measure the capacity of a container. Standard containers were defined by the weight of a particular item, such as wheat or beer, that they could carry. This custom led to several standard units. These included the barrel, the hogshead, and the peck. The gallon was originally the volume of eight pounds of wheat.

The situation was still confused during the American colonial period. The Americans chose two of the many gallons. These two were the most common. For dry commodities, the Americans were familiar with the Winchester bushel. The corresponding gallon is one-eighth of this bushel.

For liquids, Americans used the traditional British wine gallon. As a result, the US volume system includes both dry and fluid units. The dry units are about one-sixth larger than the corresponding liquid units.

In 1824, the British established a new system based on the Imperial gallon. The Imperial gallon was designed to hold exactly 10 pounds of water. Unfortunately, Americans did not adopt this new, larger gallon. So the traditional English system actually includes three different gallons: US liquid, US dry, and British Imperial.

1.1.9 Multiplying Fractions

Multiplying and dividing fractions is very different from adding and subtracting fractions. You do not need to find a common denominator. Say you have ¾ of a bag of mortar mix. You need to make three even batches. How much mix is in each batch? You want to know how much is ⅓ of ¾. The word *of* lets you know to multiply.

Step 1 Multiply the numerators together to get a new numerator. Multiply the denominators together to get a new denominator.

$$\frac{3}{4} \times \frac{1}{3} = \frac{3}{12}$$

Step 2 Reduce the fraction to its lowest terms.

$$\frac{3}{12} = \frac{1}{4}$$

1.1.10 Dividing Fractions

Dividing fractions is similar to multiplying fractions with one added step. You must invert or flip the fraction you are dividing by. Use ½ ÷ ¾.

Step 1 Invert the fraction you are dividing by.

$$\frac{3}{4} \text{ becomes } \frac{4}{3}$$

Step 2 Change the division sign to a multiplication sign. Multiply as instructed earlier.

$$\frac{1}{2} \div \frac{3}{4} \text{ becomes}$$

$$\frac{1}{2} \times \frac{4}{3} = \frac{4}{6}$$

Step 3 Reduce the fraction to its lowest terms.

$$\frac{4}{6} = \frac{2}{3}$$

1.2.0 Reading Other Measuring Devices

In addition to the six-foot folding rule, masons may be called on to use other types of measuring devices in the field. These may include measuring tapes, ranging from 25 to 200 feet in length, and engineer's rules. It is important that masons are able to recognize and to use these measuring devices as skillfully as they use folding rules. Measuring tapes (*Figure 6*) are similar in appearance to retractable steel tapes, which you were introduced to in the module *Masonry Tools and Equipment*. Measuring tapes are used for laying out buildings and other large structures.

Use the following steps to read a measuring tape:

Step 1 Place the end of the tape at the starting point.

Step 2 Reading from the starting end of the tape, measure the distance to the point being measured. Notice that the measuring tape includes the foot measurement as a smaller number between the inch increments. This is to help you keep track of the distance without having to go back to the last foot increment.

Step 3 Read the feet, inches, and fractions of an inch just as you would for a steel tape or folding rule.

The tape in *Figure 7* illustrates a measurement of 17 feet, 2½ inches from the first reference point (the starting end of the tape) to the second reference point (the point being measured). You can see that the foot measurements in between the inch increments make it easier to read the distance being measured.

The graduations in an engineer's rule are different from those used in other types of rulers. Instead of being based on fractions of an inch, they are divided by tenths of a foot (*Figure 8*). On an engineer's rule, 10 hundredths equal ⅒ of a foot, and 10 tenths equals 1 foot. Likewise, 25 hundredths (0.25) is the same as ¼ foot or 3 inches, and 50 hundredths (0.50) is the same as ½ foot or 6 inches.

Use the following steps to read an engineer's rule:

Step 1 Place the end of the tape at the starting point.

> **Think About It**
>
> **Dividing Fractions**
>
> You want to divide two-thirds of a bag of cement mix in half. What is the mathematical equation and correct answer?

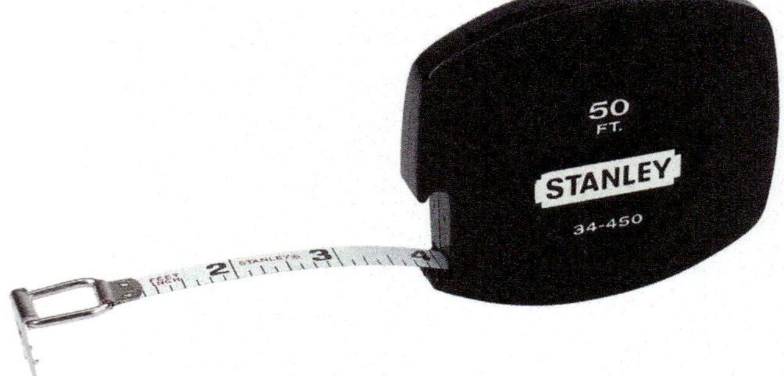

28103-13_F06.EPS

Figure 6 Measuring tape.

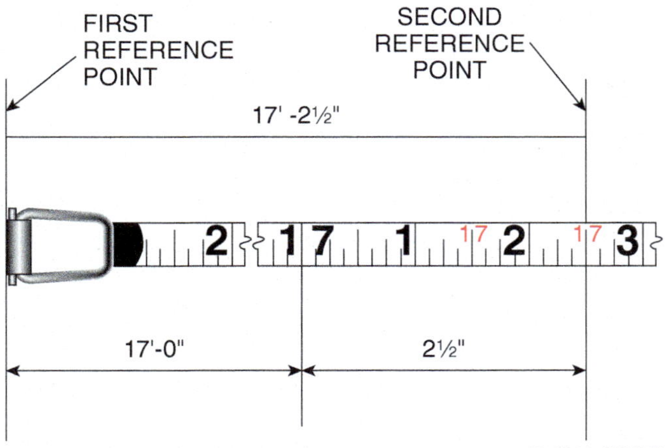

Figure 7 Reading a measuring tape.

Step 2 Reading from the starting end of the rule, measure the distance to the point being measured. Determine the number of whole units between the starting point and the point being measured. Remember, these are in tenths of a foot, not fractions of an inch.

Step 3 Use the inch and increments of tenths of an inch to determine the distance. If necessary, use a conversion table or your calculator to convert decimal units into fractions of an inch.

1.3.0 Reading Mason's Rules

On the back of mason's rules are measures that masons use to determine the distances between brick courses. The course system, also called the brick spacing rule or brick spacing system, predates the modular system of measurement. The standard brick spacing rule numbers the courses of different sizes of brick that will fill a vertical space (*Figure 9*). Standard and oversize rules are used to lay out and space standard brick courses to nonmodular dimensions. The rule has inches on the other side, marked in sixteenths of an inch.

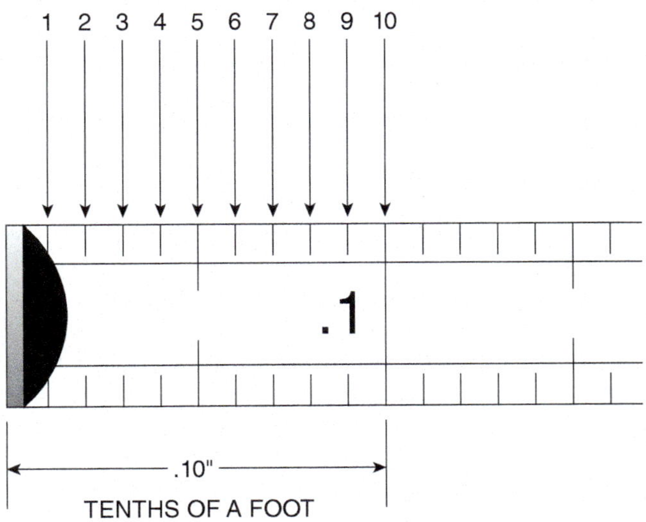

Figure 8 Reading an engineer's rule.

Figure 9 shows the standard, oversize, and modular brick spacing rules that masons commonly use on the job. The standard brick spacing rule has a gauge at the beginning that measures the size of one brick. It is used to identify the size of the brick so you will know which scale to read. On the standard brick spacing rule, all the reference measures fall between 2⅜ and 3 inches. For larger brick, you need to use the oversize brick spacing rule.

The large figures on the rule are references for the nominal sizes of standard brick and mortar thickness. The small figures at right angles to the size references count the number of courses for that size brick. The number of courses is marked for reference in *Figure 10*. On many rules, the large and small figures are shown in different colors to help you distinguish between them.

The number of courses refers to the number of times the block will need to be set in order to establish the correct height of the wall. If you are told to erect a 10-foot tall wall using nominal 8 inch × 8 inch × 16 inch concrete block, for example, the wall would be 15 courses:

10 feet × 12 inches (the number of inches in a foot) = 120 inches

Nominal block height = 8 inches

Height of wall (120 inches) ÷ Nominal height of block (8 inches) = 15

Number of courses = 15

Use the standard spacing rule to adjust course height or bed joints to meet benchmarks. Then transfer the correct markings from the spacing rule to the corner pole.

Oversize brick spacing rules have scales to measure nonmodular brick spacing. Instead of numbers, oversize brick spacing rules use letters (*A* through *K*) to indicate the number of courses. You will learn more about how to use mason's rules in later levels of the *Masonry* curriculum.

1.4.0 Recognizing Modular Increments

Today, brick is made for use on the modular grid system. The dimensions are based on a 4-inch unit called a module. The grid system makes it easier to combine different materials in a con-

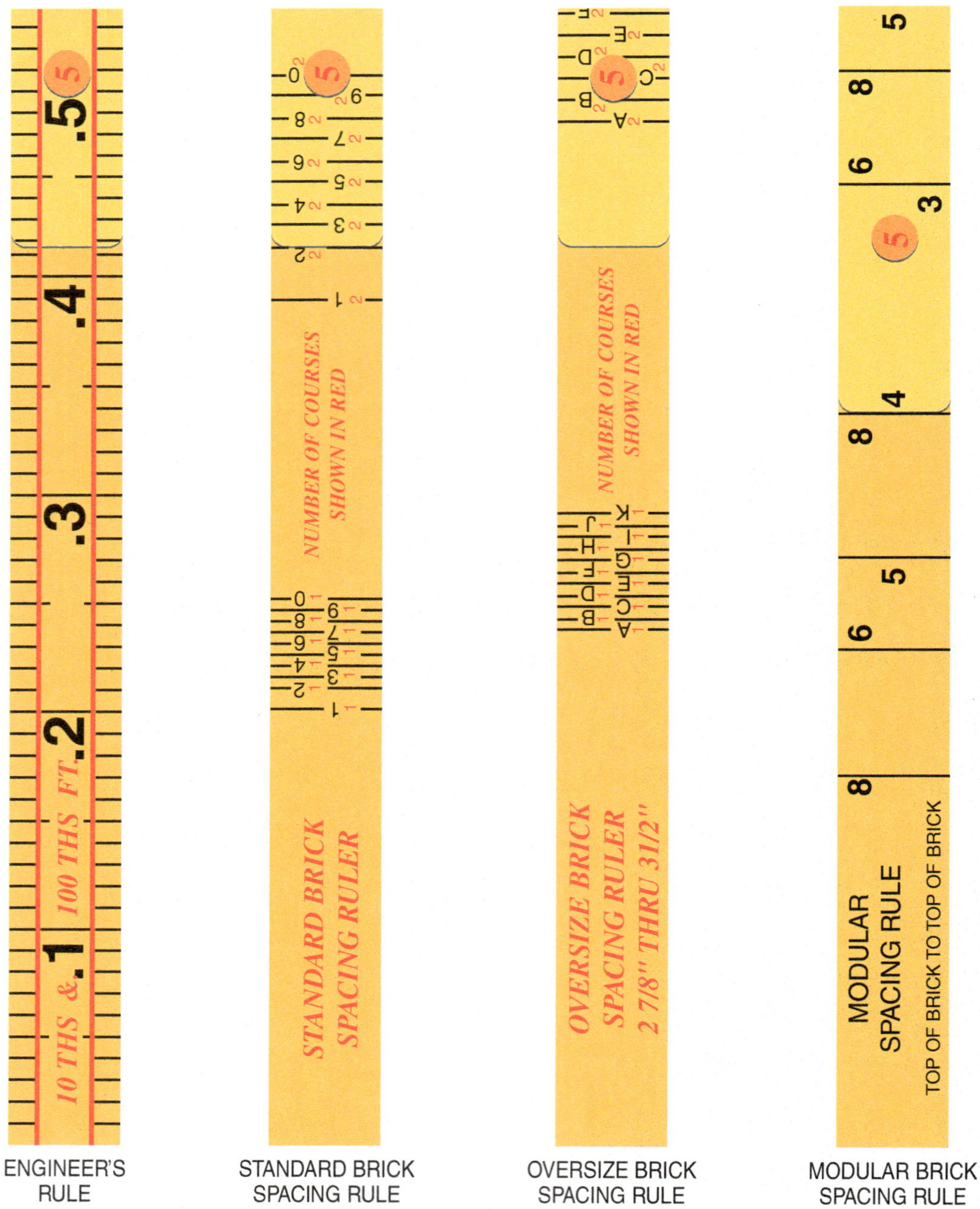

Figure 9 Standard, oversize, and modular brick spacing rules compared to an engineer's rule.

struction job. It creates a standard measurement, so different materials can be easily measured or calculated.

In modular design, the **nominal dimension** of a masonry unit is the manufactured dimension plus the thickness of the mortar joint. The nominal dimension is a multiple of 4 inches. A modular brick with a nominal length of 8 inches will have a manufactured dimension of 7½ inches if it is designed to be laid with a ½-inch mortar joint. It will have a manufactured dimension of 7⅝ inches if it is designed to be laid with a ⅜-inch mortar joint. The brick and mortar will fit in two 4-inch modules.

Table 2 shows nominal and actual manufactured dimensions for nominal brick sizes and actual dimensions for nonmodular brick sizes. It includes the planned joint thickness of ⅜ of an inch or ½ of an inch. The last column shows the number of courses required for each size of brick to equal a 4-inch modular unit or a multiple of a 4-inch unit. Remember, nominal dimensions in modular design are a multiple of 4.

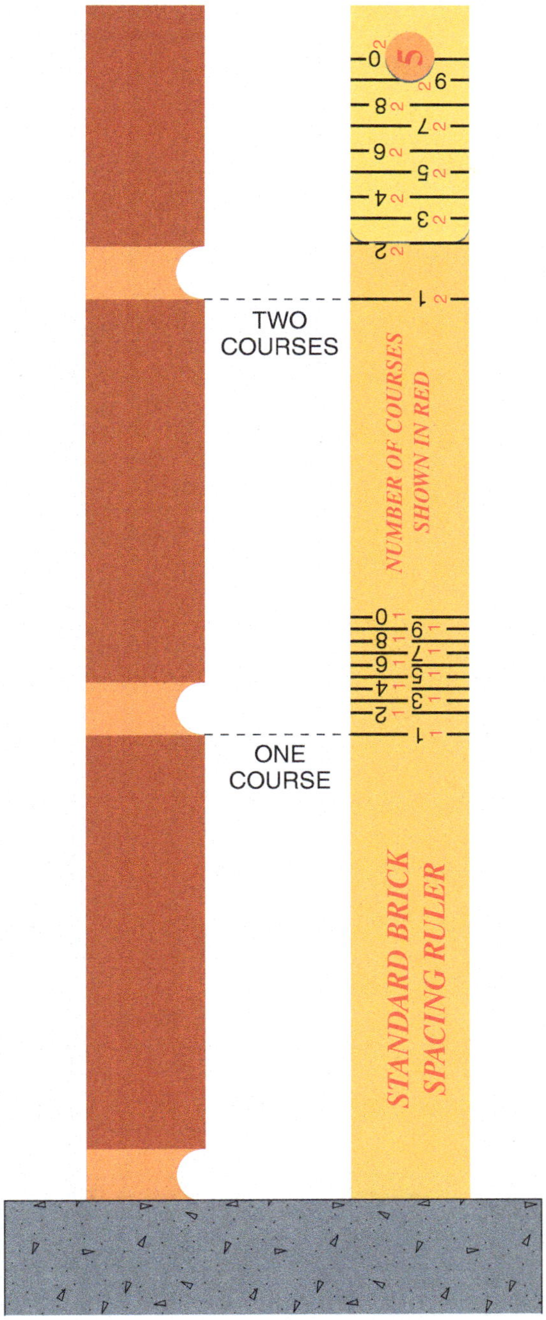

Figure 10 Reading the standard brick spacing rule.

Most masonry materials will tie and level off at a height of 16 inches vertically. Two courses of block with mortar will be 16 inches high. Six courses of modular brick with mortar will also be 16 inches vertically. As a result, the wythes can be tied together at 16-inch intervals or at multiples of 16 inches.

The modular spacing rule (*Figure 11*) has a modular scale on one side and inches marked into sixteenths on the other side.

The black figures are the references for the nominal sizes of modular brick and block. The modular markings give course numbers for different sizes. Scale 2 is for regular block or any brick with two courses equal to 16 inches in height. Scale 3 measures three courses in 16 inches and so on. Again, the specifications are the place to find the brick size or planned course height.

1.4.1 Determining the Number of Modular Brick Courses

The following example will show you how to determine how many courses of brick are needed to build a wall 8 feet high using modular brick.

Step 1 Determine the number of 16-inch sections.

$$8 \text{ ft} \times 12 \text{ in/ft} = 96 \text{ in}$$

$$96 \text{ in} \div 16 \text{ in} = 6 \text{ (16-inch sections)}$$

Step 2 Use *Table 2* to find the number of courses in a 16-inch section for modular brick.

Step 3 Multiply the number of courses per 16-inch section by the number of sections to find the total number of courses.

$$6 \times 6 = 36 \text{ courses}$$

1.4.2 Modular Brick Course Practice Exercises

You are building a wall 6 feet high. How many courses will you need for each type of brick? Refer to *Table 2*. You can check your work by looking up the answers in *Appendix A*.

1. Modular brick ____________
2. Roman brick ____________
3. Utility brick ____________
4. Norman brick ____________

1.5.0 Determining Areas and Circumferences

Knowledge of geometry is useful on a construction site. For example, if a wall is to have two windows in it, it will not need brick in these areas. Geometry allows the mason to calculate how much brick will be needed. This knowledge saves time and money when ordering materials. It can also be used to save steps when carrying brick to the workstation.

The next sections review the steps for calculating the areas of common geometric shapes. The

Table 2 Sizes of Brick

Unit Designation	Nominal Dimenions Inches w	h	l	Joint Thickness Inches	Specified Dimenions Inches w	h	l	# of Courses in 16"
Modular	4	2⅔	8	⅜	3⅝	2¼	7⅝	6
				½	3½	2¼	7½	
Engineer Modular	4	3⅕	8	⅜	3⅝	2¾	7⅝	5
				½	3½	$2\frac{13}{16}$	7½	
Closure Modular	4	4	8	⅜	3⅝	3⅝	7⅝	4
				½	3½	3½	7½	
Roman	4	2	12	⅜	3⅝	1⅝	11⅝	8
				½	3½	1½	11½	
Norman	4	2⅔	12	⅜	3⅝	2¼	11⅝	6
				½	3½	2¼	11½	
Engineer Norman	4	3⅕	12	⅜	3⅝	2¾	11⅝	5
				½	3½	$2\frac{13}{16}$	11½	
Utility	4	4	12	⅜	3⅝	3⅝	11⅝	4
				½	3½	3½	11½	
NONMODULAR BRICK SIZES								
Standard				⅜	3⅝	2¼	8	6
				½	3½	2¼	8	
Engineer Standard				⅜	3⅝	2¾	8	5
				½	3½	$2\frac{13}{16}$	8	
Closure Standard				⅜	3⅝	3⅝	8	4
				½	3½	3½	8	
King				⅜	3	2¾	9⅝	5
					3	2⅝	9⅝	
Queen				⅜	3	2¾	8	5

28103-13_T02.EPS

measurements of these shapes are typically in square feet. Plane figures are figures drawn in only two dimensions. Rectangles, triangles, and circles are common plane figures. The area of a plane figure is expressed in square units of the appropriate denomination.

1.5.1 Four-Sided Figures

Squares, rectangles, and parallelograms are four-sided regular polygons (*Figure 12*). They are figures with opposite parallel sides of the same length.

A rectangle is a polygon that has four sides of two different lengths that meet at right angles. The formula for finding the area of a rectangle is length × width, or:

$$A = lw$$

A square has four sides of the same length that meet at right angles. The formula for finding the area of a square is also length × width, simply expressed as *side times side*, or:

$$A = s^2 \text{ or } A = ss$$

A parallelogram has four sides that do not meet at right angles. The formula for finding the area of a parallelogram is base × height. The base (b) is the longest side, and the height (h) is the shortest

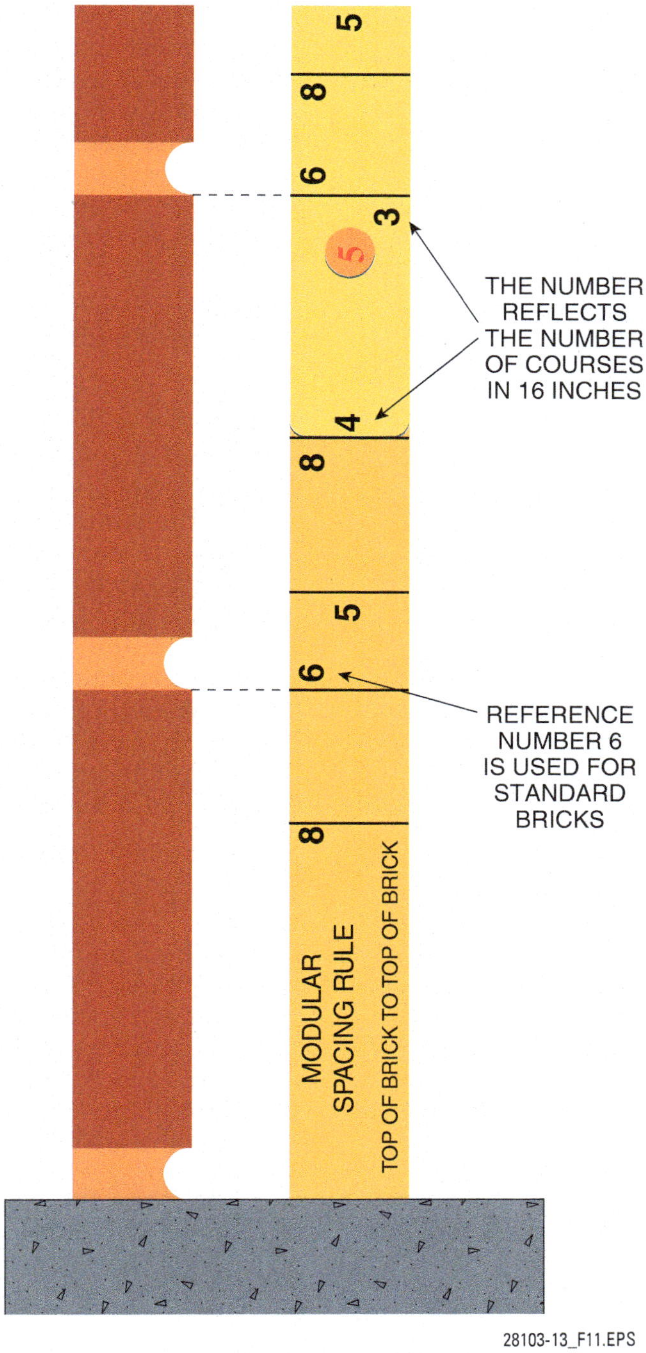

Figure 11 Reading the modular spacing rule.

distance between the upper and lower bases. This formula is expressed as:

$$A = bh$$

For example, a drawing shows a wall to be built that is 4 feet high and 10 feet long. What is the surface area of the wall?

Step 1 Find the formula for the calculation. The wall is a rectangle. The formula for the surface area of a rectangle is A = lw.

Step 2 Calculate the answer using the data from the drawing.

$$A = lw$$

$$A = 4 \text{ feet} \times 10 \text{ feet}$$
$$A = 40 \text{ square feet}$$

> **NOTE**
> Before performing a calculation, make sure that the two numbers have the same units. If the units are different, you must convert them to the same units.

1.5.2 Three-Sided Figures

Triangles are three-sided figures. They take many shapes. In all triangles, the three internal angles add up to 180 degrees. This is useful to know. If you know two of the angles, you can calculate the third. For example, if two angles of a triangle are 25 degrees and 75 degrees, the unknown angle is 80 degrees:

$$180 \text{ degrees} - 25 \text{ degrees} - 75 \text{ degrees} = 80 \text{ degrees}$$

Triangles can be identified by the relationships of the sides. Three examples are shown in *Figure 13*. All three sides of an equilateral triangle are the same size. Only two sides of an isosceles triangle are the same size. None of the sides of a scalene triangle are the same size.

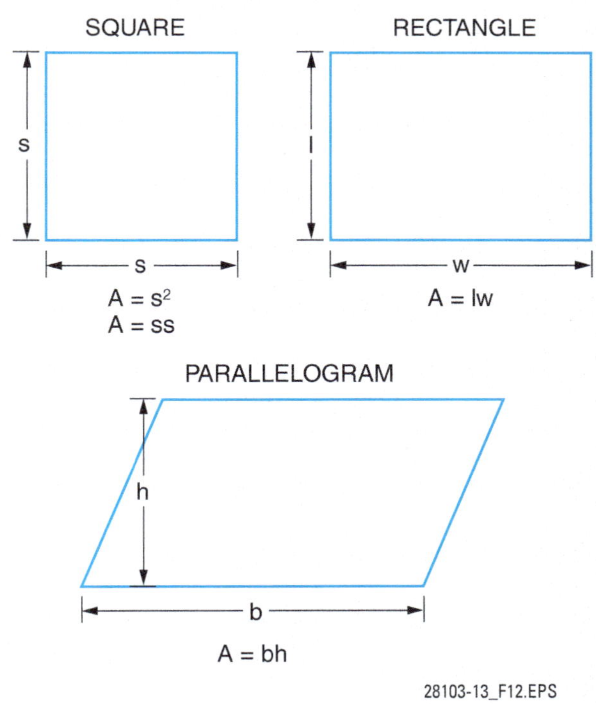

Figure 12 Four-sided figures.

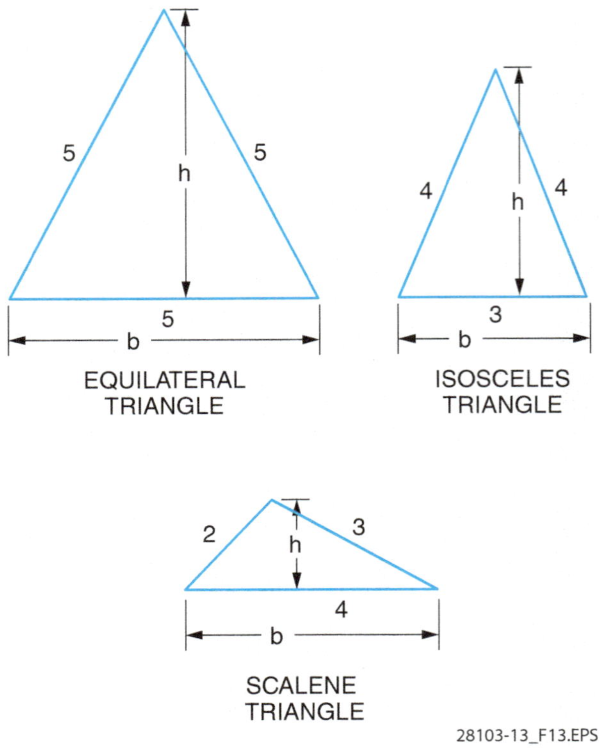

Figure 13 Triangles named by sides.

Triangles are also classified according to their interior angles (*Figure 14*). If one of the angles is 90 degrees, it is called a right triangle. If one of the angles is greater than 90 degrees, it is called an obtuse triangle. If each of the interior angles is less than 90 degrees, it is called an acute triangle.

Every triangle is really an exact half of a parallelogram. Remembering this makes it easy to calculate the area of a triangle. The height of a triangle is the length of a line drawn from one angle to the side opposite, or base, that meets the base at a right angle. In a right triangle, the base is also the shortest leg. In the case of an obtuse triangle, it will be necessary to extend the base line hypothetically outside the triangle in order to make the intersection occur at a right angle. In an acute triangle, the right angle intersection occurs inside the triangle itself.

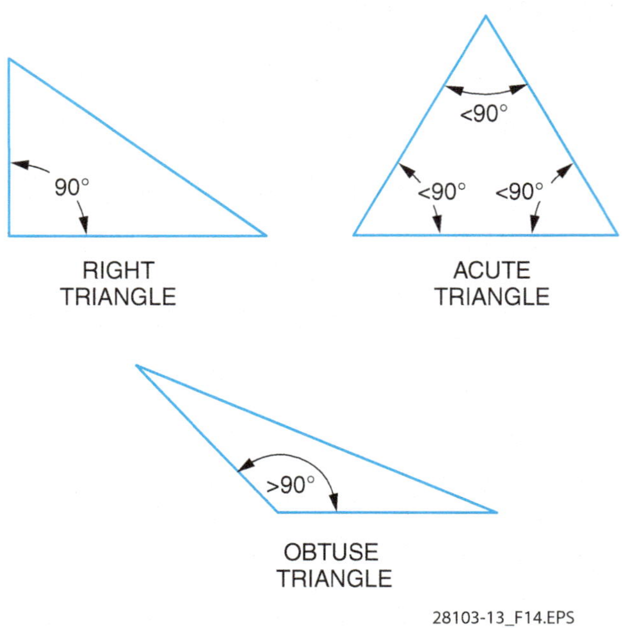

Figure 14 Triangles named by angles.

The area of a triangle is half the area of a parallelogram with the same base and height. The area of a triangle is expressed as:

$$A = \frac{1}{2}\, b \times h \text{ or } A = bh \div 2$$

Some areas are a combination of shapes. The house in *Figure 15* contains several shapes. The base of the house is a rectangle. The top portion is a triangle. The vent is a square. If you know the dimensions, you can calculate the area for the rectangle and the triangle to find the total surface area. You would then need to calculate the area of the two windows and vent and subtract that from the total surface area. Then you can find out how much brick you would need for the side of the house. You will learn how to use right triangles to square a corner in the section *Using the 3-4-5 Ratio to Square a Corner.*

1.5.3 One-Sided Figures

Circles are single, closed lines with all points the same distance from the center. *Figure 16* shows the parts of a circle. Note that the radius is half of the diameter. Either one of these dimensions may be given on a plan or drawing to indicate the size of the circle. The circumference is the outside line that defines the circle; the area is the space within it.

Figure 15 Divide figures to find the area.

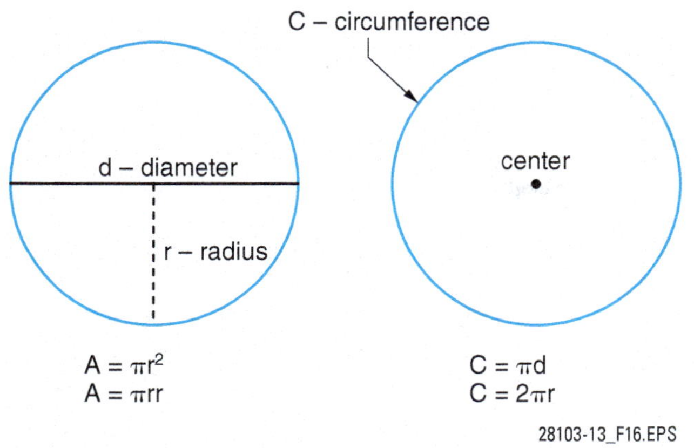

Figure 16 Parts of a circle.

> **NOTE**
> Remember to include the units, such as feet or inches, in your calculations.

If you are building a circular garden wall, you will need to know its circumference to figure out how much material will be needed. The circumference of a circle is its diameter (d) times the constant pi (π), or twice its radius (r) times the constant π. The rounded value of π is 3.14, or $^{22}/_{7}$. The formula for the circumference of a circle is expressed as:

$$C = \pi d \text{ or } C = 2\pi r$$

You may need to know the area of a circular patio to be paved with brick. The area of a circle is expressed in square units. The formula for the area of a circle is expressed as:

$$A = \pi r^2 \text{ or } A = \pi rr$$

1.5.4 Many-Sided Figures

Many-sided figures with all sides the same size and the same distance from the center are called regular polygons. They are also named after their number of sides. A five-sided figure is a pentagon, a six-sided figure is a hexagon, and an eight-sided figure is an octagon. *Figure 17* shows some regular polygons.

Occasionally, a plan will incorporate a hexagonal window over a door or an octagon in a bathroom. Sometimes a structure will be in the shape

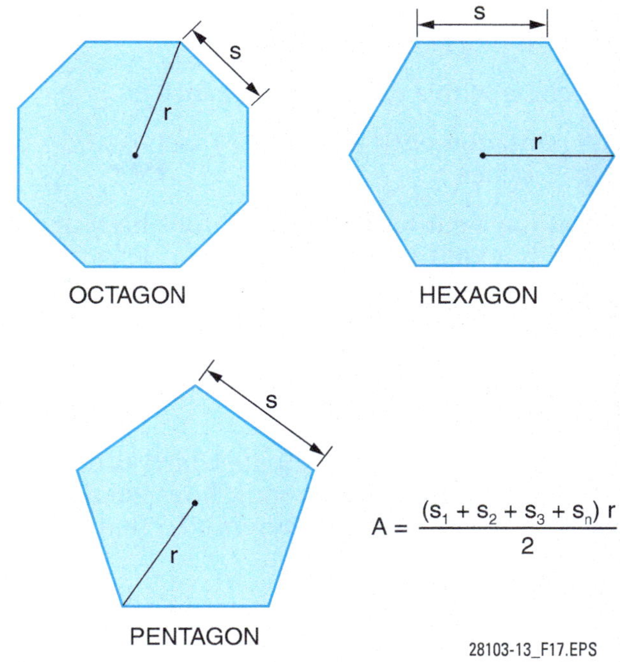

Figure 17 Regular polygons.

of an octagon instead of a square. To calculate the circumference of any regular polygon, you need to know the number and the length of each side.

The formula for the area of a regular polygon is the sum of the lengths of the sides divided by 2 then multiplied by r. The r is the distance from the center to any one angle. You may have to use your ruler on the plan to approximate that distance. The formula for the area of a regular polygon is written as:

$$A = (S_1 + S_2 + S_3 + S_n)\ r/2$$

The small n indicates that the sides continue to the required number.

1.5.5 Area and Circumference Practice Exercises

Calculate the areas of the following plane figures. You can check your work by looking up the answers in *Appendix A*.

1. A square with a height of 3 feet
2. A rectangle twice as wide as it is high, with a height of 3 feet, 8 inches

> ## Circles
> If a circle has a radius of 10 feet, is the circumference 62.8 feet or 62.8 square feet? What about the area?

3. A triangle with a base of 35 inches and a height of 2 feet
4. A circle with a diameter of 52 inches
5. A pentagon with a side of 24 inches and r of 28½ inches
6. You have a 4-foot-wide by 8-foot-high opening in a wall that was originally a door. The plans changed, and now you have to fill that space with standard block with an 8 inch × 16 inch face. How much block do you need to carry? If you add 5% for breakage, how much block does that make?
7. If you know that 1.125 standard block are needed to fill 1 square foot, how would you calculate Question 6 differently? What would the answer be?
8. A wall (*Figure 18*) that must be covered in brick veneer is 4 feet high and 6 feet long. There is a circular window with a diameter of 1 foot. What is the surface area that will be covered by brick veneer on this wall?

28103-13_F18.EPS

Figure 18 Wall with circular window.

1.6.0 Using the 3-4-5 Ratio to Square a Corner

Perhaps the most used shape in construction is the right triangle (*Figure 19*). Any vertical object or structure that forms a 90-degree angle with the ground is part of a right triangle. This can be the wall of a building, for example, or a telephone pole. If you draw an imaginary line from a point on the ground to the top of the structure, you have a right triangle. The line forms the hypotenuse of the right triangle. The base of the triangle extends from the bottom of the pole or structure to the starting point on the ground of your triangle.

Since the right triangle has one right angle, the other two angles are acute angles. They are also complementary angles. The sum of these two angles is 90 degrees. The right triangle has two sides perpendicular with each other, thus forming the right angle. To aid in writing equations, the sides and angles of a right triangle are labeled and shown in *Figure 19*. Normally, capital letters are used to label the angles, and lowercase letters are used to label the sides. The sides can be remembered as *a* for altitude and *b* for base. The third side (side *c*), opposite the right angle, is the hypotenuse. It is always longer than the other two sides.

If you know the length of any two sides of a right triangle, you can calculate the length of the third side. The equation used to calculate the length of the third side is called the Pythagorean

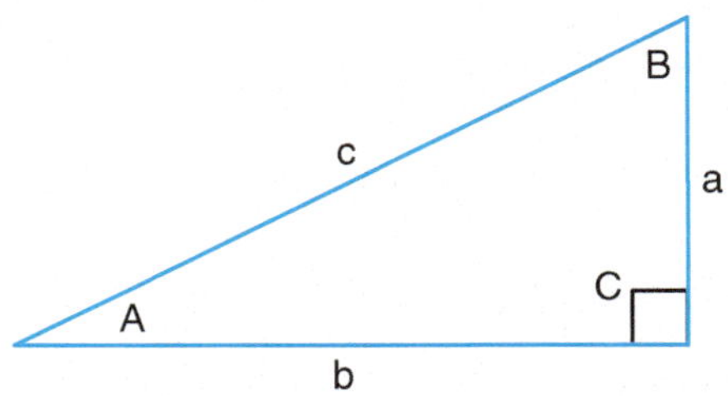

28103-13_F19.EPS

Figure 19 Right triangle.

Solid Figures and Volumes

We calculate areas by measuring two-dimensional figures. The measures are usually of length and width. The calculations are written as square measurements, such as square feet or square meters. We calculate volumes by measuring three-dimensional figures. The measures are of height, width, and depth. The calculations are written as cubic measurements, such as cubic feet or cubic meters. *Appendix B* shows examples of the most common solid figures in construction and the formulas for their volume calculations.

Notice that the cylinder, prism, and rectangular solid have the same shape from top to bottom. Solid figures with the same shape from top to bottom are easy to measure. Calculate the volume by first getting the area of the top surface. Then, multiply that area by the depth or height to get the volume. Conical, triangular, and spherical figures are more complicated, but masons rarely need to calculate the volume of these figures.

theorem, because it was first explained by the ancient Greek mathematician Pythagoras. It states that the square of the hypotenuse is equal to the sum of the squares of the remaining two sides. Expressed mathematically:

$$c^2 = a^2 + b^2$$

where c is the length of the hypotenuse and a and b are the lengths of the other two sides.

You may rearrange the equation to solve for the unknown side as follows:

$$a = \sqrt{(c^2 - b^2)}$$

$$b = \sqrt{(c^2 - a^2)}$$

$$c = \sqrt{(a^2 - b^2)}$$

For example, say you are trying to determine how long of a ladder to use to reach the top of a brick wall (*Figure 20*). If you know the height of the wall and the distance from the wall where the ladder will be placed, you can easily calculate the length of the ladder required using the Pythagorean theorem. In this example, the ladder is the hypotenuse of the right triangle formed by the wall and the level ground.

In the construction trades, the Pythagorean theorem is also called the 3-4-5 ratio. That's because whenever the longest leg of a triangle is 5 feet and the shorter legs are 3 feet and 4 feet, respectively, then the triangle is always a right triangle and contains a 90-degree angle. This is true for any triangle that has sides with a 3-4-5 ratio.

This theorem also applies if you multiply each number in the ratio (3, 4, and 5) by the same number. For example, if multiplied by the constant 3, it becomes a 9-12-15 triangle. For most construction layout and checking, right triangles that are multiples of the 3-4-5 ratio are used (such as 9-12-15, 12-16-20, 15-20-25, and 30-40-50). The specific multiple used is determined mainly by the relative distances involved in the job being laid out or checked. It is best to use the highest multiple that is practical because when smaller multiples are used, any error made in measurement will result in a much greater angular error.

So how does this work in practical terms? Let's say that C = 5 feet, A = 3 feet, and B = 4 feet. Plug these numbers into the formula as follows:

$$5^2 = 3^2 + 4^2$$

$$25 = 9 + 16$$

In practice, you can use the 3-4-5 ratio to verify that a building corner is square. To do this, follow these steps:

Step 1 Measure from the corner 3 feet along one wall, and place a mark at that point. Measure 4 feet along the other wall and mark that point.

Step 2 Use a ruler to measure the straight-line distance between the two points. If that distance is 5 feet, the corner is square. See *Figure 21*.

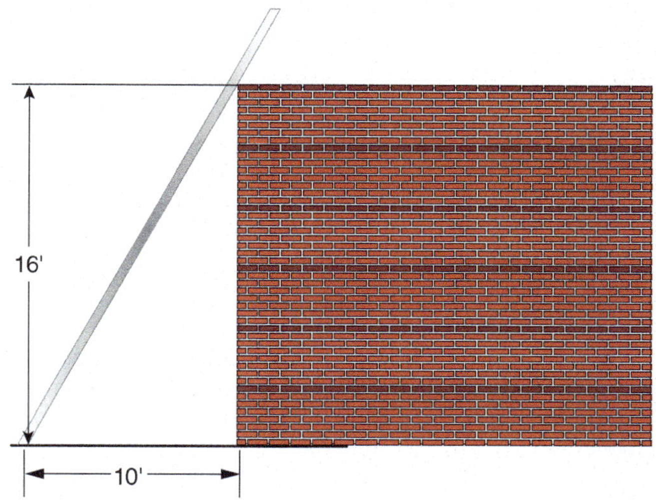

28103-13_F20.EPS

Figure 20 Brick wall and ladder.

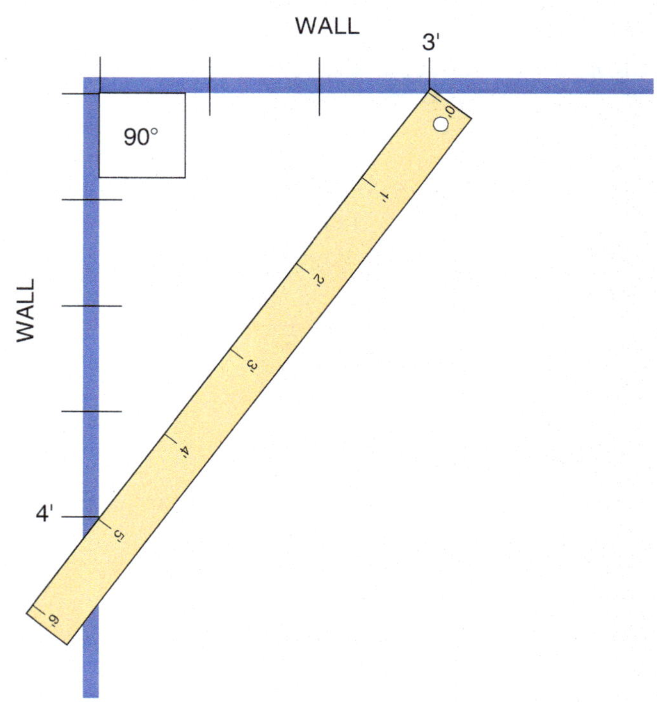

28103-13_F21.EPS

Figure 21 Checking the squareness of a corner.

Another Way to Think about Square Numbers

You know that the word square refers to shape, but in mathematical terms, it also refers to the product of a number multiplied by itself. For example, 25 is the square of 5, and 16 is the square of 4. Another way to say this is that 25 is 5 squared, or 5 times itself, and 16 is 4 squared, or 4 times itself.

In a mathematical equation this might appear as $5^2 = 25$ or $4^2 = 16$. In these examples, the numbers 5 and 4 are called the square roots, because you have to square them—or multiply them by themselves—to arrive at the squares.

The Metric System

The metric system has been used worldwide since 1875. The General Services Administration, which oversees all federal building projects, began requiring bids with metric specifications. In September 1996, all federally assisted highway construction projects were required to use metric standards.

The official name of the metric system is *Système International d'Unités* (SI). SI is a very convenient and logical system of measurements and weights. It is based on the number 10. This is similar to our system of currency: 10 pennies are equal to a dime, and 10 dimes are equal to a dollar. Terms such as kilometer, centimeter, and millimeter are units of metric measure. The base units of measure include the meter for length, the gram for weight, and the liter for liquid volume. Smaller or larger units are noted by adding a prefix to the base unit. The prefix expresses a multiple of 10. The same prefixes are used with all base units. A kilometer is 1,000 meters, while a millimeter is $\frac{1}{1,000}$ of a meter. Likewise, a kilogram is 1,000 grams, while a milligram is $\frac{1}{1,000}$ of a gram. This makes the metric system very easy to use. Refer to *Appendix B* for a list of the common SI prefixes, common metric measurements, and common US-to-metric and metric-to-US conversions,

Measuring the World

The metric system was designed in a scientific manner. Earth itself was selected as the measuring stick. The meter was defined as one ten-millionth of the distance from the equator to the North Pole. The liter was the volume of one cubic decimeter. The kilogram was the weight of a liter of pure water.

Scientific methods in 1795 were not accurate enough to measure the world. Modern measurements have shown that the world is slightly larger than was first estimated. However, the difference does not change the basic metric system.

The Yottameter

What happens when we want a number larger than the largest prefix? Every so often, new prefixes are added by the International Bureau of Weights and Measures in Paris. Added in 1991, the largest prefix is the yotta. It is one septillion or 10^{24}. That is, 1 with 24 zeros after it.

How far is a yottameter? It is about 10 billion light-years. Only the Hubble Space Telescope can see that far.

Additional Resources

Bricklaying: Brick and Block Masonry. 1988. Reston, VA: The Brick Institute of America.

Bricklaying Curriculum: Advanced Bricklaying Techniques. 1992. Raymond J. Turcotte and Laborn J. Hendrix. Stillwater, OK: Oklahoma Department of Vocational and Technical Education.

Building Block Walls: A Basic Guide for Students in Masonry Vocational Training. 1988. Herndon, VA: National Concrete Masonry Association.

1.0.0 Section Review

1. The two numbering systems that masons use, for which they have two kinds of rulers, are _____.
   a. the standard and metric systems
   b. the decimals and fraction systems
   c. the course and modular systems
   d. the standard and engineer's rules

2. The graduations in an engineer's rule are divided into _____.
   a. the letters *A* through *K*
   b. fractions of an inch
   c. percentages of a foot
   d. tenths of a foot

3. The three types of spacing rules commonly used by masons are _____.
   a. small, medium, and large
   b. residential, commercial, and specialty
   c. standard, oversize, and modular
   d. brick, block, and CMU

4. In modular design, the nominal dimension of a masonry unit is a multiple of _____.
   a. 4 inches
   b. 8 inches
   c. 16 inches
   d. 20 inches

5. A polygon that has four sides of two different lengths that meet at right angles is called a(n) _____.
   a. triangle
   b. rectangle
   c. tetrahedron
   d. pentagon

6. The formula for the Pythagorean theorem is _____.
   a. $c^2 = a^2 \times b^2$
   b. $c^2 = a^2 \div b^2$
   c. $c^2 = a^2 - b^2$
   d. $c^2 = a^2 + b^2$

Section Two

2.0.0 Masonry Drawings

Objective

Identify the basic parts of a set of drawings and list the information found on each type.

a. Identify lines, symbols, and abbreviations used on drawings.
b. Identify scales and dimensions used on drawings.
c. Identify types of construction drawings.

Performance Task 3

Locate information on construction drawings.

Trade Term

Legend: A table, list, or chart used in construction drawings to explain the meanings of the various lines, symbols, and abbreviations used in that particular set of drawings.

Construction drawings are always part of a project's documentation. Along with the specifications, they form the written guidelines for the builder. In order to do well in your work, you must be able to read and understand the information on the project drawings.

This section is a review of construction drawing material introduced in the *Core Curriculum*. It also presents some additional information that is relevant to masons about drawings, their organization, and the symbols for construction materials.

2.1.0 Identifying Lines, Symbols, and Abbreviations Used on Drawings

In order to read drawings, you have to recognize the various lines, symbols, and abbreviations used in their preparation. *Figure 22* shows typical details from construction drawings, showing a variety of brick masonry walls. You will encounter many drawings like this on the job. This section reviews the keys to lines, symbols, and scales and the information they carry. Because lines, symbols, and abbreviations can vary from drawing to drawing, you should always consult the legend, which explains the meanings of the various lines, symbols, and abbreviations used in that particular set of drawings. You should always consult the legend before attempting to read a set of drawings.

Each line symbol used on drawings means something different. *Figure 23* shows the most common types of lines. However, these lines can vary. Always consult the legend or symbol list when referring to any drawing. *Figure 24* shows an example of a construction drawing using several different types of lines.

The drafting lines shown in *Figure 23* are used as follows:

- *Light full line* – This line is used for section lines, building background (outlines), and similar uses where the object to be drawn is secondary to the system being shown, such as heating, ventilating, and air conditioning (HVAC) or electrical.
- *Medium full line* – This type of line is frequently used for hand lettering on drawings. It is further used for some drawing symbols and circuit lines.
- *Heavy full line* – This line is used for borders around title block, schedules, and for hand-lettering drawing titles. Some types of symbols are frequently drawn with a heavy full line.
- *Center line* – A center line is a broken line made up of alternately spaced long and short dashes. It indicates the centers of objects, such as holes, pillars, or fixtures. Sometimes, the center line indicates the dimensions of a finished floor.
- *Hidden line* – A hidden line consists of a series of short dashes that are closely and evenly spaced. It shows the edges of objects that are not visible in a particular view. The object outlined by hidden lines in one drawing is often fully pictured in another drawing.
- *Dimension line* – These are thin lines used to show the extent and direction of dimensions. Dimension lines have three parts: a line, a dimension, and a termination symbol. The dimension is usually placed in a break inside the dimension lines. Normal practice is to place the dimension lines outside the object's outline. However, it may sometimes be necessary to draw the dimensions inside the outline depending on the available room. *Figure 25* shows some common dimension line styles.

Other uses of the lines just mentioned include the following:

- *Extension lines* – Extension lines are lightweight lines that start about 1⁄16 inch away from the edge of an object and extend out. A common use of extension lines is to create a boundary for dimension lines. Dimension lines meet extension lines with arrowheads, slashes, or dots.

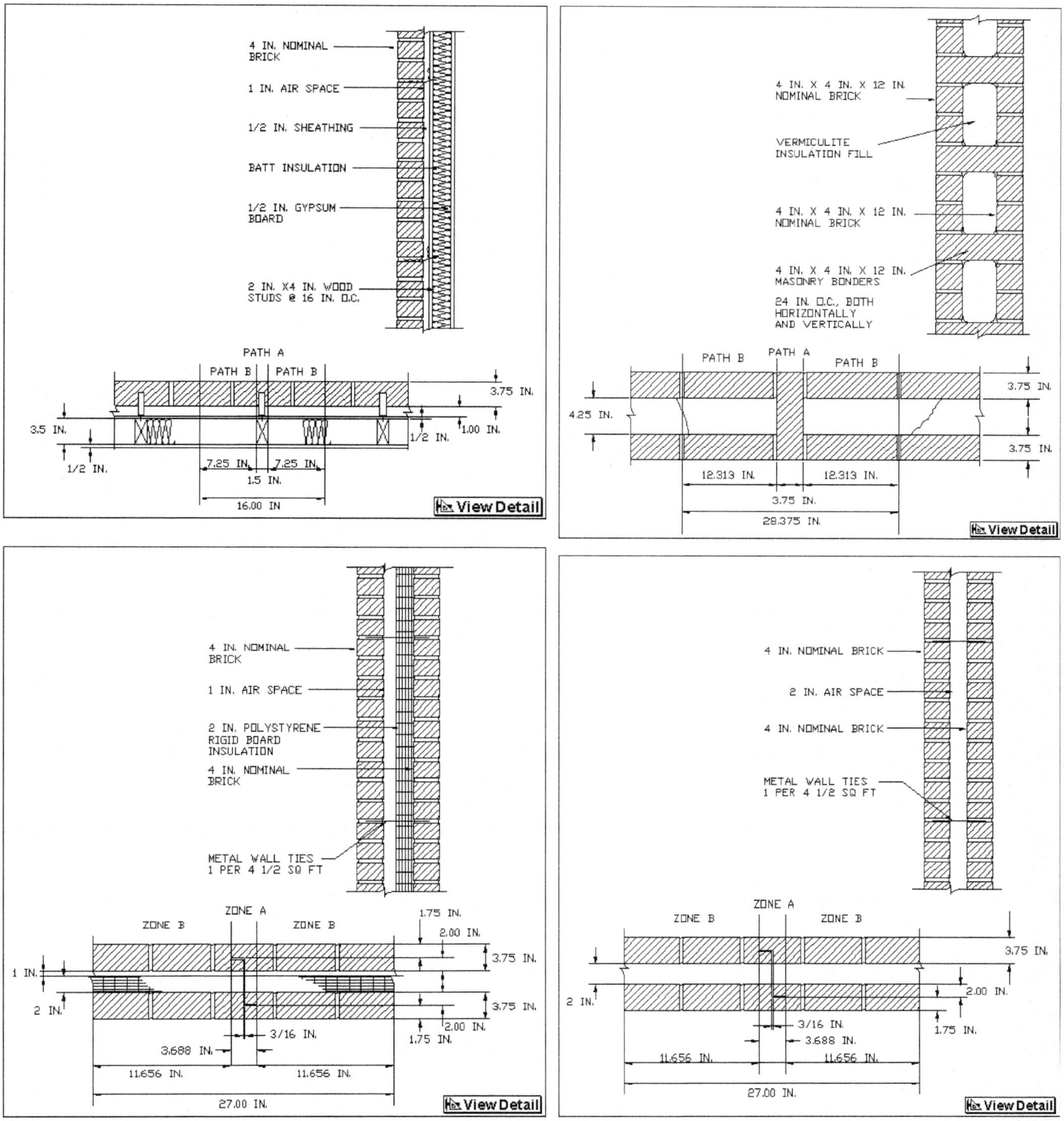

28103-13_F22.EPS

Figure 22 Details from construction drawings showing brick masonry walls.

Extension lines that point from a note or other reference to a particular feature on a drawing are called leaders. They usually end in either an arrowhead or a dot and may include an explanatory note at the end.

- *Section lines* – These are often referred to as crosshatch lines. Drawn at a 45-degree angle, these lines show where an object has been cut away to reveal the inside.
- *Phantom lines* – Phantom lines are solid, light lines that show where an object will be installed. A future door opening or a future piece of equipment can be shown with phantom lines.

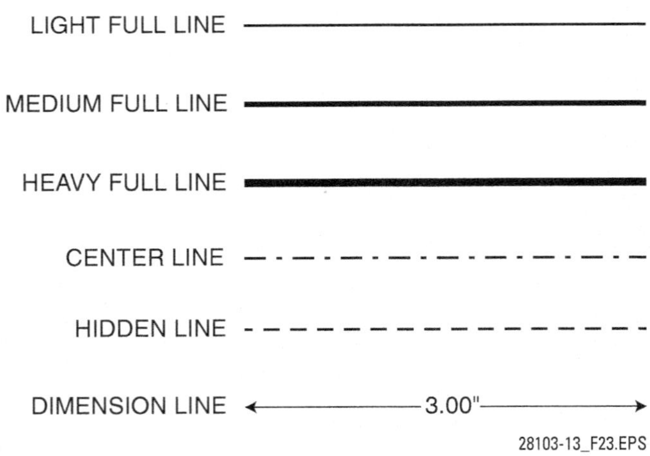

Figure 23 Lines used for construction drawings.

- *Geometric lines* – These are usually found in section drawings. They provide information about the shape and dimension of objects.

Symbols and abbreviations are used throughout a set of drawings. They convey accurate, concise, and specific information about a certain object without using much space. Some symbols will look like the object they represent while others will not. You will tend to memorize those you use frequently.

Figure 26 shows architectural symbols for common building materials. Some symbols look alike, so it is important to check where they are and what is next to them. As with other symbols used in construction drawings, the symbols used for architectural materials may vary, and you should always consult the legend for the particular set of drawings you are using.

Table 3 is a list of common architectural abbreviations. These abbreviations are used to clarify details on drawings, especially section drawings. They also appear in specification lists. Remember that these, too, may vary from drawing to drawing.

2.2.0 Identifying Scales and Dimensions Used on Drawings

Drawings are normally made using a certain scale. The inches or fractions of inches on the drawing represent real distances on the project. For residential drawings, the scale is usually ¼ of an inch equals 1 foot (1/4" = 1'-0"). This means that ¼ of an inch on the drawing equals 1 foot on the ground. For larger commercial projects, a smaller scale of ⅛ of an inch to 1 foot is normally used. For detail or sectional drawings, a larger scale is used. Commonly, these detail scales are ½ inch to 1 foot, or 1 inch to 1 foot. These scales are referred to as the ¼-, ⅛-, ½-, and 1-inch scales. When reading a drawing, it is a good idea to first check the scale.

Dimensions tell the builder how large an object is and its specific location in relation to other objects. Dimensions are usually shown in feet and inches (12'-6"), while dimensions less than 1 foot are shown in inches (6½"). Dimensioning interior walls is generally done in one of three ways: they are dimensioned to the center of the stud, to the outside of the stud, or to the outside of the finished walls. Dimensioning to the center of the studs is the most common.

Exterior masonry walls are usually dimensioned to the outside of the walls rather than to the center. *Figure 27* shows some typical dimensioning practices. Note that windows and doors in wood frame walls are usually dimensioned to the outside edge of the exterior wall. Windows and doors in brick veneer and frame walls are usually dimensioned to the edge of the framing, not the edge of the brick. It is critical to note where the dimension lines originate.

When reading a set of drawings to determine a particular dimension, do not measure the drawing or attempt to scale up the dimension from the drawing itself. It is best to calculate that dimension from the information on the drawing. Do not measure drawings because they may have shrunk or stretched from the reproduction process. If you must measure a drawing to determine a dimension, do it in several places if possible. If you are using a reduced set of drawings (they should be marked as reduced), never measure a dimension, always calculate it. If, when measuring, you find an error in the dimensions, do not simply change the drawing with the revised measurement. The error must be called to the attention of the architect.

2.3.0 Identifying Types of Construction Drawings

Construction drawings, as their name suggests, are drawings that describe a construction project in sufficient detail to allow a contractor to bid the work and then build it. The drawings also show the craftworker exactly what is to be done. Just as a mason always uses a level to check for plumb, a mason always uses a drawing to check for detail, location, and measurements. These drawings are usually bound together in a package. The package contains a cover page, a table-of-contents page, and sheets showing the technical details of the project. Depending on the complexity of the project, the specifications may also be included as part of the construction drawings. The title block

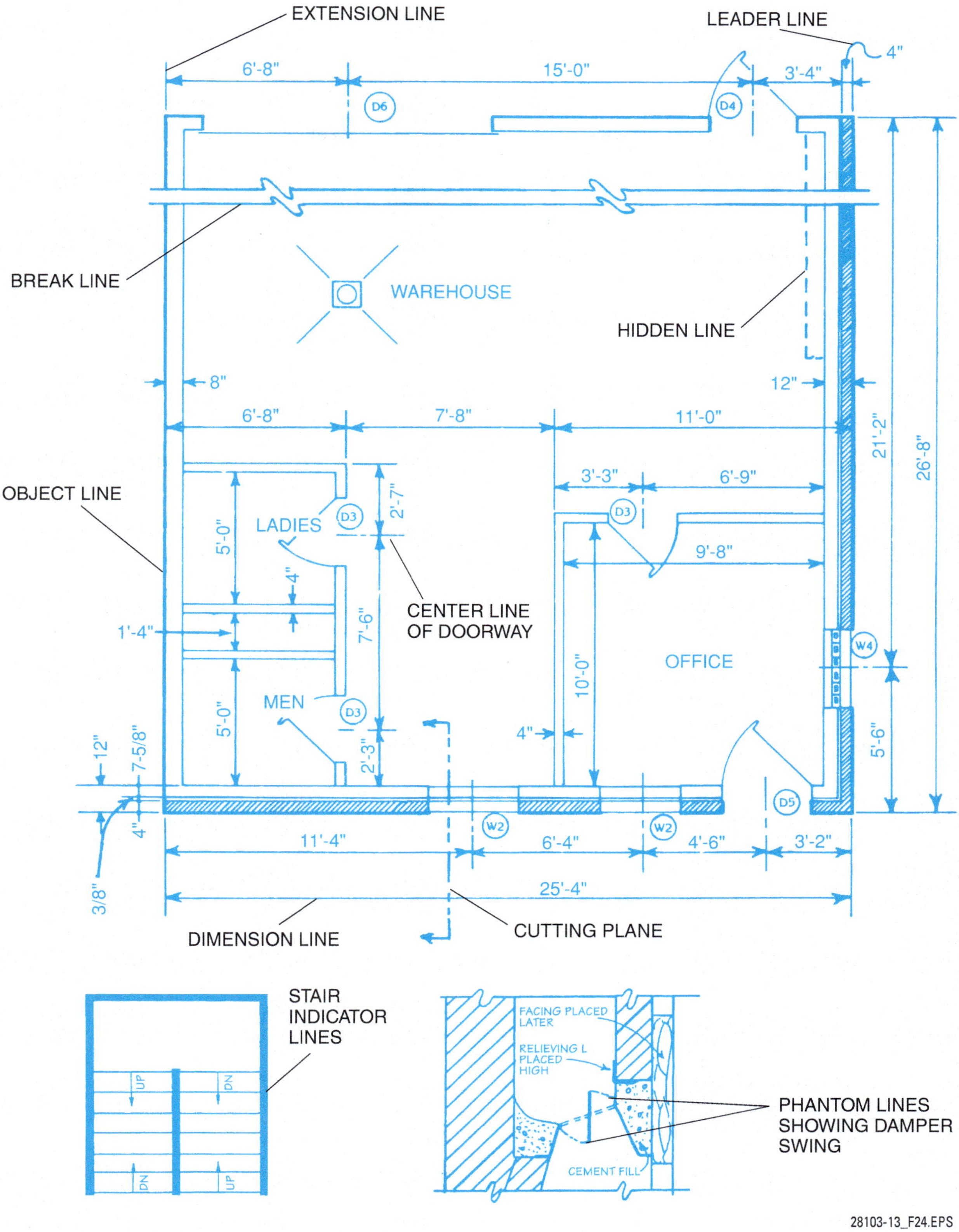

Figure 24 Sample construction drawing showing a variety of lines.

Figure 25 Dimension line styles.

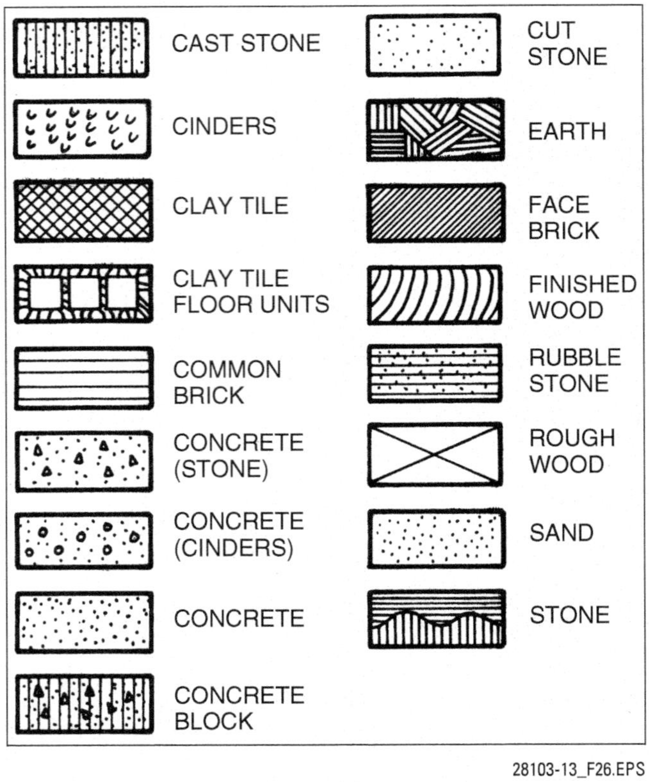

Figure 26 Architectural material symbols.

of the drawings also contains useful information about revisions to the plans.

The number of drawings in a set of plans will vary with the complexity of the project. A private home will require fewer sheets than a commercial building, and a large industrial site will require even more drawings. A large commercial project may have site plans by the surveyor, architectural drawings by the architect, structural drawings by the structural engineer, mechanical drawings by the mechanical engineer, and electrical drawings by the electrical engineer, plus detailed specification sheets that are separate from the drawings.

Construction drawings are frequently called working drawings or blueprints. Construction drawings are normally organized in the following order:

- Plot or site plans show the general layout of the land and information provided by the surveyor or site planner.
- Architectural/engineering plans include floor plans, foundation plans, elevations, section plans, roofing plans, and schedules for doors, windows, and other equipment.
- Specialty plans include mechanical, plumbing, electrical, and ductwork drawings. These may also include details for any custom features or unusual designs.

2.3.1 Plot or Site Plans

The plot or site plan shows the location of the building in relation to the property lines. It may show utilities and utility easements, contour lines, site dimensions, other buildings on the property, walks, drives, and retaining walls (*Figure 28*). This plan also shows the finished floor elevation(s) and the north direction arrow. Site plans were covered in the *Core Curriculum* module *Introduction to Construction Drawings.*

2.3.2 Floor Plans

The floor plan provides the largest amount of information, perhaps making it the most important drawing of all (*Figure 29*). The floor plan shows all exterior and interior walls, doors, windows, patios, walks, decks, fireplaces, built-in cabinets, and appliances. The floor plan is actually a cross-sectional view taken horizontally between the floor and the ceiling. The height of the cross section is usually cut about 4 feet above the floor. Sometimes this is varied to show important details of the structure.

2.3.3 Foundation Plans

The foundation plan is usually the first of the structural drawings (*Figure 30*). It shows the foundation size and materials. It shows details about excavation, waterproofing, and supporting structures, such as footings or piles. It can have sections of the footings and foundation walls. When a building has a basement, it is included here as well.

Table 3 Architectural Abbreviations

ABBREVIATIONS					
AB	ANCHOR BOLT	FDN	FOUNDATION	RM	ROOM
ADD'L	ADDITIONAL	FIN.	FINISH	SCHED	SCHEDULE
ADJ	ADJACENT	FLR	FLOOR	SECT.	SECTION
AISC	AMERICAN INSTITUTE OF	FOB	FACE OF BRICK	SHT	SHEET
	STEEL CONSTRUCTION	F.O.CONC	FACE OF CONCRETE	SIM	SIMILAR
ALT.	ALTERNATE	FOW	FACE OF WALL	SLV	SHORT LEG VERTICAL
ARCH.	ARCHITECTURAL	FS	FLAT SLAB	SPC	SPACE
ASTM	AMERICAN SOCIETY FOR	FT	FOOT	SPEC	SPECIFICATION
	TESTING & MATERIALS	FTG	FOOTING	SQ	SQUARE
BLDG	BUILDING	FW	FILLET WELD	STD	STANDARD
BM	BEAM	GA	GAUGE	STIFF	STIFFENER
B.O.	BOTTOM OF	GAL.	GALVANIZED	STL	STEEL
BOT	BOTTOM	GL	GLULAM BEAM	STOR	STORAGE
BSMT	BASEMENT	GR	GRADE	SYM	SYMMETRICAL
BTWN	BETWEEN	GR BM	GRADE BEAM	T&B	TOP AND BOTTOM
CANT.	CANTILEVER	H.A.S.	HEADED ANCHOR STUD	THK	THICKNESS
CB	CARDBOARD	HORIZ	HORIZONTAL	T.O.	TOP OF
CH	CHAMFER	HSB	HIGH-STRENGTH BOLT	TYP	TYPICAL
CJ	CONTROL/CONSTRUCTION JOINT	ID	INSIDE DIAMETER	U.N.O.	UNLESS NOTED OTHERWISE
CLR	CLEAR, CLEARANCE	IN.	INCH	VAR	VARIES
CMU	CONCRETE MASONRY UNIT	INT	INTERIOR	VERT	VERTICAL
COL	COLUMN	JNT	JOINT	VIF	VERIFY IN FIELD
CONC	CONCRETE	LB	POUND	WT	WEIGHT
CONN	CONNECTION	LIN FT	LINEAL FEET		
CONST	CONSTRUCTION	LLV	LONG LEG VERTICAL		**SYMBOLS**
CONT	CONTINUOUS	MAT'L	MATERIAL		
CONTR	CONTRACTOR	MAX.	MAXIMUM	℄	CENTER LINE
CTRD	CENTERED	MECH	MECHANICAL	ϕ	DIAMETER
DET	DETAIL	MID.	MIDDLE	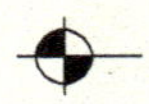	ELEVATION
DIAG	DIAGONAL	MIN	MINIMUM	&	AND
DIAM	DIAMETER	MISC	MISCELLANEOUS	W/	WITH
DIM.	DIMENSION	MTL	METAL	⅊	PLATE
DISCONT	DISCONTINUOUS	NIC	NOT IN CONTRACT	X	BY
DWG	DRAWING	NO.	NUMBER	#	NUMBER
EA	EACH	NOM	NOMINAL	@	AT
EF	EACH FACE	NTS	NOT TO SCALE	⌷	SQUARE
EL.	ELEVATION	OC	ON CENTER	L	ANGLE
ELECT.	ELECTRICAL	OD	OUTSIDE DIAMETER		
ELEV	ELEVATOR	O.H.	OPPOSITE HAND		
EQ	EQUAL	OPNG	OPENING		
EWB	END WALL BARS		PLATE		
EW	EACH WAY	PSF	POUND PER SQUARE FOOT		
EXIST.	EXISTING	PSI	POUND PER SQUARE INCH		
EXP JNT	EXPANSION JOINT	R	RADIUS		
EXT	EXTERIOR	REINF	REINFORCEMENT		
FD	FLOOR DRAIN	REQ'D	REQUIRED		

28103-13_T03.EPS

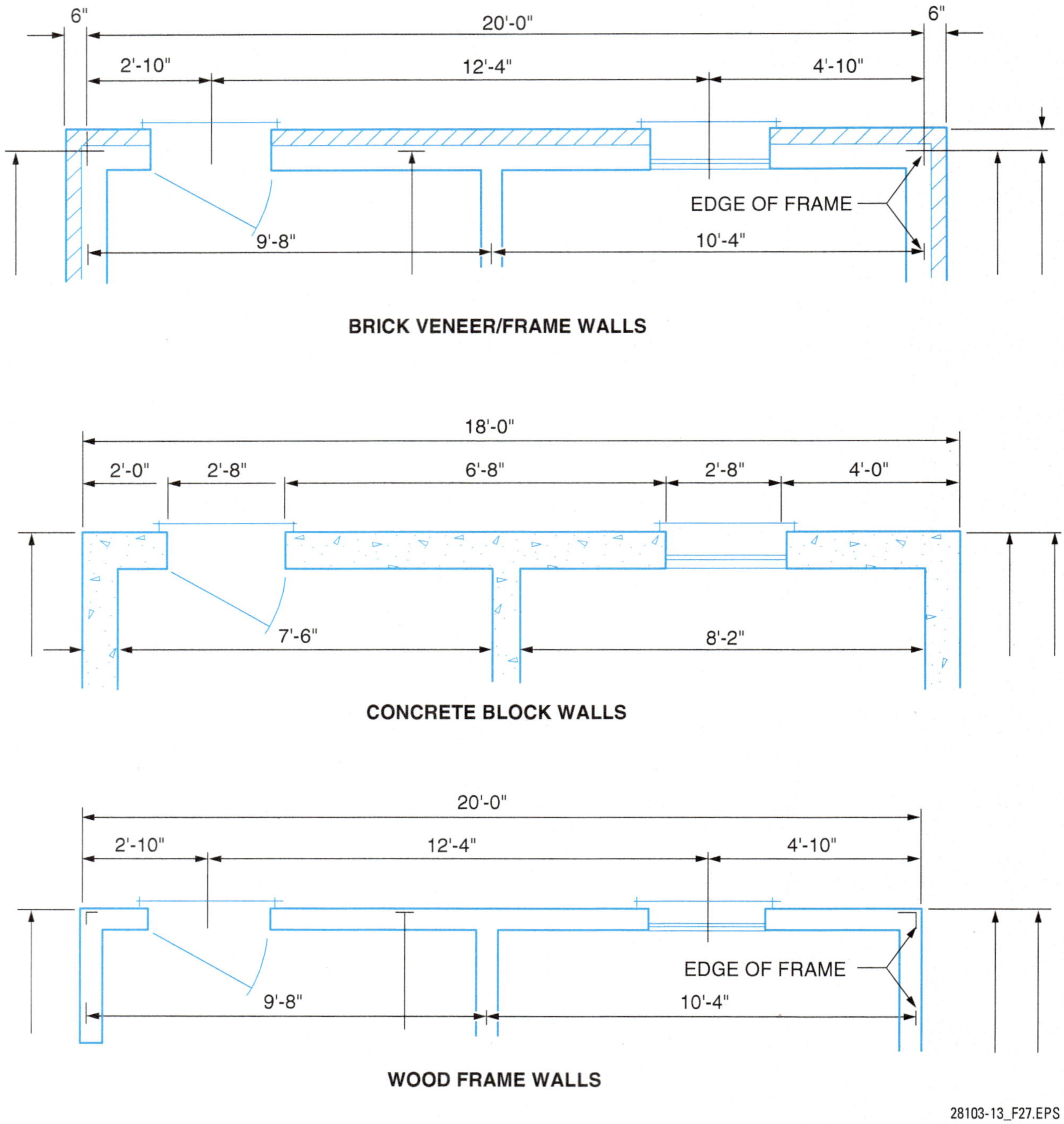

Figure 27 Dimension lines for different wall types.

Note that this detail contains alphanumeric bubble references to several construction details and section drawings that provide detailed information about the methods of construction.

2.3.4 Elevation Drawings

Elevation drawings are views of the exterior features of the building (*Figure 31*). They usually show all four sides of a building. Sometimes, for

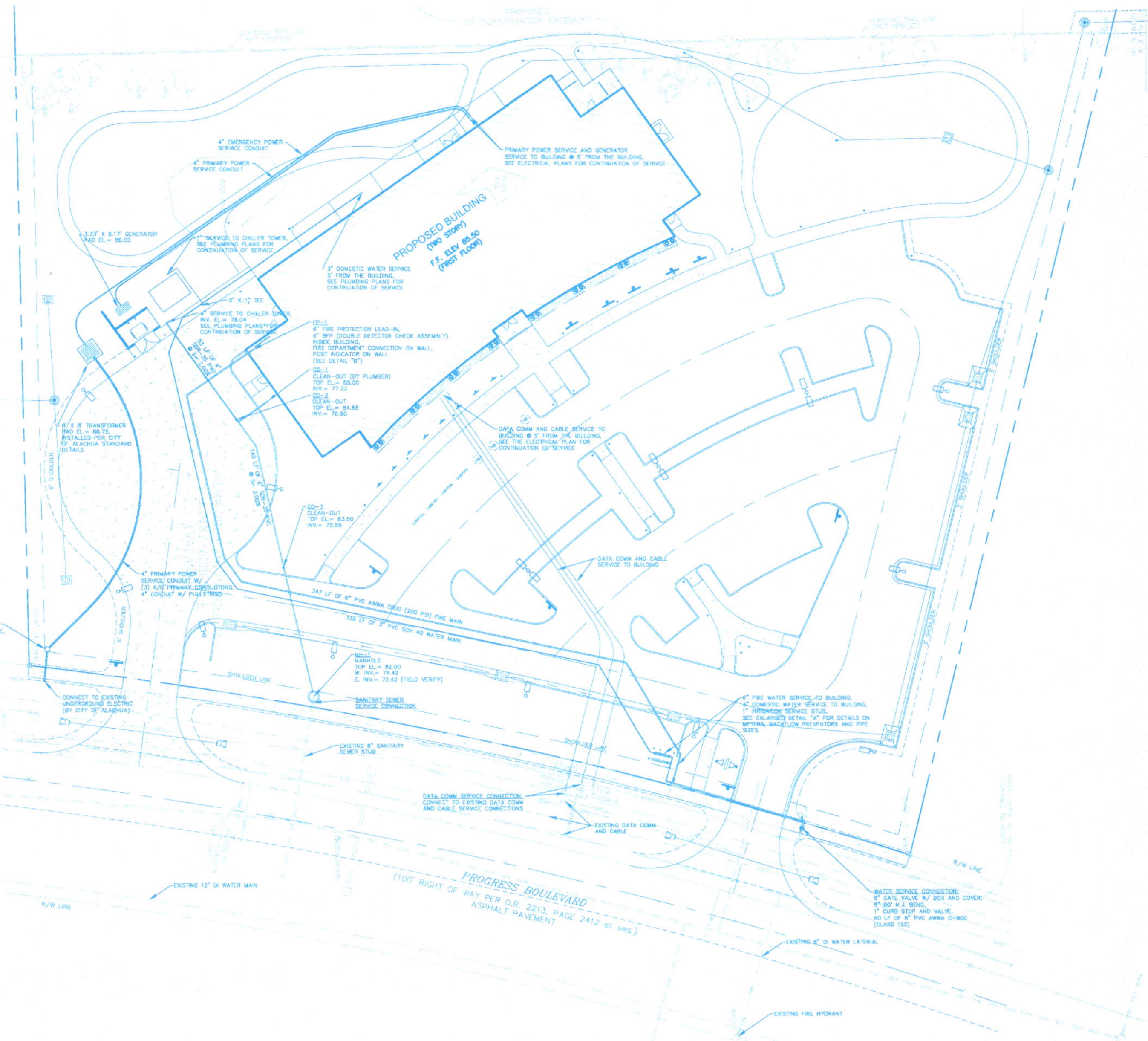

28103-13_F28.EPS

Figure 28 Plot plan.

a building of unusual design, more than four elevations may be needed. The elevation drawings show outside features, such as placement and height of windows, doors, chimneys, and roof lines. Exterior materials are indicated, as well as important vertical dimensions.

Interior elevation drawings show in greater detail the various cabinets, bookshelves, fireplaces, and other important interior features. These are sometimes called detail drawings.

2.3.5 Section Drawings

Section drawings give information about the construction of walls, stairs, or other items that cannot be easily given on the elevation or floor plan drawings (*Figure 32*). These drawings are usually drawn to a scale large enough to show the details without cluttering the drawing. A section taken through the narrow width of a building is known as a transverse section. A section taken through the entire length is known as a longitudinal section.

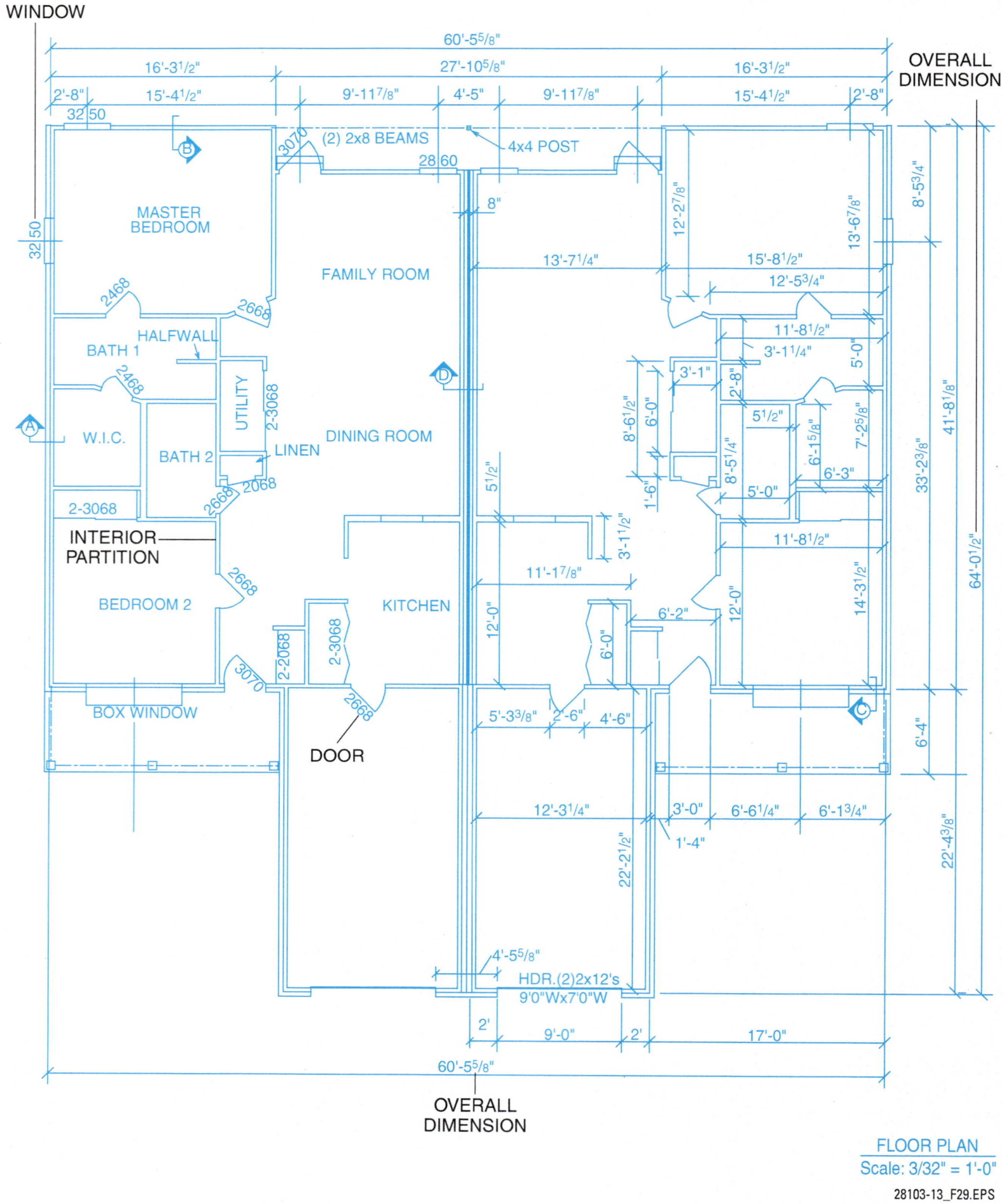

Figure 29 Basic floor plan.

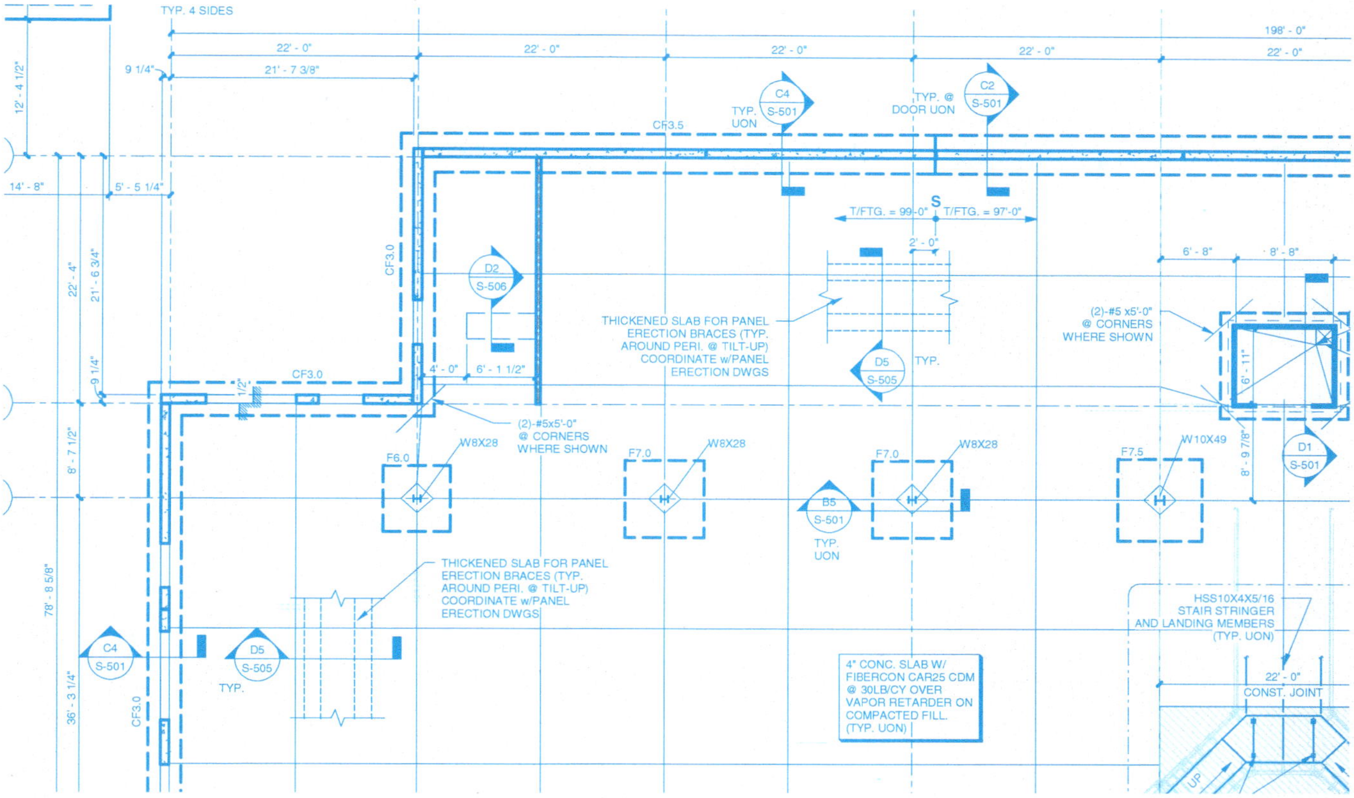

FOOTING SCHEDULE					
Mark	**Width**	**Length**	**Thickness**	**Reinforcing**	**Remarks**
CF2.0	2' - 0"	CONT.	1' - 0"	2-#5 CONT. w/ #3 @ 48" TRANS.	
CF2.5	2' - 6"	CONT.	1' - 0"	2-#5 CONT. w/ #3 @ 48" TRANS.	
CF3.0	3' - 0"	CONT.	1' - 0"	3-#5 CONT. w/ #3 @ 48" TRANS.	
CF3.5	3' - 6"	CONT.	1' - 0"	3-#5 CONT. w/ #3 @ 48" TRANS.	
F6.0	6' - 0"	6' - 0"	1' - 2"	6-#6 EA. WAY BOT.	Rectangular Footings
F6.0T	6' - 0"	6' - 0"	1' - 2"	6-#6 EA. WAY TOP & BOT.	Rectangular Footings
F7.0	7' - 0"	7' - 0"	1' - 6"	7-#6 EA. WAY BOT.	Rectangular Footings
F7.5	7' - 6"	7' - 6"	1' - 6"	8-#6 EA. WAY BOT.	Rectangular Footings
RF6x8	6' - 0"	8' - 0"	1' - 2"	#6 EA.@12"o/c EA. WAY TOP & BOT.	Rectangular Footings
RF6x9	6' - 0"	9' - 0"	1' - 2"	#6 EA.@12"o/c EA. WAY TOP & BOT.	Rectangular Footings
RF6x11.5	6' - 0"	11' - 6"	1' - 2"	#6 EA.@12"o/c EA. WAY TOP & BOT.	Rectangular Footings

28103-13_F30.EPS

Figure 30 Foundation plan.

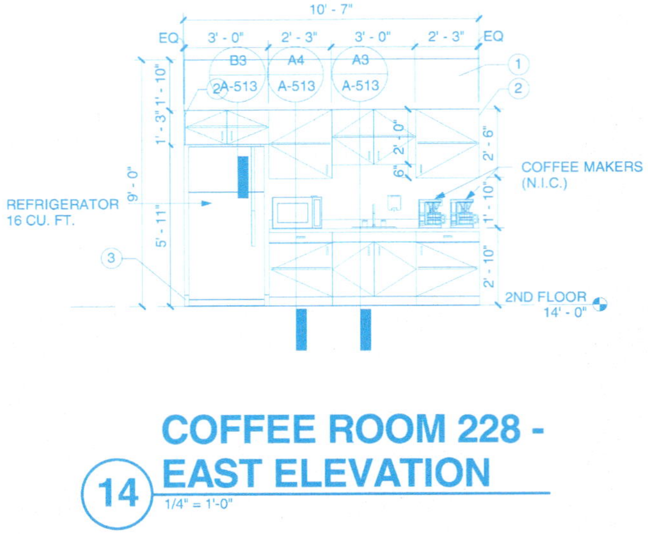

28103-13_F31.EPS

Figure 31 Detail of an elevation drawing.

When there are only a few sections, they are often incorporated onto one of the other drawings instead of a separate section drawing. The section drawings show the building from the foundation footings to the roof.

The working mason should be able to read section plans and answer questions about the work. What is the foundation wall made of? What is the distance between the brick veneer and the backing walls? How frequently is the veneer tied to the backing? What is the distance between weepholes?

2.3.6 Schedules

Schedules are an important tool in a set of plans. Although they are tables, not drawings, they give specific information that would be impractical to include on the documents. They also make it convenient for ordering material by collecting this detailed information in one place.

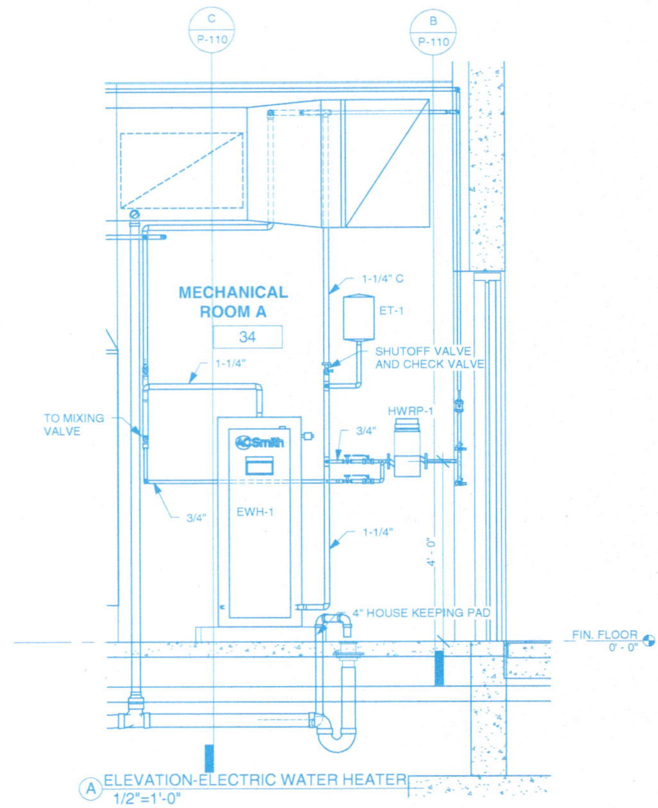

28103-13_F32.EPS

Figure 32 Section drawing.

2.3.7 Specialty Plans

Specialty plans give details about plumbing, electrical, and duct work (*Figure 33*). It might seem that a mason does not have to know about these things, but this is not the case. Masons are only one of the many kinds of craftworkers on the job. For example, masons have to cut around plumbing pipe, electrical boxes, and wall penetrations. Understanding what other workers are doing and how they affect your work is important. This information is on the drawings. Use them to help you work smarter.

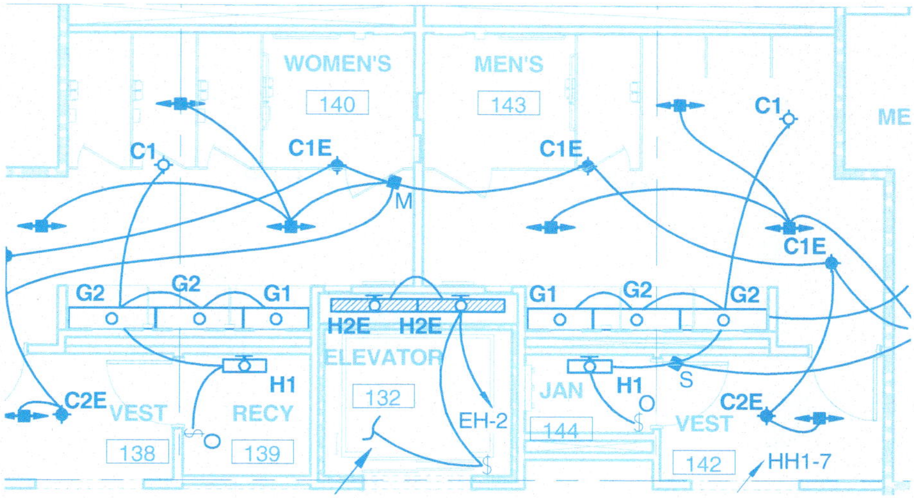

NCCER LUMINAIRE SCHEDULE										
TYPE	MFG	CATALOG NO.	DESCRIPTION	LAMPS			FIXTURE INPUT WATTS	VOLTS	MTG	REMARKS
				Qty	Watts	Desc				
INTERIOR										
A1	METALUX	AC	2x2 DUAL BASKET, LINEAR FACETED ACRYLIC REFLECTOR	2	17	T8	48	277	GRID	
A2	METALUX	AC	2x4 DUAL BASKET, LINEAR FACETED ACRYLIC REFLECTOR	2	32	T8	76	277	GRID	
A3	METALUX	RDI	2x2 DIRECT/INDIRECT, SINGLE BASKET, LINEAR FACETED ACRYLIC REFLECTOR	3	17	T8	72	277	GRID	
A4	METALUX	RDI	2x4 DIRECT/INDIRECT, SINGLE BASKET, LINEAR FACETED ACRYLIC REFLECTOR	3	32	T8	114	277	GRID	
B1	METALUX	GR8	2x2 LENSED A12.125	2	17	T8	48	277	GRID	
B2	METALUX	GR8	2x4 LENSED A12.125	2	32	T8	76	277	GRID	
C1	COOPER	PD6	6" DOWNLIGHT CLEAR SPECULAR REFLECTOR	1	26	TRT	28	277	GRID	
C2	COOPER	PD6	6" DOWNLIGHT CLEAR SPECULAR REFLECTOR	1	32	TRT	36	277	GRID	
C3	COOPER	PD6	6" DOWNLIGHT CLEAR SEMI-SPECULAR REFLECTOR WITH WHITE SPLAY, FRESNEL LENS	1	18	TRT	22	277	GRID	WET LOCATION
C4	SPECTRUM LIGHTING	SPC0812CF	8" PENDANT CLEAR SPECULAR REFLECTOR, COLOR BLACK	1	26	TRT	28	277	PENDANT	
C5	SPECTRUM LIGHTING	PROCT22	22" PENDANT, HALF REFLECTOR/HALF OPAL DIFFUSER	8	42	CFL	384	277	PENDANT	
C6	SPECTRUM LIGHTING	SPC0611CF	6" PENDANT CLEAR SPECULAR REFLECTOR , COLOR BLACK	1	32	TRT	36	277	PENDANT	
D1	METALUX	WN	SQUARE BASKET WRAPAROUND	2	32	T8	76	277	CHAIN	
D2	SPECTRUM LIGHTING	STPT3LED	3" LED TRACK LIGHT WITH 48" TRACK, COLOR BLACK	2	15	LED (INCLUDED)	30	277	CEILING	
D3	SPECTRUM LIGHTING	STPT3LED	3" LED TRACK LIGHT WITH 72" TRACK, COLOR BLACK	3	15	LED (INCLUDED)	45	277	PENDANT	
D4	SPECTRUM LIGHTING	STPT3LED	3" LED TRACK LIGHT WITH 72" TRACK, COLOR BLACK	4	15	LED (INCLUDED)	60	277	PENDANT	
F1	INCON LIGHTING	544	WIDE BAND VANITY WITH CHOME ACCENTS	2	17	T8	48	277	WALL	
G1	METALUX	SSL	STAGGERED STRIP W/ SPECULAR ASYMMETRIC REFLECTOR (3')	1	25	T8	33	277	SURFACE	
G2	METALUX	SSL	STAGGERED STRIP W/ SPECULAR ASYMMETRIC REFLECTOR (4)	1	32	T8	38	277	SURFACE	
H1	LSI INDUSTRIES	VL	2' ENCLOSED LINEAR POLYCARBONATE DIFFUSER	1	17	T8	24	277	WALL	<1>
H2	LSI INDUSTRIES	VL	4' ENCLOSED LINEAR POLYCARBONATE DIFFUSER	2	32	T8	76	277	WALL	<1>
X1	THOMAS & BETTS	LXN	EXIT SIGN, EDGE LIT, LED RED LETTER STENCIL ON MIRROR ACRYLIC PANEL, EXTRUDED AL. FINISH HOUSING	1		LED (INCLUDED)				MOUNTING TYPE SEE LIGHTING PLANS
X2	THOMAS & BETTS	ELXN400	EXIT SIGN, LED RED LETTER STENCIL, WHITE THERMOPLASTIC HOUSING	1		LED (INCLUDED)				MOUNTING TYPE SEE LIGHTING PLANS

28103-13_F33.EPS

Figure 33 Detail of an electrical drawing and schedule.

Additional Resources

The ABCs of Concrete Masonry Construction. Skokie, IL: Portland Cement Association (Video, 13 min. 34 sec.).

Masonry Design and Detailing For Architects, Engineers and Contractors, Sixth Edition. 2012. Christine Beall. New York: McGraw-Hill.

2.0.0 Section Review

1. Before attempting to read a set of drawings, you should always consult the _____.
   a. title block
   b. legend
   c. ANSI specifications
   d. plot or site plans

2. The scale usually used for residential drawings is _____.
   a. ⅛ of an inch equals 1 foot
   b. ¼ of an inch equals 1 foot
   c. ½ of an inch equals 1 foot
   d. 1 inch equals 1 foot

3. The first drawing in a set of structural plans is usually the _____.
   a. plot or site plan
   b. floor plan
   c. elevation drawing
   d. foundation plan

Section Three

3.0.0 Masonry Specifications, Standards, and Codes

Objective

Identify the purpose of specifications, standards, and codes used in the building industry and the sections that pertain to masonry.

a. Explain the purpose of specifications, standards, and codes.
b. Describe the purpose of inspections and testing.

Trade Term

MasterFormat™: A standard indexing system for construction specifications in the United States and Canada, developed by the Construction Specifications Institute (CSI) and Construction Specifications Canada (CSC) and used by project planners to prepare project specifications.

A mason must follow certain rules in addition to the building plans. These include specifications, standards, and codes. Specifications provide additional detail that is often not available on the plans. Standards provide detailed information and guidance on a single topic and are followed across the industry and even internationally. Codes are legal documents adopted by a jurisdiction that establish the minimum acceptable standards, rules, and regulations for all materials, practices, and installations used in buildings and building systems. Codes are adopted to ensure that contractors perform their work according to recognized standards. In this section, you will learn about specifications, standards, and codes, and how they apply to masonry. You will also learn about how specifications, standards, and codes govern inspections and testing, and why these activities are important for ensuring that safe construction practices are followed.

3.1.0 Understanding Specifications, Standards, and Codes

Construction plans or drawings usually include specifications. These are known as *specs*. They are written instructions or information needed to complete the work. Specs contain information not found on the drawings. The architect, the engineer, or the owner can give direction on how the job must be done. In fact, such specifications are part of the building contract. The plans cannot show all the necessary information for the mason to complete the work. Specifications add the details. These include the following:

- Quality of materials
- Quality of workmanship (minimum tolerance)
- Procedures or techniques to be used during construction
- Various responsibilities of each subcontractor

Generally, specifications are divided into two major areas: general conditions and technical specifications. The general conditions cover such legal items as insurance, permit responsibilities, and payment schedules. The mason will have little contact with these general conditions. Masons do need to read the technical specifications, though, to do their work properly.

Specifications are prepared for specific projects. They reflect the special conditions of that project. Standards apply to all projects. They describe the best practices or minimum requirements for doing certain tasks. Standards are referenced by specifications and by codes.

As explained previously, codes are legal requirements for building and construction. Model codes are set by national organizations and are adopted by local governments, which may modify them. Local codes have the force of the law, and they are enforced by building inspectors.

3.1.1 Specifications

Technical specifications directly affect the mason's work. They list how the job is to be completed. Each section includes information for the work of a single subcontractor. On every project, there will be a set of technical specifications for the masonry work.

Technical specifications are legal documents because they are part of a contract. The mason must use both the drawings and the specifications to construct the project. If there is a discrepancy between the two, the specifications have priority. The architect will address conflicts that occur during construction, such as an electrical conduit that is supposed to be positioned where a drainpipe is to be located.

In 1995, the Construction Specifications Institute (CSI) and Construction Specifications Canada (CSC) jointly developed a standard indexing system for construction specifications in the United States and Canada called **MasterFormat™**. Project planners follow the MasterFormat™ system when preparing project specifications because it makes organizing and finding information much simpler. Today, MasterFormat™ is divided into more than 50 divisions, each of which refers to a specific trade, specialty, or area of technical expertise. Masonry is covered in Division 4 of the MasterFormat™. *Figure 34* lists the sections under Division 4 that are used to organize masonry specifications on a project.

Highly detailed technical specifications spell out not only the materials, but how they are to be assembled and finished. This ensures the long-term performance of the materials. It is based on engineering studies about the properties of masonry construction.

Specifications take advantage of these engineering studies. They do this by referencing existing standards and codes. By referencing them, the specifications incorporate the provisions of the standards and codes.

Almost every type of building material has a multitude of size and performance standards and codes for it. The standards and codes guide and regulate manufacturers, designers, and builders. There are voluntary standards, national standards, international standards, national codes, local building codes, and model codes. The next sections discuss standards and codes in greater detail.

3.1.2 Standards

When you look at technical specifications, you will find items like "work to meet *ASTM C144-96*" or "as defined by *CSA A165.2*." These phrases refer to studies by the American Society for Testing and Materials (ASTM) International or the Canadian Standards Association (CSA). These are known as standards.

A standard usually has a single subject. It defines one aspect of the subject. The standards are based on studies, research, and advances in materials and construction techniques. Typical standards include the following:

- *ASTM C1072, Method for Measurement of Masonry Flexural Bond Strength*
- *ASTM C315, Specification for Clay Flue Linings*

Appendix C lists ASTM standards for masonry construction. Read through the list of titles in *Appendix C*. You can see how much study has gone into modern masonry materials and practices.

The ASTM and CSA are only two of the organizations publishing standards for the masonry industry. Consensus standards and codes are developed by other independent organizations, including the following:

- The American Society of Civil Engineers (ASCE)
- The American Institute of Steel Construction (AISC)
- The Masonry Society (TMS)
- The American Concrete Institute (ACI)

Standards are usually updated on a three-year cycle. ASTM and the other organizations gather data on new materials and techniques and include them in the standards. For ASTM publications, the year of publication follows a dash after the title number. Always make sure you are using the most current standard as adopted by your local jurisdiction.

DIVISION 04—MASONRY

- 04 00 00 - MASONRY
- 04 10 00 - *UNASSIGNED*
- 04 20 00 - UNIT MASONRY
- 04 30 00 - *UNASSIGNED*
- 04 40 00 - STONE ASSEMBLIES
- 04 50 00 - REFRACTORY MASONRY
- 04 60 00 - CORROSION-RESISTANT MASONRY
- 04 70 00 - MANUFACTURED MASONRY
- 04 80 00 - *UNASSIGNED*
- 04 90 00 - *UNASSIGNED*

28103-13_F34.EPS

Figure 34 CSI MasterFormat™ specification divisions for masonry.

3.1.3 Codes

Building codes are enforceable standards. They include all aspects of building construction. Their primary purpose is to protect the public. Safety is included in all aspects of building construction. Some areas have special concerns, like earthquakes. The local codes in these areas include special conditions for local hazards. Local codes can also include local requirements, such as conformance with a zoning plan, or lot setbacks. Codes establish minimum requirements for all aspects of masonry units and masonry work. Masonry work will meet or exceed code standards.

You may have heard of the *International Building Code®*. Throughout the years, organizations have worked together to establish model building codes. Model codes are technical documents written by an organization. They are based on or incorporate standards published by ASTM and other organizations. Local communities or states can adopt a model code. The model code then becomes a local code or law. This provides communities with sound laws without the expense of research and investigation.

ACI 530/ASCE 5/TMS 402 Building Code Requirements for Masonry Structures, consolidates several masonry codes. As its title suggests, it is published jointly by the ACI, ASCE, and TMS. *Figure 35* is an outline of the 2011 edition of *Building Code Requirements for Masonry Structures*. Each of these topics refers to many standards. It reflects many years of testing and research.

Specifications for commercial and industrial projects are based on local codes. Unlike residential projects, commercial and industrial projects are always designed and managed by architects and engineers registered by the state. They must make sure that the project complies with all local codes. Codes for commercial projects are usually more detailed. They are more stringent about fire protection, loadbearing, and parking spaces than codes for residential buildings.

3.2.0 Understanding Inspections and Testing

Every construction contract has specifications to cover inspection and acceptance of the finished work. ASTM has published many specifications about testing. Acceptance of the finished product depends on the inspection and test results.

For residential projects, inspection and testing are usually done by the local building inspector. This inspector will be looking for compliance with the local building code. The inspector works closely with the contractor to coordinate inspections with the completion of critical phases of the work. Work cannot progress on the next phase until this inspection has been made and the inspector signs off on the work. For this reason, you may see an inspector's checkoff sheet posted on the job site with the inspector's comments and signature for each phase. This may be in the job superintendent's office, or it may be posted at the next place the inspector is due to visit.

For residential projects, testing is usually limited to looking at the material tags and manufacturer's documents that come with the materials. For this reason, you should not throw these documents away until the inspector visits the site and signs off on the work.

On some projects, the architect will also inspect the work as it progresses. The architect will be looking for compliance with the drawings and specifications and will not worry about the building code. On some projects, a representative of the owner may check the craftsmanship of the work accomplished every day. Work that does not meet standards, drawings, specifications, or local codes must be done again. This can be very costly, so it is important to do it right the first time.

Inspection and testing on commercial and industrial projects is much more involved, and depending on the project size, will require a full-time staff hired by the owner. The representatives of the owner and/or architect will look for quality and compliance with the plan and specifications. The building inspector will look for compliance with the code. The contractor must pay to take out and replace work that is not up to specifications or code, just as with residential projects.

Earliest Building Code

The earliest building code was the Code of Hammurabi. The king of the Babylonian empire adopted this code in 2200 BC. It assessed severe penalties, including death, if a building was not constructed safely.

A Brief History of Codes

Prior to 2000, there were three model codes used throughout the United States:

- The *Southern Standard Building Code*, published by the Southern Building Code Congress International (SBCCI), typically used throughout the Southeast
- The *National Building Code*, published by the Building Officials and Code Administrators (BOCA) International, adopted mostly in the northeast and central states
- The *Uniform Building Code*, published by the International Conference of Building Officials (ICBO), used throughout the West

In 2000, these three organizations merged into the International Code Council. They issued the *International Building Code®* (IBC). This model code has been adopted and is now law throughout most of the United States. The National Fire Protection Association also issued a model code called *NFPA 5000®*. It is used in a few communities and has been adopted by the state of California.

Building Code Requirements for Masonry Structures (TMS 402-11/ACI 530-11/ASCE 5-11)

TABLE OF CONTENTS

28103-13_F35A.EPS

Figure 35 TMS 402-11 table of contents. (1 of 6)

28103-13_F35B.EPS

Figure 35 TMS 402-11 table of contents. (2 of 6)

28103-13_F35C.EPS

Figure 35 TMS 402-11 table of contents. (3 of 6)

28103-13_F35D.EPS

Figure 35 TMS 402-11 table of contents. (4 of 6)

28103-13_F35E.EPS

Figure 35 TMS 402-11 table of contents. (5 of 6)

28103-13_F35F.EPS

Figure 35 TMS 402-11 table of contents. (6 of 6)

Additional Resources

ACI 530/ASCE 5/TMS 402, Building Code Requirements for Masonry Structures. Latest edition. Reston, VA: American Society of Civil Engineers.

International Building Code® Latest edition. Falls Church, VA: International Code Council.

A Manual of Construction Documentation: An Illustrated Guide to Preparing Construction Drawings. 1989. Glenn E. Wiggins. New York: Whitney Library of Design.

3.0.0 Section Review

1. *Building Code Requirements for Masonry Structures,* is published jointly by _____.
   a. OSHA, EPA, and DOL
   b. NCMA, BIA, and MCAA
   c. ACI, ASCE, and TMS
   d. SBCC, BOCA, and ICBO

2. For residential projects, inspection and testing are usually done by _____.
   a. the project manager
   b. the architect
   c. the site inspector
   d. the local building inspector

SUMMARY

This module covered three topics: math, drawings, and specifications. Math skills are useful for calculating materials and supplies. They are also needed to interpret project drawings. Masons need to know how to convert fractions and measurements. They must be able to calculate areas and volumes of common geometric figures. They also need to read mason's rules.

Project drawings include several categories. A project drawing package will include plot plans, floor plans, elevations, sections, materials schedules, and structural, electrical, and mechanical drawings. They provide details by using lines, symbols, and measurement notations.

Specifications provide detail not available on the plans. Technical specifications refer to and include standards and codes. Standards are single-topic, nationally published, detailed guidelines. Model codes include references to many standards. Local codes are often based on model codes and include local matters, such as zoning.

Review Questions

1. On the job, the most common fractions a mason will deal with are ⅜ inch and _____.
   a. ¼ inch
   b. ½ inch
   c. ⅝ inch
   d. ⅞ inch

2. The common denominator of ⅔ and ¾ is _____.
   a. 3
   b. 4
   c. 6
   d. 12

3. On a six-foot folding rule, each inch is usually divided into _____.
   a. 6 parts
   b. 10 parts
   c. 16 parts
   d. 24 parts

4. A 10-foot wall built with 8 inch × 8 inch × 16 inch concrete block would be _____.
   a. 11 courses
   b. 13 courses
   c. 15 courses
   d. 17 courses

5. To find the circumference of a circle, multiply the diameter by _____.
   a. 2.25
   b. 3.14
   c. 3.66
   d. 4.12

6. A three-sided figure with an interior angle greater than 90 degrees is a(n) _____.
   a. acute triangle
   b. equilateral triangle
   c. obtuse triangle
   d. scalene triangle

7. The shortest leg of a right triangle is the _____.
   a. base
   b. orthogonal
   c. hypotenuse
   d. riser

8. The symbol shown in *Review Question Figure 1* is used on construction drawings to indicate _____.
   a. stone
   b. face brick
   c. common brick
   d. clay tile

28103-13_RQ01.EPS

Figure 1

9. If the scale on a construction drawing reads 1/4" = 1'-0" and a dimension on the drawing is 2" long, the actual length of that dimension is _____.
   a. 8'
   b. 6'
   c. 4'
   d. 2'

10. Floor plans, foundation plans, and elevations are all considered _____.
    a. specialty plans
    b. mechanical drawings
    c. architectural/engineering plans
    d. plot or site plans

11. A floor plan represents a cross section of the building, generally _____.
    a. at floor level
    b. 4 feet above the floor
    c. at ceiling level
    d. 2 feet below the top plate

12. Construction drawings show the edges of objects that are not visible in a particular view by using a(n) _____.
    a. phantom line
    b. extension line
    c. light full line
    d. hidden line

13. Details such as quality of materials and quality of workmanship are covered _____.
    a. in specifications
    b. by national standards
    c. as part of local building codes
    d. in OSHA regulations

14. If there is a discrepancy between construction drawings and specifications, _____.
    a. the building inspector must be informed
    b. drawings have priority
    c. specifications have priority
    d. the mason can choose which to follow

15. The abbreviation ASTM stands for _____.
    a. Allied School of Technical Masonry
    b. American Society for Testing and Materials
    c. Annual Survey of Testing Methods
    d. American Society of Training Managers

Trade Terms Quiz

Fill in the blank with the correct trade term that you learned from your study of this module.

1. A table, list, or chart used in construction drawings to explain the meanings of the various lines, symbols, and abbreviations used in that particular set of drawings is called a _______.
2. The _______ is the size of the masonry unit plus the thickness of one standard mortar joint, used in laying out courses.
3. The process of changing from one form of measure to another is called _______.
4. _______ is the CSI's standard indexing system for construction specifications in the United States and Canada.

Trade Terms:

Converting
Legend
MasterFormat™
Nominal dimension

Cornerstone of Craftsmanship

Steven Fechino

Engineering and Construction Manager
Mortar Net Solution

Training has been an important part of Steven Fechino's career in masonry, ever since he learned valuable skills as an apprentice. Today, as an instructor, he helps young apprentices learn the same values of craftsmanship and skill that helped make his career rewarding.

How did you get started in the construction industry?
My father and all of his friends were in the industry, and so I never even considered another career path. My father was in construction product sales, and I grew up walking on job sites with my father, measuring projects for his customers. I held the dumb end of the tape measure; I have never understood why.

What do you enjoy most about your career?
I enjoy projects that are difficult when they under construction, but that are simple to explain once they are completed!

Why do you think training and education are important in construction?
Training is the most important part of my job. I believe that training comes from everyone you meet in the industry. People I have worked with have influenced me even if they did not show me a specific trade skill. Special people have taught me to talk respectfully to customers, to ask better questions, and to listen to their requirements better. True professionals never complete their craft training.

Why do you think credentials are important in construction?
Credentials are important because you must be able to show where you have been, so that clients can decide whether they want to follow you to the next destination.

How has training/construction impacted your life and career?
Training has been a big part of my career, as I have had many mentors. The range of skills that I have developed has allowed me to support my family through this economic downturn by performing many different tasks.

Would you recommend construction as a career to others?
Yes, without a doubt! It has been a fulfilling career that has offered many challenges with exciting results.

What does craftsmanship mean to you?
Craftsmanship is your reward for hard work. It is the skill that allows you to get a job, maintain a job, and win out over another person for a job. It is what you bring to the table; it is where you leave your mark.

Trade Terms Introduced in This Module

Converting: The process of changing from one form of measure to another, for example, from feet to inches or from inches to feet.

Legend: A table, list, or chart used in construction drawings to explain the meanings of the various lines, symbols, and abbreviations used in that particular set of drawings.

MasterFormat™: A standard indexing system for construction specifications in the United States and Canada, developed by the Construction Specifications Institute (CSI) and Construction Specifications Canada (CSC) and used by project planners to prepare project specifications.

Nominal dimension: The size of the masonry unit plus the thickness of one standard (3/8 to 1/2 inch) mortar joint, used in laying out courses.

Appendix A

Answers to Practice Exercises

Answers to 1.1.2 Addition Practice Exercises

1. 1 foot 10 inches + 2 feet 5 inches = 3 feet 15 inches = 4 feet 3 inches
2. 2 feet 9 inches + 2 feet 4 inches = 4 feet 13 inches = 5 feet 1 inch
3. 2 feet 7 inches + 1 foot 6 inches = 3 feet 13 inches = 4 feet 1 inch
4. 1 foot 9 inches + 1 foot 7 inches = 2 feet 16 inches = 3 feet 4 inches

Answers to 1.1.4 Subtraction Practice Exercises

1. 2 feet 6 inches – 1 foot 8 inches = 1 foot 18 inches – 1 foot 8 inches = 10 inches
2. 2 feet 5 inches – 2 feet 7 inches = –2 inches
3. 35 feet 4 inches – 21 feet 6 inches = 34 feet 16 inches – 21 feet 6 inches = 13 feet 10 inches
4. 21 feet 3 inches – 11 feet 9 inches = 20 feet 15 inches – 11 feet 9 inches = 9 feet 6 inches

Answers to 1.4.2 Modular Brick Course Practice Exercises

6 ft × 12 in/ft = 72 in   72 in ÷ 16 in = 4.5 (16 inch sections)

1. Modular brick: 6 × 4.5 = 27 courses
2. Roman brick: 8 × 4.5 = 36 courses
3. Utility brick: 4 × 4.5 = 18 courses
4. Norman brick: 6 × 4.5 = 27 courses

Answers to 1.5.5 Area and Circumference Practice Exercises

1. $A = s^2$

   $A = (3 \text{ feet})^2$

   A = 9 square feet

2. A = lw

   w = 3 feet 8 inches = 36 inches + 8 inches = 44 inches

   l = 2w = 2(44 inches) = 88 inches

   A = (44 inches)(88 inches)

   A = 3,872 square inches ÷ 144 square inches per square foot

   A = 26.89 square feet

3. $A = bh \div 2$

   $A = (35 \text{ inches})(24 \text{ inches}) \div 2$

   $A = 840 \text{ square inches} \div 2$

   $A = 420 \text{ square inches}$

4. $A = \pi r^2$

   $r = d \div 2 = 52 \text{ inches} \div 2 = 26 \text{ inches}$

   $A = 3.14 \times (26 \text{ inches})^2$

   $A = 3.14 \times 676 \text{ square inches}$

   $A = 2{,}122.6 \text{ square inches}$

5. $A = (S_1 + S_2 + S_3 + S_4 + S_5)\ r \div 2$

   $A = (24 + 24 + 24 + 24 + 24)\ 28.5 \text{ inches} \div 2$

   $A = (120 \text{ inches})\ 28.5 \text{ inches} \div 2$

   $A = 3{,}420 \text{ square inches} \div 2$

   $A = 1{,}710 \text{ square inches} \div 144 \text{ square inches per square foot}$

   $A = 11.875 \text{ square feet}$

6. $A = lw$

   Area of hole = 4 feet × 8 feet = 32 square feet

   8 inches ÷ 12 inches per foot = 0.667 feet

   16 inches ÷ 12 inches per foot = 1.333 feet

   Area of block face = 0.667 feet × 1.333 feet = 0.889 square feet

   # brick = area of hole ÷ area of brick

   # brick = 32 square feet ÷ 0.889 square feet

   # brick = 35.99 brick, round up to 36 brick

   # brick + 5% = 37.8 brick, round up to 38 brick

7. # brick = 18 square feet × 1.25 brick per square foot

   # brick = 20.25 brick, round up to 21 brick

8. Area of wall = total area of wall – area of window

   Total area of wall = lw

   Total area of wall = 4 feet × 6 feet = 24 square feet

   Area of window = πr^2

   $r = d \div 2 = 1 \text{ foot} \div 2 = 6 \text{ inches}$

   Area of window = $\pi r^2 = 3.14 \times (6 \text{ inches})^2 = 3.14 \times 36$ square inches = 113.03 square inches = 0.785 square feet (113.03 square inches ÷ 144 square inches per square foot)

   Area of wall = 24 square feet – 0.785 square feet = 23.215 square feet

Appendix B

METRIC PREFIXES, MEASURES, AND CONVERSIONS

AREAS OF PLANE FIGURES

NAME FORMULA	SHAPE
(A = Area) Parallelogram $A = B \times h$	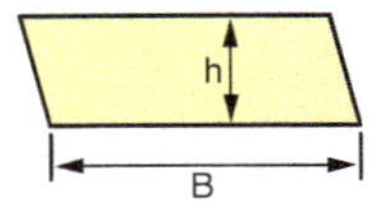
Trapezoid $A = \frac{B + C}{2} \times h$	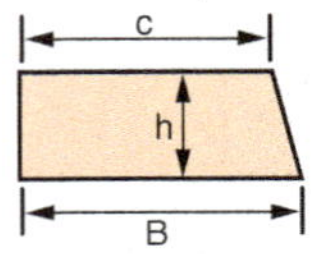
Triangle $A = \frac{B \times h}{2}$	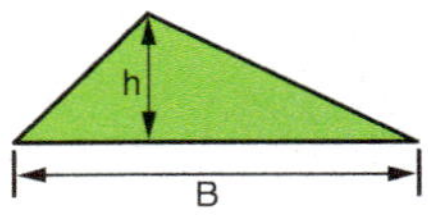
Trapezium (Divide into 2 triangles) A = Sum of the 2 triangles (See above)	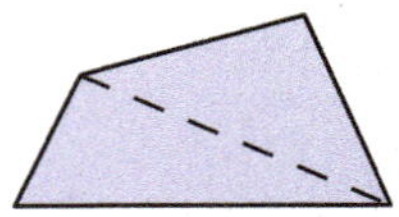
Regular Polygon $A = \frac{aP}{2}$ Where a is the length of the *apothem* (perpendicular distance from center to a side) P is the *perimeter* (sum of the sides, s)	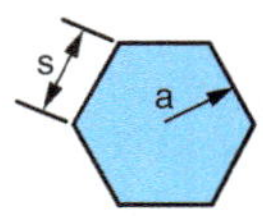
Circle $\pi = 3.14$ (1) πR^2 A = (2) $.7854 \times D^2$	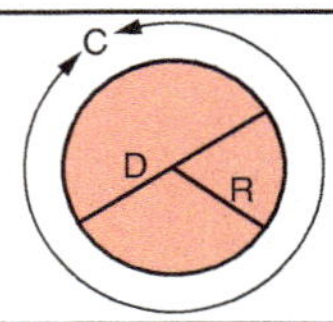
Sector (1) $\frac{a^\circ}{360^\circ} \times \pi R^2$ A = (2) Length of arc $\times \frac{R}{2}$ ($\pi = 3.14$, a = angle of sector)	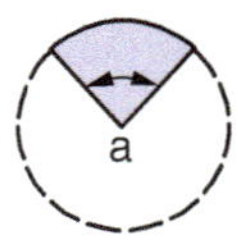
Segment A = Area of sector minus triangle (see above)	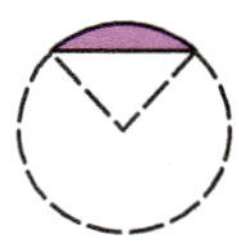
Ellipse $A = M \times m \times .7854$	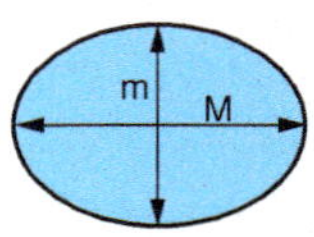
Parabola $A = B \times \frac{2h}{3}$	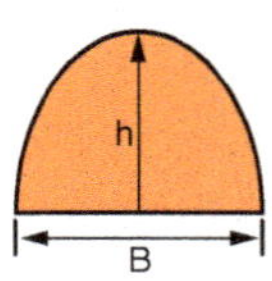

VOLUMES OF SOLID FIGURES

NAME FORMULA	SHAPE
(V - volume) Cube $V = a^3$ (in cubic units)	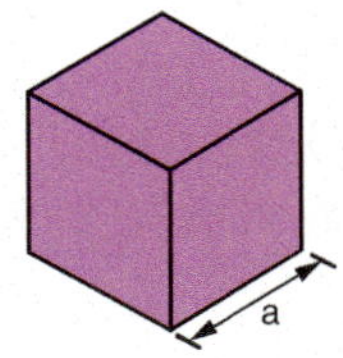
Rectangular Solids $V = L \times W \times h$	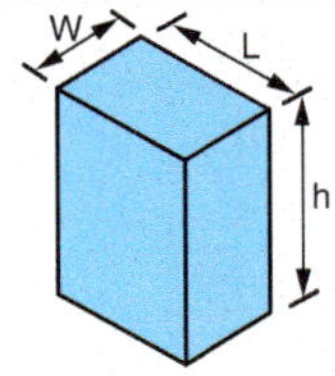
Prisms $V(1) = \frac{B \times A}{2} \times h$ $V(2) = \frac{s \times R}{2} \times n \times h$ V = Area of end × h n = Number of sides.	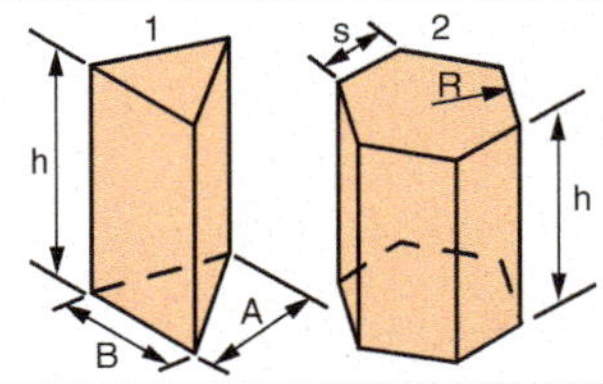
Cylinder $V = \pi R^2 \times h$ ($\pi = 3.14$)	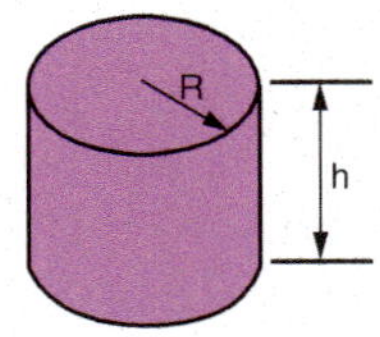
Cone $V = \frac{\pi R^2 \times h}{3}$ ($\pi = 3.14$)	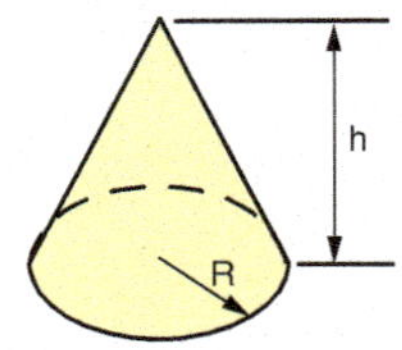
Pyramids $V(1) = L \times W \times \frac{h}{3}$ $V(2) = \frac{B \times A}{2} \times \frac{h}{3}$ V = Area of Base $\times \frac{h}{3}$	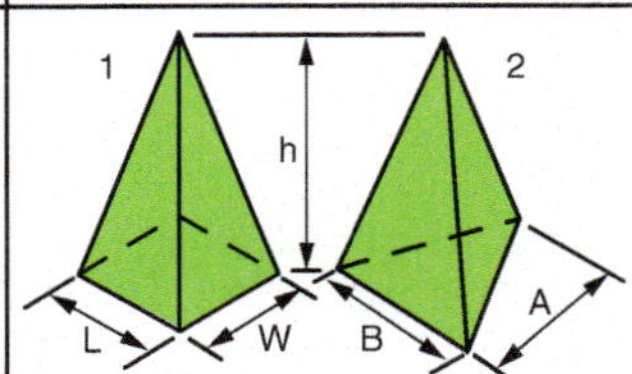
Sphere $V(1) = \frac{1}{6}\pi D^3$ $V(2) = \frac{4}{3}\pi R^3$	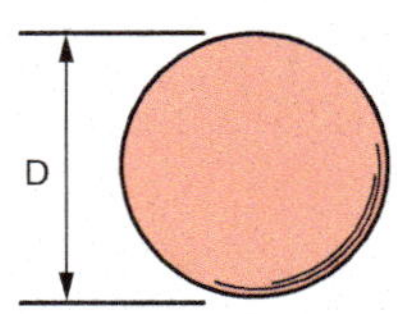
Circular Ring (Torus) $V = 2\pi^2 \times Rr^2$ V = Area of section $\times 2\pi R$ R = radius of the ring at its center, i.e., (OD – ID)/2 + ½ID; r = radius of the cross section of the ring at R.	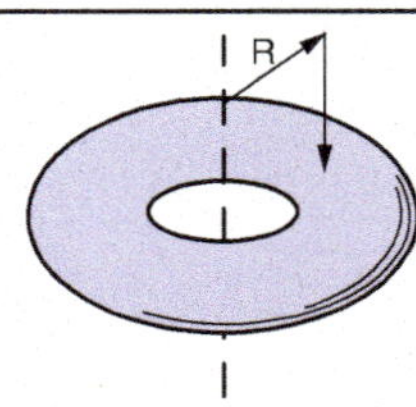

28103-13_A01.EPS

Figure A-1

Table A-1 Common SI Prefixes

PREFIX	SYMBOL	NUMBER	MULTIPLICATION FACTOR
giga	G	billion	$1,000,000,000 = 10^{9}$
mega	M	million	$1,000,000 = 10^{6}$
kilo	k	thousand	$1,000 = 10^{3}$
hecto	h	hundred	$100 = 10^{2}$
deka	da	ten	$10 = 10^{1}$
			BASE UNITS $1 = 10^{0}$
deci	d	tenth	$0.1 = 10^{-1}$
centi	c	hundredth	$0.01 = 10^{-2}$
milli	m	thousandth	$0.001 = 10^{-3}$
micro	μ	millionth	$0.000001 = 10^{-6}$
nano	n	billionth	$0.000000001 = 10^{-9}$

28103-13_TA01.EPS

Table A-2 Common Metric Measures

WEIGHT UNITS		
1 kilogram	=	1,000 grams
1 hectogram	=	100 grams
1 dekagram	=	10 grams
1 gram	=	1 gram
1 decigram	=	0.1 gram
1 centigram	=	0.01 gram
1 milligram	=	0.001 gram
LENGTH UNITS		
1 kilometer	=	1,000 meters
1 hectometer	=	100 meters
1 dekameter	=	10 meters
1 meter	=	1 meter
1 decimeter	=	0.1 meter
1 centimeter	=	0.01 meter
1 millimeter	=	0.001 meter
LIQUID VOLUME UNITS		
1 kiloliter	=	1,000 liters
1 hectoliter	=	100 liters
1 dekaliter	=	10 liters
1 liter	=	1 liter
1 deciliter	=	0.1 liter
1 centiliter	=	0.01 liter
1 milliliter	=	0.001 liter

28103-13_TA02.EPS

Table A-3 US Customary to SI Metric Conversions

U.S. CUSTOMARY		SI METRIC
WEIGHTS		
1 ounce (oz)	=	28.35 grams
1 pound (lb)	=	435.6 grams or 0.4536 kilograms
1 (short) ton	=	907.2 kilograms
LENGTHS		
1 inch (in)	=	2.540 centimeters
1 foot (ft)	=	30.48 centimeters
1 yard (yd)	=	91.44 centimeters or 0.9144 meters
1 mile	=	1.609 kilometers
AREAS		
1 square inch (in^2)	=	6.452 square centimeters
1 square foot (ft^2)	=	929.0 square centimeters or 0.0929 square meters
1 square yard (yd^2)	=	0.8361 square meters
VOLUMES		
1 cubic inch (in^3)	=	16.39 cubic centimeters
1 cubic foot (ft^3)	=	0.02832 cubic meter
1 cubic yard (yd^3)	=	0.7646 cubic meter
LIQUID MEASUREMENTS		
1 (fluid) ounce (fl oz)	=	0.095 liter or 28.35 grams
1 pint (pt)	=	473.2 cubic centimeters
1 quart (qt)	=	0.9263 liter
1 (US) gallon (gal)	=	3,785 cubic centimeters or 3.785 liters
TEMPERATURE MEASUREMENTS		
To convert degrees Fahrenheit to degrees Celsius, use the following formula: $C = 5/9 \times (F - 32)$.		

28103-13_TA03.EPS

Table A-4 SI Metric to US Customary Conversions

SI METRIC		U.S. CUSTOMARY
WEIGHTS		
1 gram (G)	=	0.03527 ounces
1 kilogram (kg)	=	2.205 pounds
1 metric ton	=	2,205 pounds
LENGTHS		
1 millimeter (mm)	=	0.03937 inches
1 centimeter (cm)	=	0.3937 inches
1 meter (m)	=	3.281 feet or 1.0937 yards
1 kilometer (km)	=	0.6214 miles
AREAS		
1 square millimeter	=	0.00155 square inches
1 square centimeter	=	0.155 square inches
1 square meter	=	10.76 square feet or 1.196 square yards
VOLUMES		
1 cubic centimeter	=	0.06102 cubic inches
1 cubic meter	=	35.31 cubic feet or 1.308 cubic yards
LIQUID MEASUREMENTS		
1 cubic centimeter (cm^3)	=	0.06102 cubic inches
1 liter (1,000 cm^3)	=	1.057 quarts, 2.113 pints, or 61.02 cubic inches
TEMPERATURE MEASUREMENTS		
To convert degrees Celsius to degrees Fahrenheit, use the following formula: $F = (9/5 \times C) + 32$.		

28103-13_TA04.EPS

Appendix C

ASTM Standards for Masonry Construction

Brick

ASTM C27, Standard Classification of Fireclay and High-Alumina Refractory Brick
ASTM C32, Standard Specification for Sewer and Manhole Brick (Made from Clay or Shale)
ASTM C34, Standard Specification for Structural Clay Load-Bearing Wall Tile
ASTM C43, Standard Terminology of Structural Clay Products (note: withdrawn 2009)
ASTM C56, Standard Specification for Structural Clay Nonloadbearing Tile
ASTM C62, Standard Specification for Building Brick (Solid Masonry Units Made from Clay or Shale)
ASTM C106, Specification for Refractories and Incinerators (note: withdrawn 1972)
ASTM C126, Standard Specification for Ceramic Glazed Structural Clay Facing Tile, Facing Brick, and Solid Masonry Units
ASTM C155, Standard Classification of Insulating Firebrick
ASTM C212, Standard Specification for Structural Clay Facing Tile
ASTM C216, Standard Specification for Facing Brick (Solid Masonry Units Made from Clay or Shale)
ASTM C279, Standard Specification for Chemical-Resistant Masonry Units
ASTM C315, Standard Specification for Clay Flue Liners and Chimney Pots
ASTM C410, Standard Specification for Industrial Floor Brick
ASTM C416, Standard Classification of Silica Refractory Brick
ASTM C530, Standard Specification for Structural Clay Nonloadbearing Screen Tile
ASTM C652, Standard Specification for Hollow Brick (Hollow Masonry Units Made from Clay or Shale)
ASTM C902, Standard Specification for Pedestrian and Light Traffic Paving Brick
ASTM C1261, Standard Specification for Firebox Brick for Residential Fireplaces
ASTM C1272, Standard Specification for Heavy Vehicular Paving Brick

Concrete Masonry Units

ASTM C55, Standard Specification for Concrete Building Brick
ASTM C73, Standard Specification for Calcium Silicate Brick (Sand-Lime Brick)
ASTM C90, Standard Specification for Loadbearing Concrete Masonry Units
ASTM C129, Standard Specification for Nonloadbearing Concrete Masonry Units
ASTM C139, Standard Specification for Concrete Masonry Units for Construction of Catch Basins and Manholes
ASTM C744, Standard Specification for Prefaced Concrete and Calcium Silicate Masonry Units
ASTM C936, Standard Specification for Solid Concrete Interlocking Paving Units
ASTM C1319, Standard Specification for Concrete Grid Paving Units

Natural Stone

ASTM C119, Standard Terminology Relating to Dimension Stone
ASTM C503, Standard Specification for Marble Dimension Stone
ASTM C568, Standard Specification for Limestone Dimension Stone
ASTM C615, Standard Specification for Granite Dimension Stone
ASTM C616, Standard Specification for Quartz-Based Dimension Stone
ASTM C629, Standard Specification for Slate Dimension Stone

Mortar and Grout

ASTM C5, Standard Specification for Quicklime for Structural Purposes
ASTM C33, Standard Specification for Concrete Aggregates
ASTM C91, Standard Specification for Masonry Cement
ASTM C144, Standard Specification for Aggregate for Masonry Mortar
ASTM C150, Standard Specification for Portland Cement
ASTM C199, Standard Test Method for Pier Test for Refractory Mortars
ASTM C207, Standard Specification for Hydrated Lime for Masonry Purposes
ASTM C270, Standard Specification for Mortar for Unit Masonry
ASTM C330, Standard Specification for Lightweight Aggregates for Structural Concrete
ASTM C331, Standard Specification for Lightweight Aggregates for Concrete Masonry Units
ASTM C404, Standard Specification for Aggregates for Masonry Grout
ASTM C476, Standard Specification for Grout for Masonry
ASTM C658, Standard Specification for Chemical-Resistant Resin Grouts for Brick or Tile
ASTM C887, Standard Specification for Packaged, Dry, Combined Materials for Surface Bonding Mortar
ASTM C1142, Standard Specification for Extended Life Mortar for Unit Masonry
ASTM C1329, Standard Specification for Mortar Cement

Reinforcement and Accessories

ASTM A82, Standard Specification for Steel Wire, Plain, for Concrete Reinforcement
ASTM A153, Standard Specification for Zinc Coating (Hot-Dip) on Iron and Steel Hardware
ASTM A167, Standard Specification for Stainless and Heat-Resisting Chromium-Nickel Steel Plate, Sheet, and Strip
ASTM A185, Standard Specification for Steel Welded Wire Reinforcement, Plain, for Concrete
ASTM A496, Standard Specification for Steel Wire, Deformed, for Concrete Reinforcement
ASTM A615, Standard Specification for Deformed and Plain Carbon-Steel Bars for Concrete Reinforcement
ASTM A641, Standard Specification for Zinc-Coated (Galvanized) Carbon Steel Wire
ASTM A951, Standard Specification for Steel Wire for Masonry Joint Reinforcement
ASTM B227, Standard Specification for Hard-Drawn Copper-Clad Steel Wire
ASTM B766, Standard Specification for Electrodeposited Coatings of Cadmium
ASTM C915, Standard Specification for Precast Reinforced Concrete Crib Wall Members
ASTM C1089, Standard Specification for Spun Cast Prestressed Concrete Poles
ASTM C1242, Standard Guide for Selection, Design, and Installation of Dimension Stone Attachment Systems

Sampling and Testing

ASTM C67, Standard Test Methods for Sampling and Testing Brick and Structural Clay Tile
ASTM C97, Standard Test Methods for Absorption and Bulk Specific Gravity of Dimension Stone
ASTM C109, Standard Test Method for Compressive Strength of Hydraulic Cement Mortars (Using 2 in. or [50 mm] Cube Specimens)
ASTM C140, Standard Test Methods for Sampling and Testing Concrete Masonry Units and Related Units
ASTM C170, Standard Test Method for Compressive Strength of Dimension Stone
ASTM C241, Standard Test Method for Abrasion Resistance of Stone Subjected to Foot Traffic
ASTM C267, Standard Test Methods for Chemical Resistance of Mortars, Grouts, and Monolithic Surfacings and Polymer Concretes
ASTM C426, Standard Test Method for Linear Drying Shrinkage of Concrete Masonry Units
ASTM C780, Standard Test Method for Preconstruction and Construction Evaluation of Mortars for Plain and Reinforced Unit Masonry
ASTM C880, Standard Test Method for Flexural Strength of Dimension Stone
ASTM C952, Standard Test Method for Bond Strength of Mortar to Masonry Units
ASTM C1006, Standard Test Method for Splitting Tensile Strength of Masonry Units
ASTM C1019, Standard Test Method for Sampling and Testing Grout
ASTM C1072, Standard Test Methods for Measurement of Masonry Flexural Bond Strength
ASTM C1093, Standard Practice for Accreditation of Testing Agencies for Masonry
ASTM C1148, Standard Test Method for Measuring the Drying Shrinkage of Masonry Mortar
ASTM C1194, Standard Test Method for Compressive Strength of Architectural Cast Stone
ASTM C1195, Standard Test Method for Absorption of Architectural Cast Stone
ASTM C1196, Standard Test Methods for In Situ Compressive Stress Within Solid Unit Masonry Estimated Using Flatjack Measurements
ASTM C1197, Standard Test Method for In Situ Measurement of Masonry Deformability Using the Flatjack Method
ASTM C1262, Standard Test Method for Evaluating the Freeze-Thaw Durability of Dry-Cast Segmental Retaining Wall Units and Related Concrete Units
ASTM C1314, Standard Test Method for Compressive Strength of Masonry Prisms
ASTM C1324, Standard Test Method for Examination and Analysis of Hardened Masonry Mortar
ASTM D75, Standard Practice for Sampling Aggregates
ASTM E72, Standard Test Methods of Conducting Strength Tests of Panels for Building Construction
ASTM E447, Test Methods for Compressive Strength of Laboratory Constructed Masonry Prisms (note: withdrawn 1988)
ASTM E488, Standard Test Methods for Strength of Anchors in Concrete Elements
ASTM E514, Standard Test Method for Water Penetration and Leakage Through Masonry
ASTM E518, Standard Test Methods for Flexural Bond Strength of Masonry
ASTM E519, Standard Test Method for Diagonal Tension (Shear) in Masonry Assemblages
ASTM E754, Standard Test Method for Pullout Resistance of Ties and Anchors Embedded in Masonry Mortar Joints

Assemblages

ASTM C901, Standard Specification for Prefabricated Masonry Panels
ASTM C946, Standard Practice for Construction of Dry-Stacked, Surface-Bonded Walls
ASTM E835, Standard Guide for Modular Coordination of Clay and Concrete Masonry Units (note: withdrawn 2011)
ASTM C1283, Standard Practice for Installing Clay Flue Lining
ASTM E1602, Standard Guide for Construction of Solid Fuel Burning Masonry Heaters

Additional Resources

This module presents thorough resources for task training. The following resource material is suggested for further study.

The ABC's of Concrete Masonry Construction, Video. Skokie, IL: Portland Cement Association.

ACI 530/ASCE 5/TMS 402, *Building Code Requirements for Masonry Structures*. Latest edition. Reston, VA: American Society of Civil Engineers.

Bricklaying: Brick and Block Masonry. Reston, VA: The Brick Institute of America.

Bricklaying Curriculum: Advanced Bricklaying Techniques. 1992. Raymond J. Turcotte and Laborn J. Hendrix. Stillwater, OK: Oklahoma Department of Vocational and Technical Education.

Building Block Walls: A Basic Guide for Students in Masonry Vocational Training. 1988. Herndon, VA: National Concrete Masonry Association.

International Building Code®. Latest edition. Falls Church, VA: International Code Council.

A Manual of Construction Documentation: An Illustrated Guide to Preparing Construction Drawings. 1989. Glenn E. Wiggins. New York: Whitney Library of Design.

Masonry Design and Detailing For Architects, Engineers and Contractors, Sixth Edition. 2012. Christine Beall. New York: McGraw-Hill.

Figure Credits

Bon Tool Company, Figure 2

Courtesy of Stanley Black + Decker, Inc., Figure 6

Courtesy of Dennis Neal, FMA&EF, Figure 15

Courtesy of the Brick Industry Association, Figure 22

Courtesy of Haskell, Figure 28, Figure 30, Figure 31, Figure 32, Figure 33

Courtesy of The Masonry Society (TMS), Figure 35

Section Review Answer Key

Answer	Section Reference	Objective
Section One		
1. c	1.0.0	1a
2. d	1.2.0	1b
3. c	1.3.0	1c
4. a	1.4.0	1d
5. b	1.5.1	1e
6. d	1.6.0	1f
Section Two		
1. b	2.1.0	2a
2. b	2.2.0	2b
3. d	2.3.3	2c
Section Three		
1. c	3.1.3	3a
2. d	3.2.0	3b

NCCER CURRICULA — USER UPDATE

NCCER makes every effort to keep its textbooks up-to-date and free of technical errors. We appreciate your help in this process. If you find an error, a typographical mistake, or an inaccuracy in NCCER's curricula, please fill out this form (or a photocopy), or complete the online form at **www.nccer.org/olf**. Be sure to include the exact module ID number, page number, a detailed description, and your recommended correction. Your input will be brought to the attention of the Authoring Team. Thank you for your assistance.

Instructors – If you have an idea for improving this textbook, or have found that additional materials were necessary to teach this module effectively, please let us know so that we may present your suggestions to the Authoring Team.

NCCER Product Development and Revision

13614 Progress Blvd., Alachua, FL 32615

Email: curriculum@nccer.org
Online: www.nccer.org/olf

❑ Trainee Guide ❑ Lesson Plans ❑ Exam ❑ PowerPoints Other ____________

Craft / Level: Copyright Date:

Module ID Number / Title:

Section Number(s):

Description:

Recommended Correction:

Your Name:

Address:

Email: Phone:

28104-13

Mortar

Overview

Mortar is one of the basic building materials used by the mason. This module describes the materials used to make mortar, their characteristics, the types of mortar used, and how to mix mortar. The purpose of mortar is to bond the units and seal the spaces between them; make up the differences in the size of the units; bond metal ties, grids, and anchors; and create a neat, attractive, and uniform appearance. As a mason, you need to know how to mix different types of mortar and how to tell the difference between good and poor mortar.

Module Five

Trainees with successful module completions may be eligible for credentialing through NCCER's National Registry. To learn more, go to **www.nccer.org** or contact us at **1.888.622.3720**. Our website has information on the latest product releases and training, as well as online versions of our *Cornerstone* magazine and Pearson's product catalog.

Your feedback is welcome. You may email your comments to **curriculum@nccer.org**, send general comments and inquiries to **info@nccer.org**, or fill in the User Update form at the back of this module.

This information is general in nature and intended for training purposes only. Actual performance of activities described in this manual requires compliance with all applicable operating, service, maintenance, and safety procedures under the direction of qualified personnel. References in this manual to patented or proprietary devices do not constitute a recommendation of their use.

28104-13
MORTAR

Objectives

When you have completed this module, you will be able to do the following:

1. Name and describe the ingredients and types of mortar.
   a. Describe the use of portland cement, hydrated lime, and sand.
   b. Describe masonry cement.
   c. Describe preblended mortars.
   d. Describe the use of water and admixtures.
   e. Describe the types of masonry mortar.
2. Describe properties of plastic and hardened mortar.
   a. Describe plastic mortar.
   b. Describe hardened mortar.
3. Identify the common problems found in mortar application and their solutions.
   a. Describe the effects of improper proportioning and poor-quality materials.
   b. Explain the effects of extreme weather and tempering.
   c. Describe efflorescence.
4. Explain how to properly set up, maintain, and dispose of mortar and use the mortar mixing area.
   a. Describe how to set up a mixing area.
   b. Describe how to maintain the mixing area.
   c. Describe how to mix mortar with a power mixer.

Performance Tasks

Under the supervision of your instructor, you should be able to do the following:

1. Properly set up a mixing area.
2. Properly mix mortar with a power mixer.

Trade Terms

Air-entraining
Extent of bond
Hydration
Masonry cement
Plasticity
Slaked lime
Water retention
Workability

Industry-Recognized Credentials

If you're training through an NCCER-accredited sponsor, you may be eligible for credentials from NCCER's Registry. The ID number for this module is 28104-13. Note that this module may have been used in other NCCER curricula and may apply to other level completions. Contact NCCER's Registry at 888.622.3720 or go to **www.nccer.org** for more information.

Contents

Topics to be presented in this module include:

Figures and Tables

SECTION ONE

1.0.0 MORTAR INGREDIENTS AND TYPES

Objective

Name and describe the ingredients and types of mortar.

a. Describe the use of portland cement, hydrated lime, and sand.
b. Describe masonry cement.
c. Describe preblended mortars.
d. Describe the use of water and admixtures.
e. Describe the types of masonry mortar.

Trade Terms

Air-entraining: A type of admixture added to mortar to increase microscopic air bubbles in mixed mortar. The air bubbles increase resistance to freeze-thaw damage.

Extent of bond: The adhesion of mortar to masonry units, as determined by its lime and moisture content.

Hydration: A chemical reaction between cement and water that hardens the mortar. Hydration requires the presence of water and optimal air temperatures.

Masonry cement: Cement that has been modified by adding lime and other materials.

Plasticity: The ability of mortar to flow like a liquid and not form cracks or break apart.

Water retention: The ability of mortar to keep sufficient water in the mix to enhance plasticity and workability and to decrease the need to temper.

Workability: The property of mortar to remain soft and plastic long enough to allow the mason to place and align masonry units and tool off the mortar joints before the mortar hardens completely.

Masonry structures are a combination of masonry units and mortar, which makes up approximately 20 percent of the surface of the structure (*Figure 1*). Mortar is made of the following basic materials:

- Portland cement
- Hydrated lime
- Sand
- Water

28104-13_F01.EPS

Figure 1 Mortar is approximately 20 percent of a masonry structure's surface.

Each material affects mortar quality. Generally, cement gives mortar strength. Sand adds bulk. Lime in the cement reacts with the water and air, and hardens in a process called hydration. In this section, you will learn how these ingredients are combined to make the various types of mortar used by masons.

1.1.0 Describing the Use of Portland Cement, Hydrated Lime, and Sand

ASTM (American Society for Testing and Materials) International created standards for each of the materials used in mortar. These standards are listed in *Table 1*. Different amounts of these materials can be used to mix different types of mortar. Other materials, known as admixtures, can be added to the mortar to change its physical or chemical properties. Color is a common admixture.

Table 1 ASTM Standards for Mortar Materials

MATERIAL	ASTM STANDARD
PORTLAND CEMENT (TYPES I, II, III)	C150
AIR-ENTRAINING PORTLAND CEMENT (TYPES IS, ISA, IP, IPA, S, SA)	C150
MASONRY CEMENT	C595
QUICKLIME	C5
HYDRATED LIME	C207
AGGREGATE	C144

28104-13_T01.EPS

1.1.1 Portland Cement

The main ingredient in mortar is portland cement. It adds to durability, high strength, and early setting of the mortar.

ASTM specifications list several types of portland cement that may be used for mortar. *Table 2* compares the five types of portland cement. Generally, only Types I, II, and III are used for mortar. Type I is most often used in mortar. Types IV and V are usually used only for concrete.

Air-entraining admixtures can be added to the various types of portland cement. These mixtures are labeled with an *A*. For example, Type IA is Type I portland cement with air-entraining admixtures.

1.1.2 Hydrated Lime

The second ingredient in mortar is hydrated lime (*Figure 2*). It sets only upon contact with the air. Thus, hardening occurs slowly over a long period of time. Lime gives the following properties to mortar:

- Bond
- **Workability**
- **Water retention**
- Elasticity

Hydrated lime in *ASTM C207, Specification for Hydrated Lime for Masonry Purposes,* is available in four types: S, N, NA, and SA. Only Type S hydrated lime should be used in mortar. Type N hydrated lime contains no limits on the quantity of unhydrated oxides. Types NA and SA are not recommended for mortar. They contain air-entraining additives that reduce the bond between the mortar and masonry units or reinforcement.

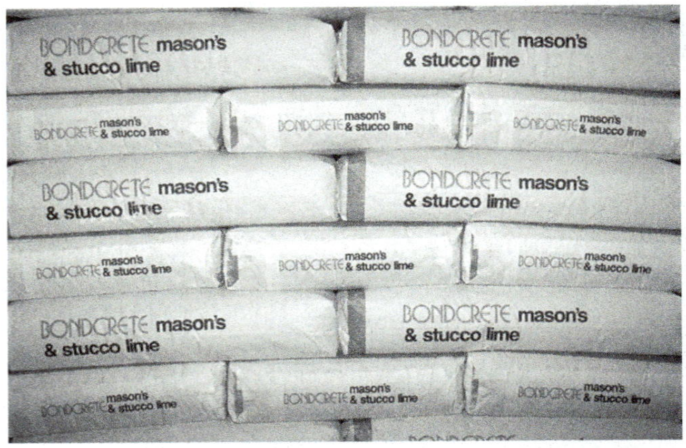

28104-13_F02.EPS

Figure 2 Commercial hydrated lime.

Type S hydrated lime adds several desirable characteristics to mortar:

- *Workability* – Mortar must be smooth and pliable. Workability is a measure of these features. Lime helps mortar to be very workable. This allows the mason to butter joints easily, results in better workmanship, and increases the ability of the wall to resist water penetration. Mortar with the correct lime content spreads easily with a trowel. Cement that is not mixed with lime is more difficult to spread.
- *Water retention* – Lime increases the water-holding capacity of the mortar, decreases the loss of water from the mortar (bleeding), and reduces the separation of sand. Mortar with a low lime content loses its moisture and sets prematurely.

Table 2 Five Types of Portland Cement

TYPE	DESCRIPTION	USE
I	General Purpose	This is a general-purpose cement and is the one masons use most often. It may be used in pavements, sidewalks, reinforced concrete bridge culverts, and masonry mortar.
II	Modified	This cement hydrates at a lower temperature than Type I and generates less heat. It has better resistance to sulfate than Type I. It is usually specified for use in such places as large piers, heavy abutments, and heavy retaining walls.
III	High Early Strength	Although this cement requires as long to set as Type I, it achieves its full strength much sooner. Generally, this cement is recommended when high strength is required in one to three days. Type III is often used in cold weather, when protection from freezing weather is important.
IV	Low Heat	This is a special cement for use where the amount and rate of heat generated must be kept to a minimum. Since the concrete cures slowly, strength also develops at a slower rate. Low heat portland cement is used in areas where there are large masses of concrete, such as dams or large bridges, and where Type I or Type II cement would generate too much heat.
V	Sulfate Resistant	This is intended for use only in construction exposed to severe sulfate actions. It also gains strength at a slower rate than normal portland cement.

28104-13_T02.EPS

When you use mortar that is too dry, you will get a poor bond. This seriously weakens the structure.

- **Extent of bond** – Extent of bond is the most important property of mortar. It is the adhesion of mortar to masonry units, as determined by its lime and moisture content. Lime decreases the strength of the cement in mortar but helps the mortar fill gaps and adhere to the structure. Greater adhesion results in a higher overall extent of bond.
- *Flexibility* – Mortar must be flexible. This will prevent cracking during strong winds, lateral pressures, and hard jolts. Tall masonry structures, such as high chimneys, are likely to sway. Lime increases the **plasticity** and flexibility of the mortar.
- *Economy* – Lime is inexpensive. In addition, lime helps make a smoother, more uniform mortar. Lime-based mortars have a slower set time. This reduces the amount of tempering and allows the mason to work efficiently.
- *Minimal shrinkage* – Very hard, high-strength mortar can crack because it shrinks after hardening. These cracks result in a loss of strength. Lime undergoes the second-least loss in volume of all the ingredients in mortar.
- *Resistance to weather* – Mortar must be able to resist strong winds, freezing temperatures, and alternating wet and dry weather. Lime-based mortar actually increases in overall strength as it ages. In fact, when small cracks appear in the mortar joints, the action of lime combined with rainwater often reseals these cracks.
- *Autogenous healing* – The ability of lime to reseal itself when hairline cracks form in the mortar is called autogenous healing. Lime reacts with water and elements in the air to reseal these minute cracks through a chemical process called recarbonation.

1.1.3 Sand

Sand is a main ingredient in mortar (*Figure 3*). Sand acts as a filler. It adds to the strength of the mix. It also decreases shrinkage during the hardening phase and therefore decreases cracking. The water, cement, and lime in the mix form a paste that coats the sand particles. This paste lubricates the sand to form a workable mix.

The amount of air contained in the mix is determined by the range of particle sizes and grades of sand used. *Table 3* shows the recommended gradation limits for sand. The percentages show how much of the sand must pass through a set of screens with progressively finer mesh sizes. Note that 35 to 70 percent of the sand must pass through the #30 screen, which has finer mesh than the #4 screen.

Gradation limits are given in *ASTM C144, Specification for Aggregates for Masonry Mortar.* Adding fine or coarse sands is an easy way to change the gradation. Sometimes it is more practical to proportion the mortar mix to suit the available sand. It is often difficult to obtain a particular gradation of sand. If the sand does not meet the grading requirement of *ASTM C144*, it can only be used if the mortar meets the property specifications of Brick Industry Association (BIA) *M1, Specification for Portland Cement-Lime Mortar for Brick Masonry, or ASTM C270.*

It is important to use good-quality sand in the production of mortar. Sand is one of the least expensive building materials on the job. Cutting costs by using poor-quality sand will lead to a variety of problems. Poor-quality sand has no place on the job, as it lowers both the quality of the final product and the mason's productivity.

1.2.0 Describing Masonry Cement

Most mortars are made from portland cement mixed with other materials. Modern **masonry cement** mixes, developed in recent years, only require the addition of sand and water. These mixes contain all the ingredients except sand and water in one bag. Some mixes also include sand; you only need to add water. Most masonry cements are a combination of portland cement, lime, gypsum, and air-entraining admixtures. Masonry cement is available in bags from 40 to 80 pounds, as shown in *Figure 4*.

New Additives for an Ancient Recipe

Mortar was used to bond masonry units in ancient times. People used naturally available materials to make mortar. Primitive mortar was made from clay, chopped straw, and sand.

Ancient Egyptians first used mud from the Nile River. In some areas, volcanic ash or ground pumice was added to mortar to increase strength. Later, burned gypsum and sand were used to make mortar.

Lime and sand were used in the United States before the 1900s. In the 20th century, portland cement and additives were developed to increase workability and strength.

LOOSE SAND PILED ON CONCRETE OR TARP

LOOSE SAND IN BAG

28104-13_F03.EPS

Figure 3 Commercial sand for mortar.

The look and quality of the mix are very consistent. The materials are mixed and ground together before packaging to make uniform mixes. Masonry cement is less subject to batch variations than mortar mixed on the job, which can vary greatly between batches. Sand must be added to most masonry cement. However, sand is not uniform. Of all the materials in mortar, sand is the least uniform from batch to batch. Even using masonry cement does not guarantee consistent mortar.

Mixing masonry cement is much faster than mixing mortar. There are fewer materials to handle and measure, which reduces mixing time.

Mortar versus Concrete

Concrete and mortar contain the same principal ingredients. This does not mean that the two are mixed and handled in the same way.

Both mortar and concrete achieve their strength by the same chemical process of hydration. Both have a limited workable life span and need similar safety precautions.

However, mortar and concrete differ in several ways. The differences are listed here. For these reasons, mortar and concrete are mixed, treated, and used in vastly different ways.

Mortar

- Mortar bonds masonry units into a single component.
- It can be mixed by hand.
- It is often placed by hand and therefore must remain pliable.
- It is placed between absorbent masonry units. Mortar loses water upon contact with the units.
- The water/cement ratio is less important than it is for concrete. Mortars have a high water/cement ratio when mixed. This ratio decreases when the mortar contacts the absorbent units.

Concrete

- Concrete is usually a structural element.
- It is usually mixed by a machine or is brought to the job site already mixed.
- It is usually poured into forms by machine.
- It is usually placed in nonabsorbent metal or wooden forms that absorb little water.
- The water/cement ratio for concrete is significant.

Table 3 Recommended Sand Gradation Limits

	PERCENT PASSING	
Sieve Size	**Natural Sand**	**Manufactured Sand**
No. 4	100	100
No. 8	95–100	95–100
No. 16	60–100	60–100
No. 30	35–70	35–70
No. 50	15–35	20–40
No. 100	2–15	10–25
No. 200	—	0–10

28104-13_T03.EPS

Masonry cement is made in grades M, S, N, and O. Masonry cement is convenient, but it must meet *ASTM C91, Standard Specification for Masonry Cement*.

WARNING!

Dry cement dust is harmful. It can enter open wounds and cause blood poisoning. When cement dust comes into contact with body fluids, it can cause chemical burns. Protect your eyes, nose, and mouth from dust. Cement dust can cause a fatal lung disease called silicosis. Repeated contact with cement or wet concrete can also cause an allergic skin reaction known as cement dermatitis.

1.3.0 Describing Preblended Mortars

Preblended mortar, also called premixed mortar, is prepared by the manufacturer prior to delivery. In preblended mortar, all of the ingredients except water are mixed and ground together according to the proportions established in *ASTM C270, Standard Specification for Mortar for Unit Masonry*. The manufacturer prepares dry batches of the cementitious materials and dried sand that have been accurately weighed and blended. One advantage of preblended mortar is that it does not need to be adjusted to account for the moisture content of the sand. Another advantage is that the mason can be assured that the proportions of the mortar materials are consistent. When using preblended mortar, the mason only needs to provide water and to mix the mortar on the job site.

1.4.0 Describing the Use of Water and Admixtures

Potable (drinkable) water makes good mortar. Water must be free of chemicals and impurities. Impurities can weaken or discolor the mix. Impurities include alkalis, salts, acids, and organic matter. Generally, in mortar, you can use any water that is safe to drink. Water purification chemicals found in city water supplies usually do not affect mortar. Water should be used at a cool ambient temperature.

Admixtures are additives put into mortar to change its appearance or properties (*Figure 5*). They can be used in mortar to increase workabil-

How Portland Cement Is Made

To make portland cement, first limestone, marl, shale, iron ore, clay, and fly ash are crushed and screened. These materials contain the right amounts of calcium compounds, silica, alumina, and iron oxide. They are then placed in a rotating cement kiln.

The kiln looks like a large horizontal pipe. It has a diameter of 10 to 15 feet and a length of 300 feet or more. One end is slightly raised. The raw mix is placed in the high end. As the kiln rotates, the materials move slowly toward the lower end.

Flame jets are at the lower end. All the materials in the kiln are heated to temperatures between 2,700°F and 3,000°F. This high heat drives off the water and carbon dioxide from the raw materials. This process forms new compounds (tricalcium silicate, dicalcium silicate, tricalcium aluminate, and tetracalcium aluminoferrite).

Marble-sized pellets, called clinker, come out of the lower end of the kiln. The clinker is then very finely ground to produce portland cement. A small amount of gypsum is added during the grinding process to control the cement's set, or rate of hardening.

28104-13_SA01.EPS

28104-13_F04.EPS

Figure 4 Masonry cement.

ity, strength, and weather resistance. Admixtures to add color to mortar are also widely used.

Some admixtures are harmless. Others can be harmful to mortar and the resulting brickwork.

28104-13_F05.EPS

Figure 5 Multipurpose admixture for faster set and increased strength.

The properties of mortar depend upon its ingredients. Admixtures should not be used unless you know their effect on the mortar. The effect of admixtures on the finished structure, masonry units, and items embedded in the wall must also be considered. The following are common admixtures:

- *Air-entraining admixtures* – These admixtures reduce water content in mortar. They increase workability and freeze-thaw resistance. Air-entraining admixtures can reduce the extent of bond. Use prepared cements that include an air-entraining agent.
- *Bonding admixtures* – Using bonding admixtures results in an increased bond between mortar and masonry units. Some organic modifiers provide an air-cure adhesive. This increases the extent of bond of dry masonry.

Slaking Quicklime

Hydrated lime is made from limestone. The limestone, as shown here, is processed through two chemical reactions to create calcium hydroxide. This process is known as slaking quicklime. The lime is mixed with water to form a putty. It takes several weeks to slake quicklime. Processing is very dangerous, as substantial heat is released by the lime/water reaction. Manufacturers slake the lime as part of producing hydrated lime for mortar.

28104-13_SA02.EPS

- *Plasticizers* – Using plasticizers adds workability to mortar. Clay, clay-shale, and finely ground limestone are inorganic plasticizers. These promote workability and water release for cement hydration. Organic plasticizers also increase workability; however, mortar may stick to your tools more.
- *Set accelerators* – By increasing cement hydration, set accelerators decrease the time needed for the mortar to harden. They can increase the compressive strength of mortar. Nonchloride accelerators should be used. Chloride salts will deteriorate the cement in masonry.
- *Set retarders* – By delaying cement hydration, set retarders allow more working time for the mortar. This is useful during hot weather. Set retarders are also used in commercial ready-mixed mortars. The effect can last for up to 72 hours, but it dissipates when the mortar contacts masonry units.
- *Water reducers* – Used to lower the amount of water in the mortar, water reducers can increase the strength of the mortar. However, mortar rapidly loses water when it contacts absorptive masonry units. Water reducers may lower the water below the level needed for cement hydration. You must be sure to allow enough water for the curing process to take place.
- *Water repellents* – Water repellents modify masonry mortar during the early period after construction. However, they can reduce the rain penetration of the mortar for the long term.
- *Antifreeze admixtures* – These admixtures are not recommended for use in mortar. The amounts needed to prevent freezing also significantly lower the compressive and bonding strength of mortar.

NOTE

The architect or project engineer should approve the addition of any admixture, as it may affect structural stability and strength, as well as color.

Table 4 lists admixtures commonly used with cement and their effects on mortar.

Design effects can be created by using colored mortar to contrast or blend with the masonry units (*Figure 6*). White mortar is made by using white portland cement with lime and white sand. Light colors, such as cream, ivory, pink, and rose, are made using a white mortar base and pigments. Dark colors can be created by using colorants in normal gray mortar. Colors may not be as bright with a gray mortar base, but gray mortar is much less expensive. The manufacturer adds color pigments to mortar during the manufacturing process. Do not attempt to add color pigments to mortar on site.

1.5.0 Describing the Types of Masonry Mortar

There are five types of mortar. Each type is mixed differently, has different uses, and is assigned a letter: M, S, N, O, or K. These are the same as those used for masonry cement. Several different types of mortar have been developed to meet the needs of different job requirements. Selecting the right mortar type will depend on the type of masonry work and its location in the structure. *Table 5* shows the average compressive strength, water retention, and air content for each type of mortar.

Table 4 Admixtures and Their Effects

ADMIXTURE	WORKABILITY	STRENGTH	WEATHER RESISTANCE
Air-entraining agents	Increase	Decrease slightly	Increase
Bonding agents	—	Increase	—
Plasticizers	Increase	—	—
Set accelerators	Decrease	Increase	—
Set retarders	Increase	—	—
Water reducers	Decrease	Increase	Increase
Water repellants	—	—	Increase

28104-13_T04.EPS

28104-13_F06.EPS

Figure 6 Design effects using white mortar.

- Type M mortar has the highest compressive strength. It is somewhat more durable than the other mortars. In laboratory tests, ASTM has found that a 2-inch cube of Type M mortar will stand up to at least 2,500 pounds per square inch (psi) of compression after it has cured for 28 days. Type M mortar is used for applications that are subjected to high compressive loads, severe frost action, or high lateral loads from earth pressures, hurricane winds, or earthquakes, and in structures below grade, such as manholes and catch basins.
- Type S mortar has a medium compressive strength, good workability, and excellent durability. This type is often interchanged with Type M because it offers not only a relatively high compressive strength but also better workability, water retention, and flexibility. The ASTM specifications state that Type S mortar must stand up to at least 1,800 psi of compression. Type S mortar is used in structures requiring a high flexural extent of bond that are subjected only to normal compressive loads.
- Type N mortar is a medium-strength mortar with a psi rating of at least 750. It has excellent workability. It does not have the strength of Types M or S. Type N is the most frequently used mortar because it is highly resistant to

Table 5 Mortar Types for Classes of Construction[1]

Mortar	Type	Average Compressive Strength at 28 Days, Min. psi (MPa)	Water Retention, Min. %	Air Content, Max. %	Aggregate Ratio (Measured in Damp, Loose Conditions)
Cement Lime	M	2500 (17.2)	75	12	Not less than 2¼ and not more than 3½ times the sum of the separate volumes of cementitious materials
	S	1800 (12.4)	75	12	
	N	750 (5.2)	75	14[2]	
	O	350 (2.4)	75	14[2]	
Mortar Cement	M	2500 (17.2)	75	12	
	S	1800 (12.4)	75	12	
	N	750 (5.2)	75	14[2]	
	O	350 (2.4)	75	14[2]	
Masonry Cement	M	2500 (17.2)	75	18	
	S	1800 (12.4)	75	18	
	N	750 (5.2)	75	20[3]	
	O	350 (2.4)	75	20[3]	

1. Laboratory-prepared mortar only.
2. When structural reinforcement is incorporated in cement-lime or mortar-cement mortar, the maximum air content shall be 12 percent.
3. When structural reinforcement is incorporated in masonry-cement mortar, the maximum air content shall be 18 percent.

28104-13_T05.EPS

weather. Type N mortar is used in veneer applications only.

- Type O mortar is a low-strength mortar with very good workability. Because of its low durability, this type is not recommended for use where bad weather conditions exist. Type O mortar should have a compressive strength of at least 350 psi. Type O mortar is used on nonloadbearing masonry of allowable compressive strength not exceeding 100 psi. Type O mortar is widely used for tuckpointing.
- Type K mortar is so low in compressive strength and extent of bond that it is seldom used. ASTM requires that the testing strength be only 75 psi after 28 days of curing. Type K mortar made without cement is sometimes used for restoration of historic buildings. Type K mortar is used on interior nonloadbearing partitions where low compressive strength and extent of bond are permitted by building codes.

Types of Mortar

You can remember the five types of mortar (M, S, N, O, and K) by taking every other letter from the words MaSoN wOrK.

Additional Resources

Masonry Construction, First Edition. 1996. David L. Hunter, Sr. Upper Saddle River, NJ: Prentice Hall.

Technical Note TN8, *Mortars for Brickwork.* 2008. Reston, VA: The Brick Industry Association. **www.gobrick.com/Portals/25/docs/Technical%20Notes/TN8.pdf**

1.0.0 Section Review

1. Of the five types of portland cement, the types that are generally only used for mortar are _____.
   a. Types I, III, and V
   b. Types I, II, and III
   c. Types II and IV
   d. Types III, IV, and V

2. Masonry cement is made in grades _____.
   a. M, S, N, O, and K
   b. I, II, III, IV, and V
   c. M, S, N, and O
   d. A, O, W, and R

3. The ASTM standard that applies to preblended mortar is _____.
   a. C144
   b. C270
   c. C207
   d. C91

4. Additives put into mortar to change its appearance or properties are called _____.
   a. entrainments
   b. aerations
   c. preblends
   d. admixtures

5. The type of mortar that is often interchanged with Type M is _____.
   a. Type K
   b. Type O
   c. Type S
   d. Type M

Section Two

2.0.0 Mortar Properties

Objective

Describe properties of plastic and hardened mortar.

a. Describe plastic mortar.
b. Describe hardened mortar.

On the basic level, a mechanical bond is formed between the masonry unit and the mortar. This bond ties the masonry in a wythe into a single unit.

For the majority of masonry construction, the most important property of mortar is extent of bond. The extent of bond depends on the properties of the mortar and the bonding surface:

- The mortar must have the right proportions of ingredients for its use. It must stay wet enough to lay and level the masonry.
- The masonry surface should be irregular to provide strength of bonding. It should be absorptive enough to draw the mortar into its irregularities.
- The masonry surface should not be so dry that it dries out the mortar. Slow, moist curing improves mortar bond and compressive strength.

The second most important property of mortar is bond integrity. The work of the mason defines the amount of bond between masonry units. Bond integrity depends on the mason, who does the following:

- Keeps tools and masonry units clean
- Butters every joint fully
- Does not move the masonry unit after it is leveled
- Levels units shortly after they are laid
- Uses mortar at a consistency that is appropriate for the masonry unit
- Keeps the mortar tempered
- Mixes fresh mortar after two hours

2.1.0 Describing Plastic Mortar

Plastic mortar is freshly mixed and ready for use. Properly mixed mortar is soft enough to spread easily but thick enough to retain its shape. The plastic properties of mortar include the following:

- Workability
- Water retention
- Water content
- Rate of hardening

2.1.1 Workability

Several factors influence the workability of mortar. These include consistency, setting time, adhesion, and cohesion.

Consistency, or uniformity, measures how similar mortar is from one batch to the next or from one place to another in the same batch. Consistency is controlled and improved by closely following the mix recipe and the mixing procedure. The order in which the ingredients are added is important. Mixing the mortar for the same amount of time is also a key to consistency.

Setting time is the time the mason has to use the mortar before it hardens and is no longer workable. It is determined by the type of cement used in the mix. Setting times can be extended by using the maximum water allowable in the mix, by wetting the mortarboard before placing mortar on it, and by tempering the mortar during use.

Adhesion is how well the mortar sticks to the masonry units. Greater adhesion means greater bonding strength.

Cohesion is how well mortar sticks to itself. Mortar with good cohesion will extrude between masonry units without smearing or dropping away (*Figure 7*).

2.1.2 Water Retention

Water retention is related to workability. Mortar with good water retention remains soft and plastic. It should remain plastic long enough for the mason to align, level, and plumb the masonry units. Poor water retention may break the bond between the mortar and the units.

Good water retention is important to developing adhesion between the mortar and the masonry units. This will create strong, watertight joints. Water in the mortar allows the cement to hydrate. This is the chemical process that makes mortar harden.

Mortar Consistency for Block and Brick

Mortar used for block requires a different consistency from mortar used for brick. Be sure to temper mortar to the proper consistency for the type of masonry unit being used.

28104-13_F07.EPS

Figure 7 Mortar with good cohesion.

Water retention is improved by adding more lime, finer sand, entrained air, and more water. These also increase the workability of the mortar. However, the proportions of the ingredients must continue to stay within the limits specified in the mix formula.

2.1.3 Water Content

Water content is possibly the most misunderstood aspect of masonry mortar. Mortar and concrete are made from the same materials. Some designers mistakenly assume that mortar requirements are similar to those for concrete. This is false. Mortar and concrete have very different water-cement ratios. Mortar, concrete, and grout have different amounts of water.

2.1.4 Consistent Rate of Hardening

Mortar hardens when cement reacts chemically with water. This process is called hydration. The rate of hardening is the speed at which it develops resistance to an applied load. If it hardens too quickly, the mason may have difficulty placing it. Very slow hardening may impede the progress of the work. The mortar will flow from the completed masonry. During winter construction, slow hardening may also subject mortar to early damage from frost action. A well-defined, consistent rate of hardening assists the mason in laying the masonry units and in tooling the joints at the same degree of hardness. Uniform joint color of masonry reflects consistent tooling times.

2.2.0 Describing Hardened Mortar

The qualities needed in good mortar are different after it has hardened. You can control the final appearance and performance of the mortar. By using good-quality materials in the mortar mix, you are halfway there. Your workmanship provides the other half. The hardened properties of mortar include the following:

- Durability
- Compressive strength
- Strength of bond
- Volume change
- Appearance

2.2.1 Durability

Durability of mortar is measured by its ability to resist weather damage. Mortar must withstand repeated cycles of freezing and thawing. Freezing temperatures cause the water in mortar to expand. This causes cracking. Mortars with high-compressive strength are denser. They contain less water and are more durable. Air-entrained mortars withstand weathering better. Mortars made with masonry cement have higher air content than mortars made with portland cement and lime. Therefore, masonry-cement mortars resist freeze-thaw cycles well.

2.2.2 Compressive Strength

Compressive strength is the ability of the mortar not to crumble under pressure. High-strength mortar is needed for loadbearing walls or supports (*Figure 8*), masonry below grade, or masonry for paving. Compressive strength is measured in pounds per square inch, or psi. The compressive strength of mortar increases as the cement content increases. It decreases as more lime, air entrainment, and water are added. Type M mortar is the strongest. But there is more to the wall than the mortar. Masonry structures gain their strength from several factors, including the following:

- The compressive strength of the masonry units
- The compressive strength of the mortar
- The design of the structure
- The workmanship of the mason
- The amount of curing

28104-13_F08.EPS

Figure 8 Mortar must be able to support loads.

If Type M mortar is used when Type S is specified, the structure may not be stronger. Brick walls are designed to move. They expand and contract with temperature changes. If the mortar is too hard, it will not move properly. This will weaken the final structure.

2.2.3 Extent of Bond

Extent of bond is the ability of the mortar to hold the masonry units together. This depends on the ability of the mortar to stick to the surface of the masonry units. Adhesion improves with greater contact between the mortar and the masonry unit.

The mortar must be completely spread over the unit face or shell. The mortar must have enough flow to wet the contact surfaces. Masonry units have surface irregularities, or micropores. This increases strength of bond by increasing the area of the surface to be bonded. Both clay and concrete units will absorb wet mortar into their irregularities. *Figure 9* shows an enlargement of this process.

Mechanical extent of bond in cured mortar depends on the mechanical interlocking of cement hydration crystals. Voids, smooth masonry units, and contamination can hinder this process. Voids are air holes in the mortar joint. These weaken the strength of bond. Voids also allow water into the joint. Smooth masonry units such as molded brick or smooth stone do not have micropores. They provide less strength of bond with the mortar. Loose sand particles, dirt, and other contamination also weaken the strength of bond.

The strength of the bond depends upon the strength of the mortar itself. The main factors affecting mortar strength include the following:

- Ingredients, such as types and amounts of cement, retained water, and air content
- Surface texture and moisture content of the masonry units used
- Curing conditions, such as temperature, wind, relative humidity, and amount of retained water

Mortar strength is directly related to cement content. Both extent of bond and mortar stiffness increase as cement content increases. Extent of bond also increases as water content increases. The optimum extent of bond is obtained by using a mortar with the highest water content compatible with workability.

2.2.4 Volume Change

As mortar cures, the volume changes as free water is lost. Shrinkage results from the volume changes in mortar. Shrinkage causes cracks to appear in joints.

A mortar joint made with properly proportioned mortar will shrink. However, the resulting cracks are usually very small and not of concern. Larger shrinkage cracks will appear when there is a large amount of water loss from the mortar.

The rate of water loss from mortar is determined by several factors. First is the amount of water in the mortar. Second is the rate at which it is absorbed into the masonry units. Air-entrained mortars need less water. However, workability and extent of bond are directly related to the flow of the

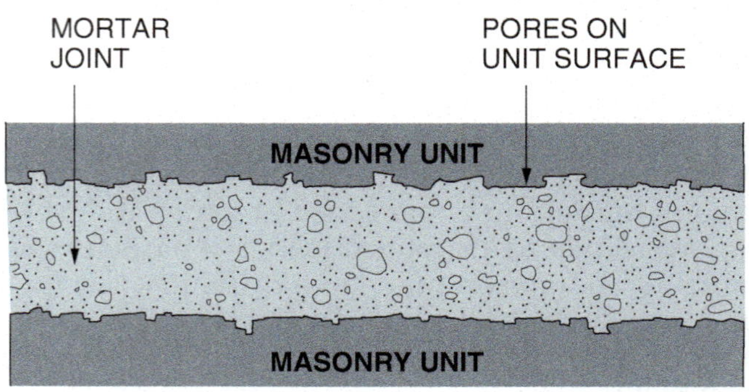

28104-13_F09.EPS

Figure 9 Mortar bonding to a porous masonry surface.

mortar and should be given priority when determining water content in the mixing operation.

2.2.5 Appearance

Mortar makes up approximately 20 percent of the face of a masonry structure. The mortar should have a uniform color and shade. This affects the overall appearance of the finished structure. Several factors affect the color and shade of mortar joints:

- Atmospheric conditions
- Admixtures
- Moisture content of the masonry units
- Uniformity of proportions in the mortar mix
- Water content
- When the mortar joints are tooled

Sulfate Resistance

Sulfates in the soil or water can cause mortar to expand. The damage is similar to freeze-thaw cycles. Mortars that will contact soil, groundwater, seawater, or industrial processes must also withstand sulfate expansion. The use of sulfate-resistant materials can minimize the damage.

Masonry-cement mortars and Type V portland cement are sulfate resistant. You can use either one, but Type V cement is slow to gain strength. Applying a protective coating to the finished work is another good option.

Additional Resources

Concrete Masonry Handbook for Architects, Engineers, and Builders, Fifth Edition. 1991. William C. Panarese, Steven H. Kosmatka, and F. A. Randall, Jr. Skokie, IL: Portland Cement Association.

Masonry Cement Mortar: Its Proper Use in Construction. 1993. Pat Howley. Murray Hill, NJ: ESSROC Materials.

2.0.0 Section Review

1. The term used to describe how well the mortar sticks to the masonry units is _____.
   a. adhesion
   b. admixture
   c. cohesion
   d. retention

2. The ability of mortar to withstand crumbling under pressure is called _____.
   a. durability
   b. strength of bond
   c. cohesion
   d. compressive strength

SECTION THREE

3.0.0 MORTAR APPLICATION PROBLEMS AND SOLUTIONS

Objective

Identify the common problems found in mortar application and their solutions.

a. Describe the effects of improper proportioning and poor-quality materials.
b. Explain the effects of extreme weather and tempering.
c. Describe efflorescence.

Trade Term

Slaked lime: Lime reduced by mixing with water to a safe form that can be used in the production of mortar.

Typically, the mason faces three general types of problems when mixing mortar:

- Improper proportioning of materials
- Poor quality of materials
- Working in extreme weather

Efflorescence is another mortar problem in which salts seep out of the mortar and stain the surface. It does not show up until after the mortar has hardened.

3.1.0 Understanding the Effects of Improper Proportioning and Poor-Quality Materials

Improper proportioning of materials is a common problem in mixing mortar. It is caused primarily by poor work techniques. The first step in preparing the mix is to plan adequately. Ensure that the mix ratios are completely understood and the formula written down. The mason or helper preparing the mix should go through a mental checklist of the things needed and the steps required for the mixing operation. Planning ahead will help prevent any missing pieces or shortages that would lead to problems during the mixing.

The four most common problems are as follows:

- *Adding too much water* – This problem is caused by using a hose to add water. Adding excessive amounts of water is known as drowning the mortar. This lowers the water-cement ratio, which leads to lower mortar strength. This problem can be corrected by adding dry ingredients in their proper proportions until the desired consistency is reached. The best way to avoid adding too much water is to premeasure the water in a bucket. This is important because trying to correct this error can lead to new problems.
- *Adding too much sand* – This problem is caused by not measuring the materials. If you use a shovel to initially fill the power mixer or mortar box, calibrate it ahead of time. You should calibrate the shovel at least twice a day, once before you start in the morning and once midway through the day. If you continue to have problems with measuring the sand, using the cubic foot box may be the best answer. This is actually a more efficient way to load sand, but you will need a helper to help lift the box and empty it into the drum.
- *Adding too much sand to counteract the effects of too much water* – This practice results in weaker mortar and should be avoided. Overly sanded mortar is harsh, difficult to use, and forms weak bonds with the masonry units. The color of the mortar will also vary. Never add sand to a mix without adding the other ingredients in proportion.
- *Adding too much or too little cement* – Mortar that is too high in cement content is known as fat mortar or rich mortar. It is sticky and hard to remove from the drum. Mortar that is low in cement content is called lean mortar. It will display all the characteristics of weak mortar, such as poor bonding, lack of cohesion and adhesion, and lack of compressive and tensile strength.

All of these problems can be avoided by proper planning, thinking ahead to anticipate problems, using calibrated containers, and approaching the mixing process in a methodical manner.

Problems with poor-quality or spoiled materials can be solved by properly storing, inspecting, and testing materials prior to use. Do not use cement that has knots or lumps; it will not mix well, and taking the lumps out will waste time. You can avoid spoiling materials by planning to minimize spills onto raw materials.

Lay out the mixing area in an organized way. This will prevent good material from being spoiled. It will also increase your efficiency. For example, place the mixer downhill from the cement pallets, to avoid getting them wet from possible water spills. Leave a pathway through the mixing site for wheelbarrows or forklifts. Mortar can be easily spilled during transfers. A spill into your materials piles will ruin future batches.

3.2.0 Understanding the Effects of Extreme Weather and Tempering

Extreme weather presents special problems in mixing mortar. Certain admixtures can be added to improve the performance of mortar in hot and cold weather. The Brick Industry Association (BIA) defines *cold* as any temperature below 40°F and *hot* as any temperature above 100°F, or 90°F with an 8-mile-per-hour wind. Winds above 10 miles per hour (mph) are also considered extreme weather for masons working outdoors.

Air-entraining admixtures will increase the workability and freeze-thaw durability of mortar. Adding too much will reduce the strength of the mortar. Adding an air-entraining admixture separately at the mixer is not recommended. This is due to the lack of control over the final air content. If air-entraining is needed, use factory-prepared cements and lime that include an air-entraining agent.

At low temperatures, performance can be improved by using Type III (high early strength) cement. Also, mortar made with lime in dry form is preferred for winter use over **slaked lime** or lime putty because it requires less water.

Another option for cold weather is to heat the material used to make the mortar. The sand and the water can both be heated. It is easier and safer to heat the water. Consider heating the materials when the temperature falls below 40°F. After mixing the heated ingredients, the temperature should be within the range considered optimal for hydration as specified by the manufacturer or the applicable standard.

Some admixtures are not recommended due to their negative effect on mortar. Any antifreeze admixtures, including several types of alcohol, would have to be used in such large quantities that the strength of the mortar would be greatly reduced. Do not use calcium chloride. It is commonly used as an accelerator in concrete, but it produces increased shrinkage, efflorescence, and metal corrosion in mortar.

Fresh mortar should be prepared as it is needed. That way its workability will remain about the same throughout the day. Mortar that has been mixed but not used immediately tends to dry out and stiffen. However, loss of water on a dry day can be reduced by wetting the mortarboard. Covering the mortar in the mortar box, wheelbarrow, or tub will also help keep it fresh.

Mortar should be mixed with the maximum amount of water consistent with workability. This will provide maximum extent of bond. If necessary to restore workability, mortar may be tempered by adding water (*Figure 10*). Tempering is permitted only to replace water lost by evaporation. This can usually be controlled by requiring that mortar be used within two hours after mixing, depending on conditions. If used within this time, it will need little tempering.

If not used within this time, thorough remixing is then necessary. Remixing can be done in the wheelbarrow or on a mortarboard. Small additions of water may slightly reduce the compressive strength of the mortar; however, the end effect is acceptable. Masonry built using a workable plastic mortar has a better extent of bond than masonry built using dry, stiff mortar. Mortar used within one hour after mixing should not need tempering unless the weather is very hot and evaporative conditions prevail.

Colored mortar is very sensitive to tempering; additional water may cause a noticeable lightening of the color. Mortar should be tempered with caution to avoid variations in the color of the hardened mortar. Mixing smaller batches can lessen the need to temper.

CAUTION

Do not heat the water over 160°F. If the water is too hot, the mix may flash-set when the cement is added. A flash set occurs when the cement sets prematurely due to excessive heat.

3.3.0 Understanding Efflorescence

Efflorescence is a deposit of water-soluble salts on the surface of a masonry wall (*Figure 11*). It is usually white and generally appears soon after the wall has been built. Of course, this is when the owner and architect tend to be most concerned with the structure's appearance.

28104-13_F10.EPS

Figure 10 Tempering mortar.

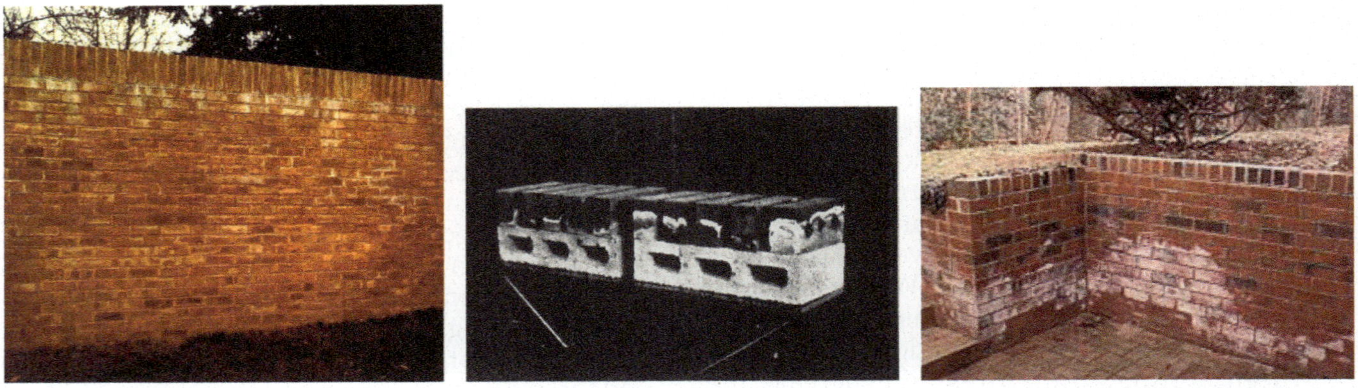

Figure 11 Types of efflorescence.

Efflorescence occurs when soluble salts and moisture are both present in the wall in sufficient concentrations. If either of those elements is missing or below a certain concentration, efflorescence will not occur.

The salts usually originate in the walls, in either the masonry units or the mortar itself. In some cases, the salts may come from ground moisture behind a basement wall. For this reason, basement walls should be protected with a moisture barrier.

Efflorescence can be prevented in several ways:

- Reduce salt content in the mortar materials by using washed sand.
- Use potable water for making the mortar.
- Keep materials properly stored and off the ground prior to use.
- Protect newly built masonry walls with a canvas or suitable waterproofing material.
- Use quality workmanship to ensure strong, tight bonds between mortar and masonry units.

Mortar Hydration Hardening

Mortar that has stiffened because of hydration hardening should be discarded. It is difficult to tell by sight or feel whether mortar stiffening is due to evaporation or hydration. The most practical method is to consider the time elapsed after mixing. To avoid the need for extensive tempering and to avoid hydration hardening, a batch of mixed mortar should be used within 2 hours.

Efflorescence

Efflorescence can be removed by washing the wall with water or a muriatic acid / water solution. Use a diluted solution of one part muriatic acid to nine parts water. This washing will remove the efflorescence, but it may return after a period of time. The best solution is to reduce the amount of moisture entering the wall and to lower the amount of water-soluble salts in the mortar materials.

Additional Resources

Technical Note TN1, *Cold and Hot Weather Construction.* 2006. Reston, VA: The Brick Industry Association. **www.gobrick.com/Portals/25/docs/Technical%20Notes/TN1.pdf**

3.0.0 Section Review

1. Mortar that is too high in cement content is commonly referred to as _____.
   a. fat or rich
   b. hard or stiff
   c. thick or slow
   d. buttered or soft

2. The Brick Industry Association defines *cold* as any temperature below _____.
   a. 25°F
   b. 30°F
   c. 35°F
   d. 40°F

3. In order for efflorescence to occur, two things must be present in a wall: _____.
   a. air-entrained mortar and moisture
   b. soluble salts and acidic soil
   c. soluble salts and moisture
   d. acidic soil and moisture

SECTION FOUR

4.0.0 THE MORTAR MIXING AREA

Objective

Explain how to properly set up, maintain, and dispose of mortar and use the mortar mixing area.

a. Describe how to set up a mixing area.
b. Describe how to maintain the mixing area.
c. Describe how to mix mortar with a power mixer.

Performance Tasks 1 and 2

Properly set up a mixing area.
Properly mix mortar with a power mixer.

Masons should be able to mix their own mortar, or mud, as it is known in the trade. They need to know how to modify the mix for different requirements of the work. On large jobs, mortar may not be mixed on the site but brought in from an off-site mixing plant. On small jobs, however, it may be practical for masons to mix their own mortar either by hand or with a power mixer. The following sections present the steps needed to prepare for mixing mortar.

4.1.0 Setting Up the Mixing Area

Efficient placement of the materials and the mixer can save time and work. Because they need to be close together, the mixer site and the material stockpile sites should be set at the same time. Avoid a site that is downhill from the work area if the mortar will be moved in wheelbarrows.

The mixer should be located as close as possible to the main section of the work with mix ingredients arranged around it. The stockpiles should be on the side of the mixer away from the work so that the mortar-filled light trucks and other equipment do not have to pass around the stockpiles. If the ground is soft or damp, set up a plywood runway for equipment.

Use a hose to bring the water to a barrel near the mixer or to fill the water measure directly. Make sure the hose is not in the path of traffic. When setting up for a good-sized job, the mortar mixing area needs to be as follows:

- Away from materials and equipment movement paths
- Close to stockpiles, so materials will not have to be moved a long distance
- Close to the work site, so mortar will not have to be moved a long distance, but not so close as to interfere with the work
- Positioned so that mortar-filled wheelbarrows will not have to be pushed uphill to the work area

If the job requires only a small amount of mortar, the mortar can be mixed in a wheelbarrow and rolled directly to the work site.

4.2.0 Maintaining the Mixing Area

The materials used in making mortar should be stored on the job site in an area convenient to the mixing site but not where they will interfere with the work. Cement and lime are generally sold and delivered by the bag. Sand is ordered by weight, but it is usually delivered in loose form.

The cement and lime should be kept in a storage shed. If kept outside, they must be covered with plastic sheets or canvas tarpaulins (tarps) (see *Figure 12*). Any other materials stored outside should be covered in a similar manner. In either case, the bags should be kept off the ground by using wooden pallets to keep them from ground moisture.

Sand can be dumped on the ground, and in most cases, does not need to be protected from moisture in warm weather (refer to *Figure 3*). In

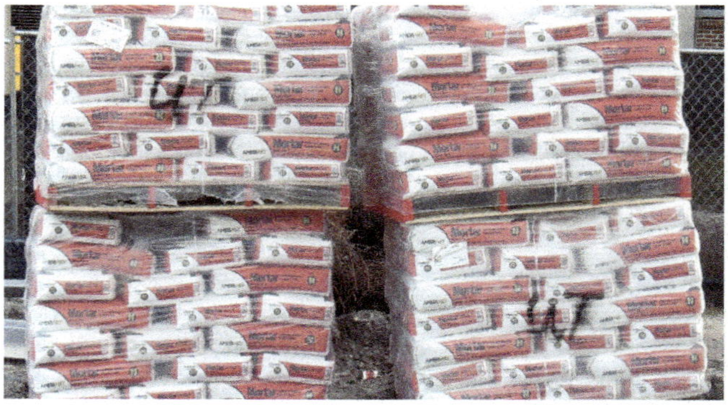

28104-13_F12.EPS

Figure 12 Storing cement.

Mixing Area

Keep the mixing area clean by measuring and loading materials carefully. Clear away spills with a shovel or brush. A clean mixing site is safer as well as easier to use.

winter, cover the stockpile to prevent wet sand from freezing. Another method to keep sand from freezing is to run pipes through the stockpile, and then pump heated air or water through the pipes (see *Figure 13*). If the sand does freeze, it must be thawed before it can be used.

Keep foreign material out of the stockpile. If the only location for sand at the job site is muddy or vegetation-covered ground, the sand can be dumped onto plastic sheets or tarpaulins.

To ensure quality control of a mortar mix and consistency between batches, start with the proper measuring of mortar ingredients. Proper measuring will ensure that the properties of the mortar, such as workability, color, and strength, will be the same from mix to mix. This ability to produce the same results in each batch will increase your productivity and lead to a quality, good-looking job. *Table 6* gives the specifications for mixing each type of mortar.

28104-13_F13.EPS

Figure 13 Sand pile warmed by heated pipe.

The proportion specifications list the recipe for mixing mortar materials. A certain amount of each material is added to the mix. If more mortar is needed, each material must be added in proportion to the other materials. Mortar specified by this method does not need laboratory testing.

Mortar mix specifications are expressed in terms of a proportion of cement, lime, and sand volumes. For example, for a Type N mortar, the proportions might be expressed as 1:1:6, or 1 part portland cement, 1 part lime, and 6 parts sand. For a Type O cement-lime mortar with a mix ratio of 1:2:9, the proportions would be 1 part portland cement, 2 parts lime, and 9 parts sand.

Portland cement is available in 94-pound bags containing 1 cubic foot of material (*Figure 14*). Hydrated lime comes in 50-pound bags containing 1 cubic foot. Masonry cement comes in 70-pound bags containing 1 cubic foot. Because sand comes in bulk form, it needs a practical and consistent measuring method. Many masons measure sand by the shovel. The number of shovels of sand in a cubic foot will vary, however, depending on the amount of water in the sand. One method that works well is to use a cubic foot box (*Figure 15*). This box measures 1 foot long × 1 foot wide × 1 foot deep and holds exactly 1 cubic foot of material.

The cubic foot box can be used to measure all the sand used in the mix. It is more commonly used to calibrate the shovel for loading the sand into the mixer. The mason determines the number of shovels of sand required for a cubic foot

Measuring and Mixing

Careful measurement of mortar materials and thorough mixing are the first steps to making a good-looking masonry structure.

Mortar Box Safety

Some mortar boxes have special fittings, so they can be moved by a forklift. Always place these boxes for easy access by the forklift operator. It is much easier to adjust the box before it is full of mortar. Never use a forklift to move a mortar box that does not have fittings for doing so. The wet mortar could shift, and the box will fall. This can cause serious injury and/or damage.

28104-13_SA03.EPS

Table 6 Proportion Specifications for Mortar

Mortar	Type	Proportions by Volume (Cementitious Materials)								Aggregate Ratio (Measured in Damp, Loose Conditions)
		Portland or Blended Cement	Mortar Cement			Masonry Cement			Hydrated Lime or Lime Putty	
			M	S	N	M	S	N		
Cement Lime	M	1	—	—	—	—	—	—	¼	Not less than 2¼ and not more than 3 times the sum of the separate volumes of cementitious materials
	S	1	—	—	—	—	—	—	over ¼ to ½	
	N	1	—	—	—	—	—	—	over ½ to 1¼	
	O	1	—	—	—	—	—	—	over 1¼ to 2½	
Mortar Cement	M	1	—	—	1	—	—	—	—	
	M	—	1	—	—	—	—	—	—	
	S	½	—	—	1	—	—	—	—	
	S	—	—	1	—	—	—	—	—	
	N	—	—	—	1	—	—	—	—	
	O	—	—	—	1	—	—	—	—	
Masonry Cement	M	1	—	—	—	—	—	1	—	
	M	—	—	—	—	1	—	—	—	
	S	½	—	—	—	—	—	1	—	
	S	—	—	—	—	—	1	—	—	
	N	—	—	—	—	—	—	1	—	
	O	—	—	—	—	—	—	1	—	

NOTE: Two air-entraining materials shall not be combined in mortar.

28104-13_T06.EPS

by filling the box. The box is no longer necessary for most of the mixing operations. You should recheck your shovel count twice each day to allow for factors such as moisture content of the sand, changing shovels, or changing people doing the shoveling.

Water should always be measured with a container. Never pour water directly into the mixer from a hose, as it is not possible to calculate the exact amount added. A plastic five-gallon bucket works well.

The volume of mortar obtained from a mix will only equal the volume of sand in the mix, even though you have also added a quantity of cement, lime, or masonry cement. This is because the cement, lime, and water occupy the voids between

Preventing Runoff Contamination

To protect the environment from accidental contamination, masons take steps to ensure that construction materials, leftover mortar, and contaminated wash water are not allowed to enter storm drains and pollute rivers and streams. Mixing containers and washout chutes should not be emptied in areas that may flow to a street, a public drain, or a waterway. Do not wash concrete or brick areas unless the wash water can flow into a bermed area that prevents further runoff. Place tarps and drop cloths under mixers to protect against spills.

A company called Eco-Pan® (**www.eco-pan.com**) has developed a washout containment system that prevents mortar and other cementitious materials from contaminating the environment. The company delivers large washout basins to your job site. These basins collect wastewater rather than letting it run into the ground. The company then comes to collect the basins when full and disposes of the contaminated water in accordance with United States Environmental Protection Agency (EPA) standards.

28104-13_F14.EPS

Figure 14 Bag of portland cement.

28104-13_F15.EPS

Figure 15 Cubic foot box.

the grains of sand. The mix simply becomes denser as more cement and lime are added. For this reason, consider only the amount of sand added when you want to know how much mortar will be produced in your mix.

4.3.0 Mixing Mortar with a Power Mixer

You learned how to mix mortar by hand in the module *Introduction to Masonry*. In this section, you will learn how to mix mortar by machine. Machine mixing is the preferred method for preparing a batch of mortar. For larger jobs, a power mixer (*Figure 16*) can produce 4 to 7 cubic feet of mortar at one time with the same quality and the same properties batch after batch. Machine mixing requires less human energy, which frees masons to use their energy where it counts most—in the construction of their project.

When using a power mixer, place the materials near the mixing area. The mixer should be at the center of these materials in such a way that all materials can be easily reached. Leave a clear path for the wheelbarrow or other equipment used to move the finished mortar. The mixer should be securely supported and the wheels blocked to avoid tire wear.

WARNING!

Wear eye protection and other appropriate personal protective equipment when using a power mixer. Never place any part of your body in the mixer.

4.3.1 Machine Mixing Steps

Review the operation manual for the machine used, and review safety procedures for working with power equipment. Have the mix formula written down, including the number of cubic feet of cement and lime as well as the cubic feet or shovelsful of sand needed for the size of the batch to be mixed.

28104-13_F16.EPS

Figure 16 Mortar mixer.

Cement Storage

In the United States, portland cement is normally sold in paper bags by volume. Bags must be stored off the ground and covered. This prevents the cement from absorbing moisture, which can cause lumps to form in the cement powder. Cement powder should be free flowing. Do not use any cement with lumps that cannot be broken up easily into powder by squeezing them in your hand.

Follow these steps for mixing mortar:

Step 1 With the blades turning, add a small amount of water (*Figure 17*). Add enough to wet the inside of the mixing drum to prevent the mortar from caking on the mixing paddles or the sides of the drum.

Step 2 Add one-third to one-half of the sand needed for the batch. Keep the paddles turning to prevent stress on the turning mechanism.

Step 3 Add all the necessary cement and lime or the masonry cement to the mixer. This is best done by the bag. Place the bag on the safety grate, and open the bag by cutting with a small knife or trowel or by pulling the bag opening. Some mixers have a metal tooth on the safety grate, specifically designed for this purpose.

Step 4 Add the remaining sand at this time. Then add more water to bring the mixture to the desired consistency. Allow the mixing process to continue for three to five minutes in order to completely blend the materials. Do not mix any longer, however, as this will allow extra air to be trapped in the mix. This will cause the mortar to be spongy and weak.

Step 5 When the mixing is complete, with the drum still turning, grasp the drum handle and dump the mortar into the wheelbarrow or mortar box (*Figure 18*). The turning blades will clear most of the mortar out of the mixer. Next, take the blades out of gear and turn off the mixer. The remaining mortar can now be removed by hand.

Step 6 If more mortar is needed immediately, start the process over again by adding enough water to the drum to clean the blades as they are turning. If no more mortar will be needed for some time, leave the mixer turned off, and clean the inside and the blades with a water hose and a stiff bristle brush.

4.3.2 Safety Tips for Power Mixers

As with any piece of power equipment, a power mortar mixer should be treated with respect and caution. You should read the manufacturer's operation manual. Follow the procedures listed in the manual. Here are some basic safety tips:

- Wear appropriate personal protective equipment when working with mortar mixers. This includes shatterproof safety glasses or goggles to protect your eyes and a nose or face mask to prevent the inhalation of cement or lime dust. Hearing protection and steel-toed boots may also be required, depending on the manufacturer's specifications.
- When operating a mixer's electric motor or gasoline engine, follow all safety precautions in the operator's manual. Always ensure that the motor or engine is off and cool before attempting to service it.

28104-13_F17.EPS

Figure 17 Adding water to wet the mortar mixer.

28104-13_F18.EPS

Figure 18 Dumping mortar into wheelbarrow.

- Before operating the mixer, ensure that the drive belt connecting the engine to the drum is free of cracks, fraying, cuts, or wear. Check the mixing paddles to ensure they are properly adjusted according to the manufacturer's specifications.
- When adding water, shoveling sand, or pouring cement or lime into the drum, be extremely careful not to place the shovel or bucket into the mouth of the mixer where the blades are turning. Serious injury could result if equipment is caught by the mixing blades.
- Do not place your hands or arms into the turning mixer. Do not wear loose clothing; it can be caught in the turning mixer blades. Stand clear of the dump handle while the mixer is operating to avoid injury in the event of kickback.
- If the mixer does not have a safety grate over the mixing chamber, consider adding one.
- Do not scrape the last mortar out of the drum while the blades are still turning. Disengage the blades and turn off the mixer first.
- Turn off the mixer before doing final cleanup. Follow the manufacturer's specifications for turning off the mixer to ensure that all safety steps have been followed correctly.

WARNING!

Keep your hands out of the mouth of the mixer. If a torn bag falls into the mixer, do not try to remove it while the mixer is turning. Turn the machine off and allow its blades to stop completely before reaching inside.

Mortar Mixing

For large jobs, the dry mix is delivered to the site in a large hopper. The mixing machine is placed below the hopper. Batches of mortar can be mixed easily and rapidly. The materials must still be carefully measured to obtain a consistent mortar.

28104-13_SA04.EPS

Additional Resources

Technical Note TN8B, *Mortars for Brickwork: Selection and Quality Assurance.* 2006. Reston, VA: The Brick Industry Association. **www.gobrick.com/Portals/25/docs/Technical%20Notes/TN8B.pdf**

Technical Note TN20, *Cleaning Brickwork.* 2006. Reston, VA: The Brick Industry Association. **www.gobrick.com/Portals/25/docs/Technical%20Notes/TN20.pdf**

4.0.0 Section Review

1. The mixer should be located as close as possible to _____.
   a. the entrance/exit to the work site
   b. secondary work areas
   c. materials and equipment movement paths
   d. the main section of the work

2. In most cases, during warm weather sand does *not* need to be protected from _____.
   a. wind
   b. contact with other mortar ingredients
   c. moisture
   d. sunlight

3. When using a mortar mixer and the mixing is complete, grasp the drum handle and dump the mortar into the wheelbarrow or mortar box _____.
   a. while at the same time removing the blades
   b. after removing the blades
   c. after the drum has stopped turning
   d. while the drum is still turning

SUMMARY

Mixing quality mortar is one of the primary skills a mason needs. Good mortar adds to the strength and stability of the structure. Although you can purchase premixed mortar, a mason must know the ingredients in mortar. Each of the materials affects the overall properties and appearance of the final product. Knowing these properties will help you mix strong, effective mortar under different conditions.

Review Questions

1. The ingredient of cement that causes hydration of mortar by reacting with water and air is _____.
   a. sand
   b. lime
   c. admixtures
   d. portland cement

2. The main ingredient in mortar is _____.
   a. sand
   b. portland cement
   c. slaked lime
   d. water-soluble colorant

3. Flexibility is a mortar property that is provided by the ingredient _____.
   a. lime
   b. sand
   c. water
   d. portland cement

4. The only type of hydrated lime that should be used in mortar is Type _____.
   a. N
   b. NA
   c. S
   d. SA

5. Portland cement types that are used for concrete, rather than mortar, are Types _____.
   a. I and II
   b. II and IV
   c. III and V
   d. IV and V

6. The amount of air contained in the mortar mix is determined by _____.
   a. the amount of sand in the mix
   b. the range of sand particle sizes and grades
   c. the range of lime particle sizes and grades
   d. the amount of portland cement in the mix

7. An admixture that adds workability to mortar is called a(n) _____.
   a. plasticizer
   b. set accelerator
   c. set retarder
   d. antifreeze

8. Mortar that is sometimes used for reconstruction of historic buildings is _____.
   a. Type S
   b. Type M
   c. Type K
   d. Type O

9. For the majority of masonry construction, the most important property of mortar is _____.
   a. plasticity
   b. extent of bond
   c. durability
   d. bond integrity

10. The compressive strength of mortar increases as _____.
    a. the water content increases
    b. the lime content increases
    c. the sand content increases
    d. the cement content increases

11. The water-cement ratios for mortar and concrete are _____.
    a. inversely proportional
    b. fairly similar
    c. identical
    d. very different

12. A mortar joint made with properly proportioned mortar will _____.
    a. expand
    b. shrink
    c. widen
    d. soften

13. The major cause of improper proportioning of materials when mixing mortar is _____.
    a. poor-quality material
    b. use of weather-damaged materials
    c. incorrect mixing directions
    d. poor work techniques

14. In cold weather, consider heating water for mortar mixing when the air temperature is below _____.
    a. 50°F
    b. 45°F
    c. 40°F
    d. 35°F

15. Tempering mortar is permitted only _____.
    a. to replace moisture lost through evaporation
    b. if mortar begins to stick to the trowel
    c. when the air temperature is higher than 95°F
    d. after the mortar batch is more than two hours old

16. Efflorescence on a masonry surface usually appears _____.
    a. before the mortar hardens
    b. soon after the wall is built
    c. within three months
    d. after one year

17. Sand for use in mortar is usually stored _____.
    a. in a loose pile on the ground
    b. in 50-pound bags
    c. in covered bins
    d. in a shed or other building

18. A mixing specification for mortar of 1:1:6 indicates _____.
    a. 1 part portland cement, 1 part sand, 6 parts water
    b. 1 part portland cement, 1 part lime, 6 parts sand
    c. 1 part sand, 1 part lime, 6 parts portland cement
    d. 1 part water, 1 part portland cement, 6 parts sand

19. The amount of material in a 94-pound bag of portland cement is _____.
    a. 2 cubic feet
    b. 1½ cubic feet
    c. 1 cubic foot
    d. ½ cubic foot

20. To prevent mortar from caking on the mixing paddles, _____.
    a. wet the drum with a small amount of water before you add dry material
    b. mix all the dry material first, then add the water
    c. fill the drum with water, then add the dry materials
    d. hose them down after mixing each batch

Trade Terms Quiz

Fill in the blank with the correct term that you learned from your study of this module.

1. Lime reduced by mixing with water to a safe form that can be used in the production of mortar is called _______.
2. _______ is the adhesion of mortar to masonry units as determined by its lime and moisture content.
3. The ability of mortar to keep sufficient water in the mix to enhance plasticity and workability and to decrease the need to temper is called its _______.
4. _______ is the property of mortar to remain soft and plastic long enough to allow the mason to place and align masonry units and strike off the mortar joints before the mortar hardens completely.
5. An admixture added to mortar to increase microscopic air bubbles in mixed mortar is called _______.
6. The chemical reaction between cement and water that hardens the mortar, which requires the presence of water and optimal air temperatures is called _______.
7. _______ is cement that has been modified by adding lime and other materials.
8. The ability of mortar to flow like a liquid and not form cracks or break apart is called _______.

Trade Terms:

Air-entraining
Extent of bond
Hydration
Masonry cement
Plasticity
Slaked lime
Water retention
Workability

Cornerstone of Craftsmanship

Todd B. Hartsell

Masonry Instructor
Central Cabarrus High School

How did you get started in the construction industry?
My family has been in the residential masonry business for over 80 years, starting with my grandfather, and including my father, uncles, brothers, and cousins. I also took masonry classes in the high school where I currently teach.

Who or what inspired you to enter the industry? Why?
When I was growing up, I got to see the great skill that it took my father, Carl Edward Hartsell, to create the beautiful work that he did. That's what drove me to enter the trade. I worked in the industry as a mason for 12 years, and when my former masonry instructor retired in 1997, I became the masonry teacher at Central Cabarrus High School in Concord, North Carolina.

What do you enjoy most about your career?
I enjoy teaching the skill to young people, emphasizing that masonry is an art that takes years of practice to master. I also enjoy the fact that you can look back on your work and your craftsmanship for years and decades, and share it with friends and family.

Why do you think training and education are important in construction?
Proper training is vital for ensuring quality craftsmanship as well as the future of the masonry and construction industries.

Why do you think credentials are important in construction?
Credentials are a record of training that an apprentice has received in the trade and are a good resource for an employer to use when looking for new employees. They document the knowledge and training that a potential employee has already earned.

How has training/construction impacted your life and career?
By taking masonry courses in high school, training in the apprenticeship program, and working in the field, I have learned to appreciate the history of masonry as shared by experienced masons, and in turn to share those experiences along with my own job experiences and skills with younger, less experienced masons. As an instructor for the past 15-plus years, I've have trained many young people who have made masonry their career. Several of my students have won state and national SkillsUSA competitions and I love to see young people work hard in a great trade. I feel that I'm just giving back to others what the industry did for me.

Would you recommend construction as a career to others?
Yes! It is very rewarding to help build new homes or commercial buildings while making a good living, as well as seeing the fruits of your labor when the project is complete.

What does craftsmanship mean to you?
Craftsmanship is the skill it takes to create the final product, which takes experience to perform the kind of art that can only be done by a highly skilled and trained craftsman.

Trade Terms Introduced in This Module

Air-entraining: A type of admixture added to mortar to increase microscopic air bubbles in mixed mortar. The air bubbles increase resistance to freeze-thaw damage.

Extent of bond: The adhesion of mortar to masonry units, as determined by its lime and moisture content.

Hydration: A chemical reaction between cement and water that hardens the mortar. Hydration requires the presence of water and optimal air temperatures of between 40°F and 80°F.

Masonry cement: Cement that has been modified by adding lime and other materials.

Plasticity: The ability of mortar to flow like a liquid and not form cracks or break apart.

Slaked lime: Lime reduced by mixing with water to a safe form that can be used in the production of mortar.

Water retention: The ability of mortar to keep sufficient water in the mix to enhance plasticity and workability and to decrease the need to temper.

Workability: The property of mortar to remain soft and plastic long enough to allow the mason to place and align masonry units and tool off the mortar joints before the mortar hardens completely.

Additional Resources

This module presents thorough resources for task training. The following resource material is suggested for further study.

Concrete Masonry Handbook for Architects, Engineers, and Builders, Fifth Edition. 1991. William C. Panarese, Steven H. Kosmatka, and F. A. Randall, Jr. Skokie, IL: Portland Cement Association.

Masonry Cement Mortar: Its Proper Use in Construction. 1993. Pat Howley. Murray Hill, NJ: ESSROC Materials.

Masonry Construction, First Edition. 1996. David L. Hunter, Sr. Upper Saddle River, NJ: Prentice Hall.

Technical Note TN1, *Cold and Hot Weather Construction*. 2006. Reston, VA: The Brick Industry Association. **www.gobrick.com/Portals/25/docs/Technical%20Notes/TN1.pdf**

Technical Note TN8, *Mortars for Brickwork*. 2008. Reston, VA: The Brick Industry Association. **www.gobrick.com/Portals/25/docs/Technical%20Notes/TN8.pdf**

Technical Note TN8B, *Mortars for Brickwork: Selection and Quality Assurance*. 2006. Reston, VA: The Brick Industry Association. **www.gobrick.com/Portals/25/docs/Technical%20Notes/TN8B.pdf**

Technical Note TN20, *Cleaning Brickwork*. 2006. Reston, VA: The Brick Industry Association. **www.gobrick.com/Portals/25/docs/Technical%20Notes/TN20.pdf**

Figure Credits

Courtesy of Dennis Neal, FMA&EF, CO05, Figure 1, Figure 3 (right), Figure 4 (top), Figure 6, Figure 8, SA04, Figure 17, Figure 18

Courtesy of Portland Cement Association, SA01, Figure 7 (bottom), Figure 12,

Courtesy of Graymont Inc., Figure 2, SA02

Courtesy of RLT Construction, Figure 3 (left)

Quikrete, Figure 4 (middle, bottom), Figure 5

Courtesy of the Brick Industry Association, Table 5, Figure 11, Figure 13, Table 6

Courtesy Kraft Tool Co., SA03

The QUIKRETE Companies, Figure 14

Section Review Answer Key

Answer	Section Reference	Objective
Section One		
1. b	1.1.1	1a
2. c.	1.2.0	1b
3. b	1.3.0	1c
4. d	1.4.0	1d
5. c	1.5.0	1e
Section Two		
1. a	2.1.1	2a
2. d	2.2.2	2b
Section Three		
1. a	3.1.0	3a
2. d	3.2.0	3b
3. c	3.3.0	3c
Section Four		
1. d	4.1.0	4a
2. c	4.2.0	4b
3. d	4.3.1	4c

NCCER CURRICULA — USER UPDATE

NCCER makes every effort to keep its textbooks up-to-date and free of technical errors. We appreciate your help in this process. If you find an error, a typographical mistake, or an inaccuracy in NCCER's curricula, please fill out this form (or a photocopy), or complete the online form at **www.nccer.org/olf**. Be sure to include the exact module ID number, page number, a detailed description, and your recommended correction. Your input will be brought to the attention of the Authoring Team. Thank you for your assistance.

Instructors – If you have an idea for improving this textbook, or have found that additional materials were necessary to teach this module effectively, please let us know so that we may present your suggestions to the Authoring Team.

NCCER Product Development and Revision

13614 Progress Blvd., Alachua, FL 32615

Email: curriculum@nccer.org
Online: www.nccer.org/olf

❑ Trainee Guide ❑ Lesson Plans ❑ Exam ❑ PowerPoints Other ____________

Craft / Level: Copyright Date:

Module ID Number / Title:

Section Number(s):

Description:

Recommended Correction:

Your Name:

Address:

Email: Phone:

28105-13

Masonry Units and Installation Techniques

Overview

In this module, you will apply what you have learned from the previous modules of *Masonry Level One* in order to build single- and double-wythe masonry walls from concrete masonry units (block) and brick. You will learn about the characteristics of block and brick; how to set up, lay out, and bond block and brick; how to cut block and brick; how to lay and tool block and brick; and how to clean block and brick once they have been laid. You will also learn about masonry reinforcements and accessories that you will use on the job to lay block and brick professionally and safely.

Module Six

Trainees with successful module completions may be eligible for credentialing through NCCER's National Registry. To learn more, go to **www.nccer.org** or contact us at **1.888.622.3720**. Our website has information on the latest product releases and training, as well as online versions of our *Cornerstone* magazine and Pearson's product catalog.

Your feedback is welcome. You may email your comments to **curriculum@nccer.org**, send general comments and inquiries to **info@nccer.org**, or fill in the User Update form at the back of this module.

This information is general in nature and intended for training purposes only. Actual performance of activities described in this manual requires compliance with all applicable operating, service, maintenance, and safety procedures under the direction of qualified personnel. References in this manual to patented or proprietary devices do not constitute a recommendation of their use.

28105-13

Masonry Units and Installation Techniques

Objectives

When you have completed this module, you will be able to do the following:

1. Describe how to install concrete masonry units.
   a. Identify the characteristics of concrete masonry units.
   b. Explain how to set up, lay out, and bond concrete masonry units.
   c. Explain how to lay and tool concrete masonry units.
   d. Explain how to clean concrete masonry units.
2. Describe how to install brick.
   a. Identify the characteristics of brick.
   b. Explain how to set up, lay out, and bond brick.
   c. Explain how to lay and tool brick.
   d. Explain how to clean brick.
3. Describe how to cut concrete masonry units and brick.
   a. Explain how to cut with chisels and hammers.
   b. Explain how to cut with masonry hammers.
   c. Explain how to cut with saws and splitters.
   d. Explain how to check units and cuts.
4. Describe how to install masonry reinforcement and accessories.
   a. Describe how to install masonry reinforcements.
   b. Describe how to install masonry accessories.

Performance Tasks

Under the supervision of your instructor, you should be able to do the following:

1. Lay a dry bond for block.
2. Tool a bed joint for block.
3. Lay block to the line in courses that are true for height, level, plumb, and range.
4. Build a block lead.
5. Lay a dry bond for brick.
6. Tool a bed joint for brick.
7. Lay brick to the line in courses that are true for height, level, plumb, and range.
8. Build a brick lead.
9. Accurately cut block using the following tools:
   - A masonry hammer
   - A brick set
   - A power saw
   - A splitter
10. Accurately cut brick using the following tools:
    - A masonry hammer
    - A brick set
    - A power saw
    - A splitter

Trade Terms

- Bond
- Closure unit
- Corner lead
- Crowd the line
- Dry bond
- Frogged
- Hurricane strap
- Lintel
- Racking
- Range
- Return
- Slack to the line
- Spread
- Tail
- Veneer tie

Industry-Recognized Credentials

If you're training through an NCCER-accredited sponsor, you may be eligible for credentials from NCCER's Registry. The ID number for this module is 28105-13. Note that this module may have been used in other NCCER curricula and may apply to other level completions. Contact NCCER's Registry at 888.622.3720 or go to **www.nccer.org** for more information.

Contents

Topics to be presented in this module include:

Figures and Tables

Section One

1.0.0 Concrete Masonry Unit Materials and Installation

Objective

Describe how to install concrete masonry units.

a. Identify the characteristics of concrete masonry units.
b. Explain how to set up, lay out, and bond concrete masonry units.
c. Explain how to lay and tool concrete masonry units.
d. Explain how to clean concrete masonry units.

Performance Tasks 1 through 4

Lay a dry bond for block.

Tool a bed joint for block.

Lay block to the line in courses that are true for height, level, plumb, and range.

Build a block lead.

Trade Terms

Bond: Laying out masonry units to establish spacing.

Corner lead: A lead built at a corner in which each course of masonry is shorter than the course below it.

Dry bond: Laying out masonry units without mortar to establish spacing.

Frogged: A style of brick in which a depression in the bedding surface of a brick lightens its weight.

Range: To align the masonry unit or wall along the plane or face of the wall. Walls should be ranged from one corner to another.

Spread: To lay a bed of mortar in a line for a course of masonry units.

Tail: To check the spacing of head joints by checking the diagonal edges of the courses on a lead or corner.

Most masonry materials are made of clay, concrete, or stone. Concrete masonry units (CMUs), commonly referred to as block, are classified into six types:

- Loadbearing concrete block
- Nonbearing concrete block
- Concrete brick
- Calcium silicate units
- Pre-faced or prefinished concrete facing units
- Concrete units for manholes and catch basins

WARNING!

Always wear appropriate personal protective equipment when cutting concrete masonry units.

1.1.0 Identifying the Characteristics of Concrete Masonry Units

Each of the six types of CMU has its own ASTM (American Society for Testing and Materials) International standards for performance characteristics. The standards describe the expected performance of the CMU in compressive strength, water absorption, loadbearing, and other characteristics. Block is valuable for its fire resistance, sound absorption, and insulation value.

1.1.1 Compressive Strength

The compressive strength of a CMU is a measure of how much weight it can support without collapsing. These figures are set according to ASTM test results. The tests are performed on a specific shape, size, and weight of CMU. The CMU tested then becomes the standard for that particular compressive strength. Compressive strength is measured in pounds per square inch (psi). This is the weight that the unit, and the structure made of the units, can support. A structure supports less imposed weight than the sum of its unit strength because the structure has to support itself as well.

The quality of a mason's work is an important part of how a masonry structure performs and whether it meets its compressive strength specification. Tests have shown that the compressive strength of a loaded wall is about 42 percent of the compressive strength of a single CMU when the mason uses a face-shell mortar bedding. When the mason uses a full mortar bedding, the compressive strength increases to about 53 percent. The engineer factors these components into the equation for picking the CMU. The factoring is contained in the job specifications. This is another reason that specifications are important.

1.1.2 Moisture Absorption and Content

Moisture in the CMU has an effect on shrinking and cracking in the finished structure. Generally, the lower the moisture, the less likely the units are to shrink after they are set. Acceptable moisture content and absorption rates are set by ASTM standards and local codes. Manufacturers specify that their units meet ASTM or other standards. Unit tests are made to make sure they do. For the mason, this means keeping the CMU dry. Never wet the CMUs immediately before or during the time they are to be laid. Stockpile them on planks or pallets off the ground. Use plastic or tarpaulin covers for protection against rain and snow.

When stopping work, cover the tops of masonry structures to keep rain or snow off. Be sure moisture does not get into cavities between wythes. When laying CMU for interior use, dry them before laying. They should be dried to the average condition to which the finished wall will be exposed.

1.1.3 Concrete Block Characteristics

Block is produced in four classes: uncored loadbearing, uncored nonbearing, cored loadbearing, and cored nonbearing. As you learned in *Introduction to Masonry,* block is classified as solid if no more than 25 percent of its surface is cored, which means that the brick has holes extending through it to reduce weight. If the block has no holes at all, it is called uncored. Block comes in multiple weights; two of the most common are normal and lightweight. Lightweight CMUs are made with fly ash, pumice, and scoria or other lightweight aggregate. Loadbearing and appearance qualities of the two weights are similar. The major difference is that lightweight CMUs are easier and faster to lay. *Figure 1* illustrates the various parts of three common types of block.

Most CMUs are governed by *ASTM C90, Standard Specification for Loadbearing Concrete Masonry Units,* which covers cored loadbearing CMU. At one time, *ASTM C90* specified Grades N and S. Grade S has been discontinued for this CMU, and *ASTM C90* CMU is now ungraded. The ASTM does specify a minimum compressive strength of 800 psi, however. Uncored loadbearing CMU must be either Grade N or S. Grade N has a compressive strength of 1,500 psi, and the compressive strength of Grade S is 1,000 psi.

Block comes in modular sizes, with colors determined by the cement ingredients, the aggregates, and any additives. The basic CMU is called a stretcher. It has a nominal face size of 8 to 16 inches with a standard ⅜-inch mortar joint. The most commonly used CMU has a nominal width of 8 inches, but 4-, 6-, 10-, and 12-inch widths are also common. Three modular brick have the same nominal height as one nominal 8-inch CMU. Two nominal brick lengths equal one nominal CMU length. So, laying one stretcher covers the same area as laying six modular brick.

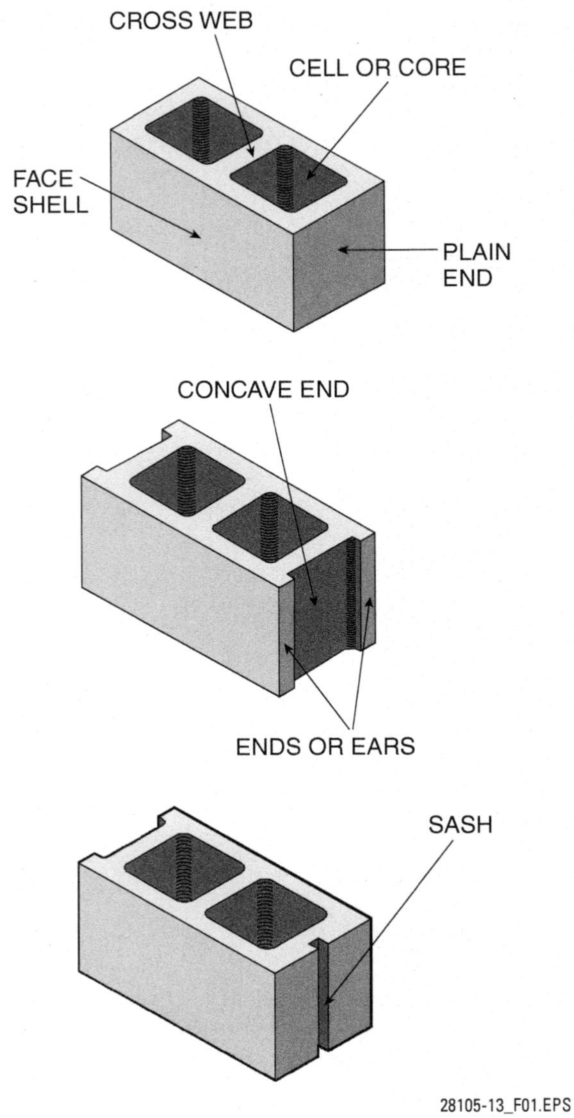

Figure 1 Parts of CMUs.

Block comes in a variety of shapes to fit common and special purposes. *Figure 2* shows a sampling of CMU sizes and shapes. Most cores are tapered slightly to provide a larger bed-joint surface. Block edges may be flanged, notched, or smooth. There are local variations as well, with some shapes available only in specific parts of the country.

Nonstructural CMU is specified under *ASTM C129,* listing a minimum compressive strength of 500 psi. This CMU is used for screening and nonbearing partition walls. Elegantly surfaced, uncored nonstructural CMU is often used as a veneer wall for wood, steel, or other backing. Cored nonstructural CMU is made with pattern cores much like clay tile. Pattern-core CMU comes in a variety of shapes and modular sizes, and is commonly used for screen walls.

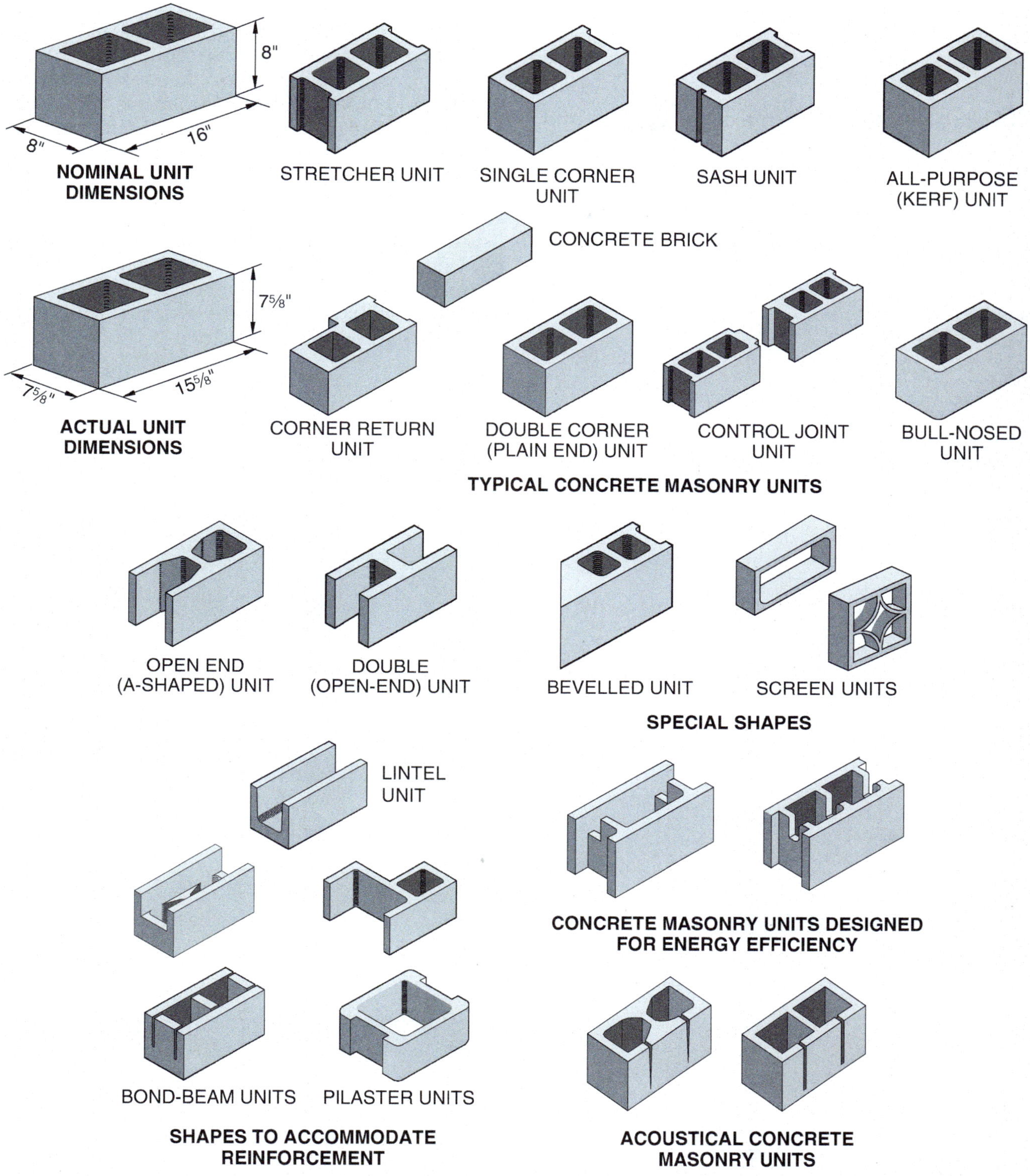

Figure 2 Common CMU shapes.

1.1.4 Concrete Brick

Concrete brick can be used in the same way as clay brick. It is available in cored, uncored, and **frogged** styles (*Figure 3*). A frog is a depression in the bedding surface of a brick that lightens the weight of the brick. It also makes for a better mortar joint by increasing the area of mortar contact.

Concrete brick is designed to be laid with a ⅜-inch mortar joint. It comes in many sizes, with the most popular nominal dimensions of 4 to 8 inches. This size gives three courses in a height of 8 inches, like standard modular brick.

ASTM C55 specifies two grades of concrete brick:

- Grade N is used for architectural veneers and facing units in exterior walls. It has high resistance to moisture and frost penetration, and has a compressive strength of 3,000 psi.
- Grade S is also used for architectural veneers and facing units. Grade S has moderate resistance to moisture and frost, and is used in the southern region of the United States. Its compressive strength rating is 2,000 psi.

Slump block has an irregular face resembling stone, as shown in *Figure 4*. In other respects, slump block meets concrete brick standards.

Uncored block is also made from a slump mixture. Because of the greater surface area, the block face is very irregular. Its height, surface texture, and appearance resemble stone.

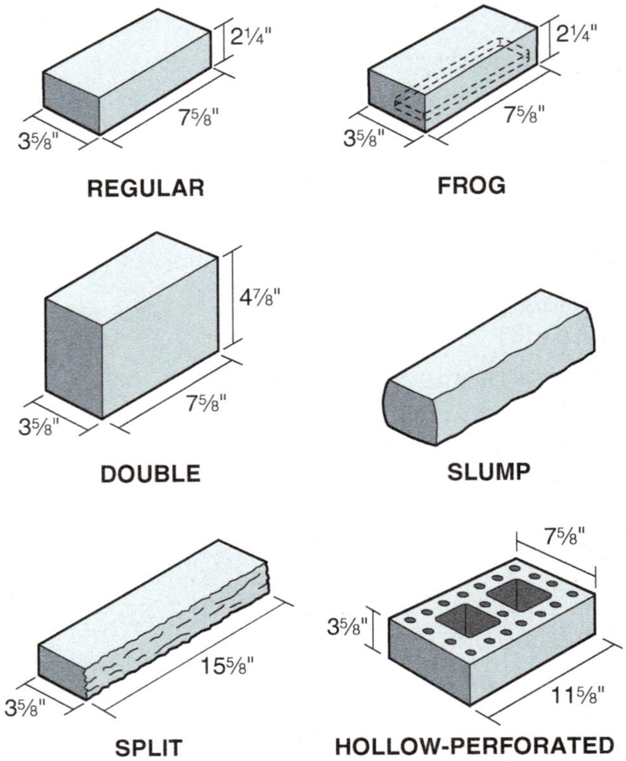

28105-13_F03.EPS

Figure 3 Concrete brick.

28105-13_F04.EPS

Figure 4 A wall made from slump block.

Reinforcing Block Walls

Grout is a mixture of cementitious material and aggregate with enough liquid content to make it flow readily. Grout is often pumped into CMU wall cavities with special pumps.

28105-13_SA01.EPS

28105-13_SA02.EPS

1.1.5 Architectural Block

As you learned in the module *Introduction to Masonry*, architectural block has a variety of surface finishes that affect the unit's texture. Architectural block is often designed to imitate brick and stone. The surface finish allows the block to be used for both structural and finish purposes without the need for veneer (*Figure 5*). Architectural block can be used on interior and exterior walls. The block may be finished on one or both faces. Common types of architectural block facings include split, scored, ribbed, ground, sandblasted, striated (raked), glazed, offset, and slump block. *Figure 6* shows several common 8-inch nominal-dimension architectural block shapes.

The standards for architectural block are covered in *ASTM C90, Standard Specification for Loadbearing Concrete Masonry Units. ASTM C90* provides the minimum thickness requirements for all loadbearing block. Split-face units that are not solidly grouted can be less than the minimum specified thickness when no more than 10 percent but not less than ¾ inch of its shell is less than the minimum thickness. Because the terminology used to describe architectural block can vary from region to region, the National Concrete Masonry Associa-

Figure 5 Varieties of architectural block.

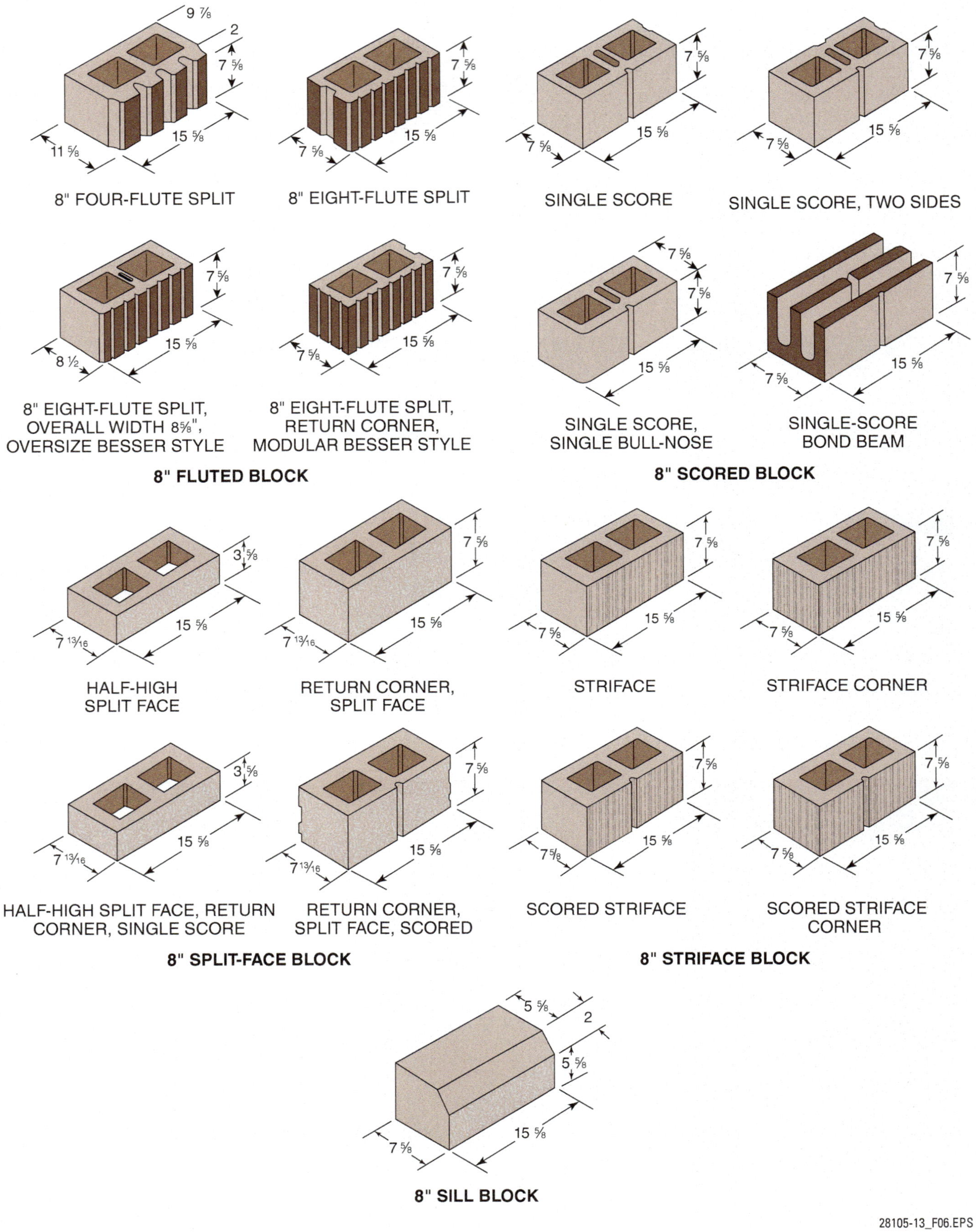

Figure 6 Dimensions (in inches) of common architectural block.

tion (NCMA) developed a standardized nomenclature to describe architectural block (see *Table 1*).

When installing scored or ribbed block, lay the block so that the scores or ribs align vertically. Note that this placement may require the use of a different **bond** pattern, which may affect the wall's ability to withstand structural loads. The projections also make it difficult to tool the mortar properly.

Color and facings can be added to architectural block by the manufacturer during manufacture. The facings are made of resins, portland cement, ceramic glazes, porcelainized glazes, or mineral glazes. The slick facing is easily cleaned. These

Table 1 Standard Nomenclature for Architectural Block

Each unit is described using a three-part code in the following format: XXYYYWWHHLL, where XX describes the number of scores or ribs, YYY describes the architectural finish, and WWHHLL describes the overall nominal unit dimensions for width, height, and length. The various codes are described below.

Scores or Ribs:	
00	No scores or ribs, applicable for any running bond
01	One score, applicable for one-half running bond (units overlap the unit above and below by one-half the unit length)
02	2 scores, applicable for one-third running bond
03	3 scores, applicable for one-half or one-quarter running bond
04	4 ribs, applicable for one-half or one-quarter running bond
05	5 scores, applicable for one-half running bond
06	6 ribs, applicable for one-half running bond
07	7 scores, applicable for one-half or one-quarter running bond
08	8 ribs, applicable for one-half or one-quarter running bond
Architectural Finish:	
BN1	Bull-nose unit with bull-nose radius of 1" (25 mm)
BN2	Bull-nose unit with bull-nose radius of 2" (51 mm)
SCV	Vertically scored unit
GRF	Ground face unit
MDC	Circular ribs, rib projects beyond the overall unit thickness
MNC	Circular ribs, rib projection included in overall unit thickness
MDR	Rectangular ribs, rib projects beyond the overall unit thickness
MNR	Rectangular ribs, rib projection included in unit thickness
STR	Striated unit
STS	Striated unit, 1" (25-mm) uniform striation pattern
STT	Striated unit, 1/16" (1.6-mm) uniform striation pattern
SPF	Split-face unit
NPF	Split-face ribbed unit, rib projections included in unit thickness
SLP	Slump block
**Q	Locally provided product

28105-13_T01.EPS

units are popular for use in gyms, hospital or school halls, swimming pools, and food processing plants. Along with texture, color and facings offer designers a wide variety of aesthetic effects (see *Figure 7*). Block can also appear to change color when lit by electric and natural light, which can give even standard gray block a warmer appearance.

1.1.6 Calcium Silicate Units

Calcium silicate units (*Figure 8*) are made of a mixture of sand, water, lime, and calcium silicate. The calcium silicate acts as a leavening agent and creates gas bubbles in the mix. The units are not fired or cured in a kiln but cured in an autoclave, or pressurized tank, with pressurized live steam. In the autoclave, the lime reacts with the silicate to bind the sand particles into a very lightweight, strong unit. ASTM performance specifications cover this type of block with grading standards identical to those for traditional products.

The units are also called sand-lime brick or aerated block. They are used extensively in Europe, Australia, Mexico, and the Middle East. In the United States, this brick is used mostly in flues, chimney stacks, and other high-temperature locations. They resist sulfates in soil, do not effloresce, and are not damaged by repeated freeze-thaw cycles. The block is now manufactured in the United States in a variety of sizes for commercial or home building.

28105-13_F07.EPS

Figure 7 Aesthetic effects of colored and glazed architectural block.

28105-13_F08.EPS

Figure 8 Calcium silicate units.

1.2.0 Setting Up, Laying Out, and Bonding Concrete Masonry Units

Setting up the job and laying out the structure are two distinct steps. Both must be complete before the mason can start to lay units. Setting up refers to preparing the materials and the site. Laying out refers to establishing the wall lines for the masonry structure.

Masons deal with four types of bonds:

- A mortar bond is made by the joining of mortar and a masonry unit. The strength of this bond depends on the mortar.
- A pattern bond is a pattern formed by masonry units and mortar joints on the face of a floor or wall. Unless it is the result of a structural bond, a pattern bond is purely decorative.
- A structural bond is made by interlocking or tying masonry units together so they act as a single structural unit.
- A structural pattern bond is the result of a structural bond that forms a pattern as well as a bond. Most traditional pattern bonds began as structural pattern bonds.

Note that it is difficult to make a distinction between a structural bond and a pattern bond. The act of overlapping or interlocking masonry to create a structural bond also creates a pattern. Defining a particular pattern as a structural bond or a structural pattern bond depends on whether it is used to simply form a veneer or to tie masonry wythes together.

1.2.1 Setting Up

Masonry setup work starts when the contract for the job is signed. The first step for the masonry contractor is to read the contract, construction drawings, and specifications. The next step is to review the schedule, plus any standards and codes cited in the contract. When these steps have been completed, the masonry contractor is ready to estimate the workers, materials, and equipment needed for that job. Review the information in the module *Measurements, Drawings, and Specifications* to get a clearer idea of what this work entails.

The next step is to estimate again, check figures, and order the masonry equipment and materials. A visit to the job site and discussion with the engineer or construction foreman will give the masonry contractor an idea of where and how to store masonry materials. The masonry contractor must specify a delivery date and location on site. Materials must be stored close to where they will be used and protected from the weather. The crew must be hired and briefed. Then, the work is ready to begin, but there is still a lot to do before laying the first masonry unit. The following checklist shows some of the preliminary procedures:

- Check that all materials are stored close to workstations and protected from moisture. Block must be laid dry in order to avoid shrinkage upon drying.
- Place mortar pans and mortarboards by workstations. If you are using scaffold, place the pans at intervals on the scaffold near the point of final use.
- Stockpile units on each side of the mortar pans and at intervals along the wall line. If you are using scaffold, place units along the top of the scaffold near the point of final use. Stockpiles should require the mason to move block as little as possible once laying starts.
- Stack block in stockpiles with the bottom side down, just as they will be laid in the wall. The top of the block has a larger shell and web. Stack faced units with the faced sides in the direction they will go, just as they will be laid in the wall. Stack all units so the mason will move or turn them as little as possible.
- Check all scaffold for proper assembly and position. Ensure that braces are attached and planks are secured at each end. Scaffold should be level and close to the wall.
- Check all mechanical equipment, power tools, and hand tools. Make sure they are clean, in good condition, and the right size for the job.

The contractor will typically assign a mason tender to keep mortar pans full and working stacks of masonry units at a comfortable height for the mason. The objective of all setup work is to make everything efficient and convenient for the masons once they begin laying.

1.2.2 Job Layout

Laying out the wall or other masonry element calls for a review of the plans and specifications. The first steps are to plan out the work, locate where it will go, and then establish bond.

1.2.3 Planning

Planning out the work means the mason needs to check the plans for wall lengths, heights, door dimensions, and window openings. What pattern or bond is specified? What is the nominal size of the masonry unit? How are openings to be treated?

The Law of Gravity Applies

Cubes of CMU and brick must be placed on level ground. If a 1,500-pound load of brick collapses, it can cause serious injuries and expensive damage.

28105-13_SA03.EPS

Are the dimensions and the masonry units on the modular scale of 4-inch increments?

After answering these questions, the mason can draw a rough layout of the wall and lay out the bond pattern. If the job is sized on the modular grid, graph paper might be handy for the spacing drawing. This drawing can show where the bond pattern will start and how it will fit around the specified openings. From this drawing, the mason can count and calculate how many masonry units to cut.

The question of whether the designer did or did not use the modular grid becomes important. *Figure 9* shows door and window openings located in a running bond. Notice how the openings are set off the modular grid in the diagram on the left. The amount of cutting is significant compared to the example on the right. Using many small units reduces wall strength as well. Sometimes the mason can persuade the designer or engineer to shift the openings slightly to avoid so much cutting. In other cases, the dimensions are critical and cannot be changed.

Note that the dimensions provided on plans are nominal dimensions. As you have learned, the nominal dimension of a masonry unit is the size of the unit plus the thickness of one standard (½ inch or ⅜ inch) mortar joint. In *Figure 9*, the window and door measurements in A and B are nominal measures, while drawing C shows the actual dimensions. *ACI 530/ASCE 5/TMS 402, Building Code Requirements for Masonry Structures*, specifies that a mortar joint cannot be thicker than ¾ inch (as you have already learned elsewhere in *Masonry Level One*, ACI is the American Concrete Institute, ASCE is the American Society of Civil Engineers, and TMS is The Masonry Society). Project specifications can be even more stringent. Therefore, a mason cannot simply use thicker joints to make up for discrepancies. Block may have to be cut in order to make the structure meet the required height or length.

1.2.4 Locating

The mason will check the location first. Masonry walls are supported by a footing or other support, usually made of concrete. The surveyor or foreman will mark the corners of the structure or slab. On some jobs, the foreman will drive nails into the wall footing to mark the building line. *Figure 10* shows a foundation layout with a footing plan for block foundation walls. At the job site, the first thing to do is to locate the footing. Next, brush it off. Remove any dried concrete particles or large aggregates to ensure a good bond between the footing and the first course.

Check that the footing is level. If the footing is not within project tolerances, it must be fixed. Do not apply a mortar joint thicker than ¾ inch to level the first course. This can result in a joint too thick to carry the load of the wall. If the footing is out of level, notify your supervisor.

The next step is to locate the walls. Take measurements from the foundation or floor plan and transfer them to the foundation, footing, or floor slab. All measurements on the plans must be followed accurately. Be sure the door openings are placed exactly and the corners are on the footings as shown on the detailed drawings. Check to see that you are not confusing the measurements for the interior and exterior walls. If it appears that the wall cannot be laid out exactly because of errors in the footing, notify your supervisor.

Weather Considerations

Mortar temperature should be between 40°F and 80°F in order for proper hydration to occur. In temperatures below 40°F, it may be necessary to heat the water and/or sand in order to keep the mortar at a high-enough temperature. One method often used to heat sand is to pile it over a large-diameter pipe, such as a culvert pipe, with a fire inside the pipe.

28105-13_SA04.EPS

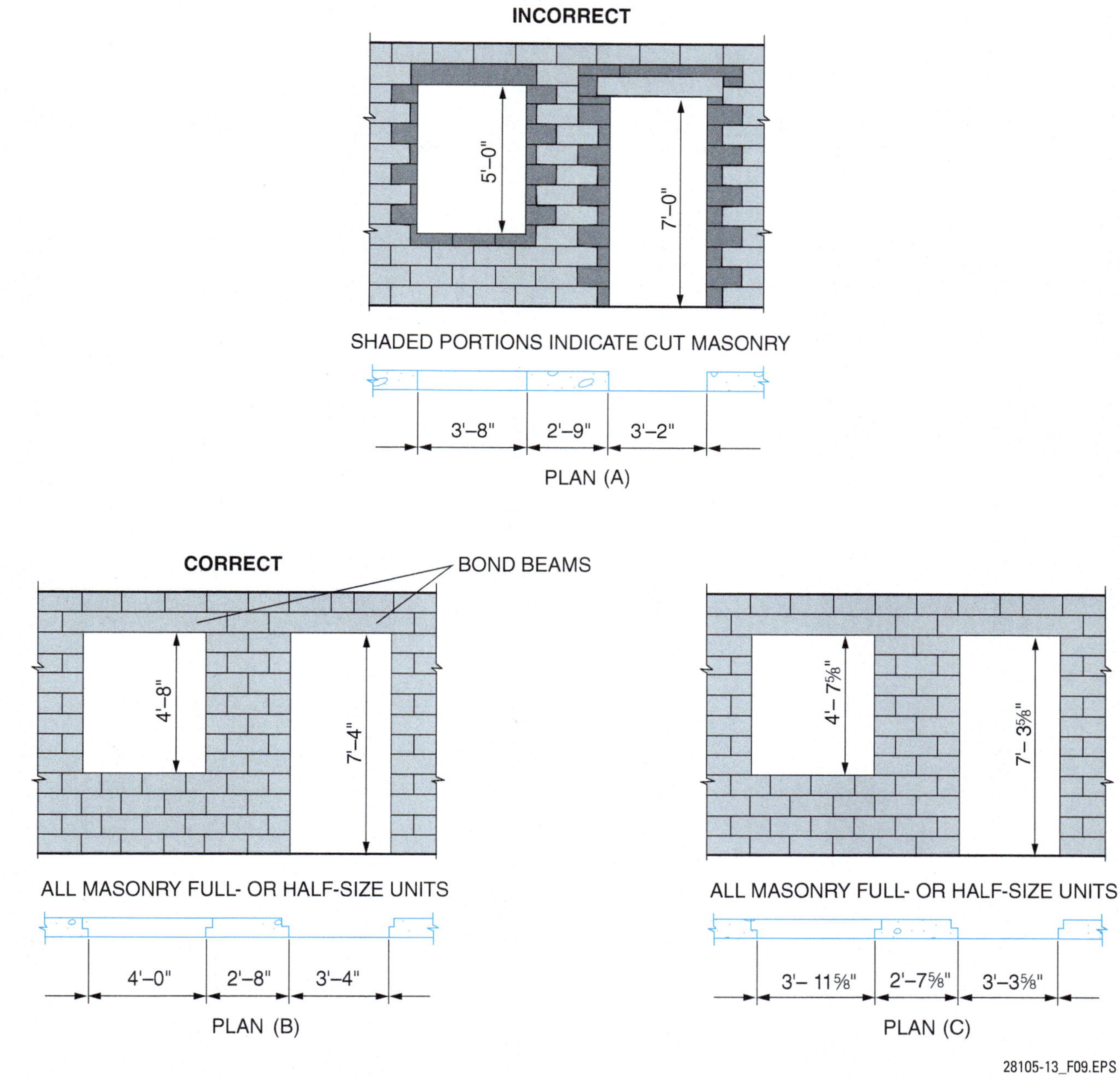

Figure 9 Door and window openings.

The next task is to establish two points, corner-to-corner or corner-to-door. Then, run a chalkline between the two points and snap it on the footing or foundation. Because a chalkline is easily erased, mark key points along the chalkline with a marking pencil, nail, or screwdriver. This will allow resnapping the chalkline without refinding the points.

Mark the entire foundation for walls, openings, and control joints. After snapping the chalkline, mark over the chalk with a marker or nail. Once you have completed all markings, check the measurements of the markings against the foundation plan. Again, be sure you are reading the correct measurements. If there is to be a veneer wall, check that you are dimensioning the veneer, not the backing wythe. If everything does not fit precisely and exactly, it must be done over. It is easier to redo measurements than to redo a masonry wall.

1.2.5 Dry Bonding

Dry bonding is an alternative to measuring to establish the positioning of the masonry units. Starting with the corners, lay the first course with no mortar. This method is also called a **dry bond**. This is a visual check of how the units will fit. It also checks the pattern bond drawing and the calculations for cut units. For block, it allows for adjustments to be made while maintaining uniform head joints. It is especially important to dry-bond

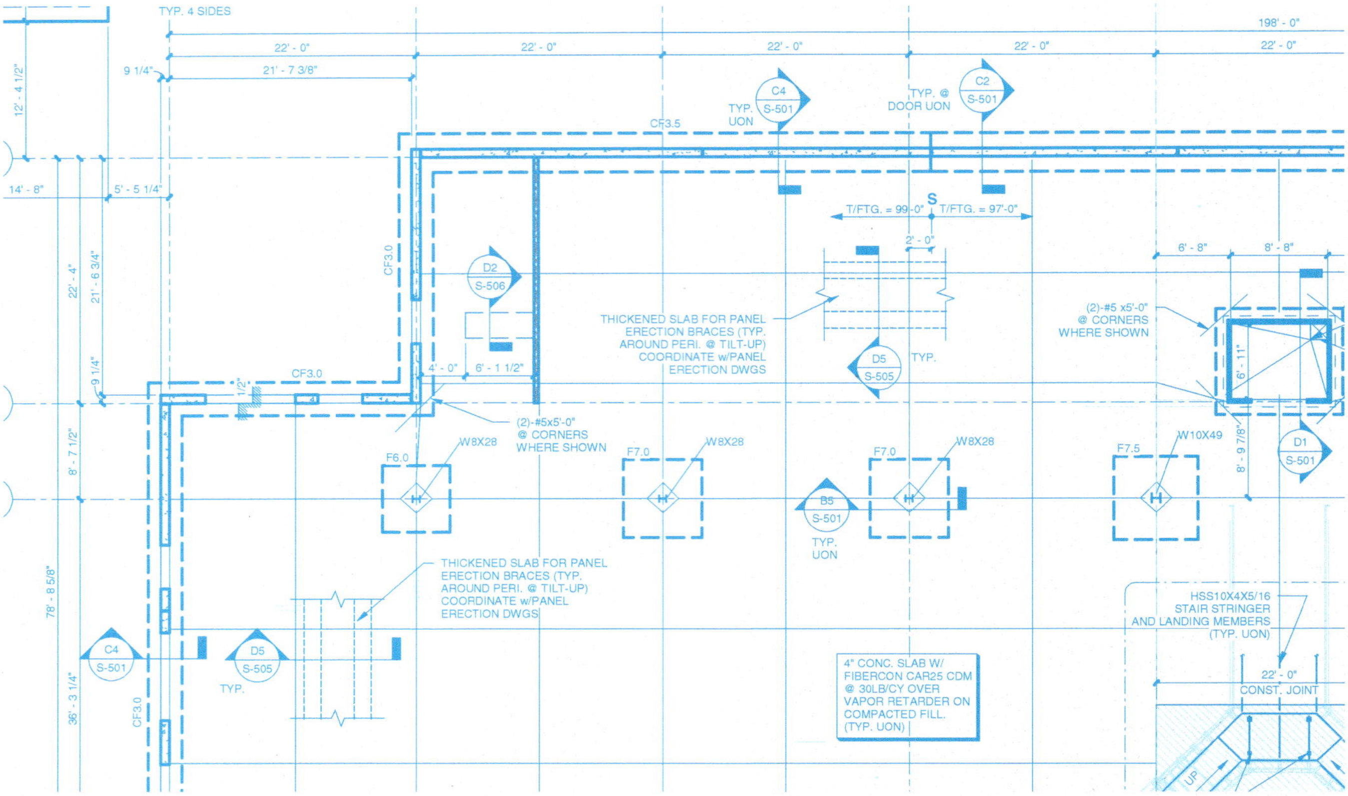

FOOTING SCHEDULE					
Mark	**Width**	**Length**	**Thickness**	**Reinforcing**	**Remarks**
CF2.0	2' - 0"	CONT.	1' - 0"	2-#5 CONT. w/ #3 @ 48" TRANS.	
CF2.5	2' - 6"	CONT.	1' - 0"	2-#5 CONT. w/ #3 @ 48" TRANS.	
CF3.0	3' - 0"	CONT.	1' - 0"	3-#5 CONT. w/ #3 @ 48" TRANS.	
CF3.5	3' - 6"	CONT.	1' - 0"	3-#5 CONT. w/ #3 @ 48" TRANS.	
F6.0	6' - 0"	6' - 0"	1' - 2"	6-#6 EA. WAY BOT.	Rectangular Footings
F6.0T	6' - 0"	6' - 0"	1' - 2"	6-#6 EA. WAY TOP & BOT.	Rectangular Footings
F7.0	7' - 0"	7' - 0"	1' - 6"	7-#6 EA. WAY BOT.	Rectangular Footings
F7.5	7' - 6"	7' - 6"	1' - 6"	8-#6 EA. WAY BOT.	Rectangular Footings
RF6x8	6' - 0"	8' - 0"	1' - 2"	#6 EA.@12"o/c EA. WAY TOP & BOT.	Rectangular Footings
RF6x9	6' - 0"	9' - 0"	1' - 2"	#6 EA.@12"o/c EA. WAY TOP & BOT.	Rectangular Footings
RF6x11.5	6' - 0"	11' - 6"	1' - 2"	#6 EA.@12"o/c EA. WAY TOP & BOT.	Rectangular Footings

28105-13_F10.EPS

Figure 10 Foundation plan.

block across openings to ensure proper layout and to make any necessary adjustments in the mortar joints.

From the corners, the mason lays units along the wall markings for the entire foundation, as in *Figure 11*. Check the specifications for the size of brick joints. Remember that all block is laid with a ⅜-inch joint. If you run into spacing problems, use the dry-bond method and mark any adjustments on the foundation or wall.

Lay the units through door openings to see how bond will be maintained above the doors. Then check spacing for openings above the first course, such as windows. Do this by taking away units from the first course and checking the spacing for the units at the higher level. These checks will show whether the joint width will work out for each course up to the top of the wall. Use the pattern bond diagram to help you. If spacing has to be adjusted slightly, mark it on the diagram and on the foundation.

28105-13_F11.EPS

Figure 11 Example of a dry bond.

After the units have been laid out correctly, mark the end of every other unit with a marking pencil, directly on the foundation. This will guide you in laying mortar when the dry units are removed.

Once all of this has been done, the mason can use the square to mark the exact location and angle of the corners. The next step is checking the corner layout on the drawings.

The layout of the corner itself is important, especially when you are working with block and modular spacing. The architect will detail the corner layout on the working drawings. Different block layouts, as shown in *Figure 12*, are possible. Each layout takes up a slightly different amount of space. This will affect the modular spacing and determine whether any block will have to be cut. Building the corners as specified is the key to maintaining modular dimensions.

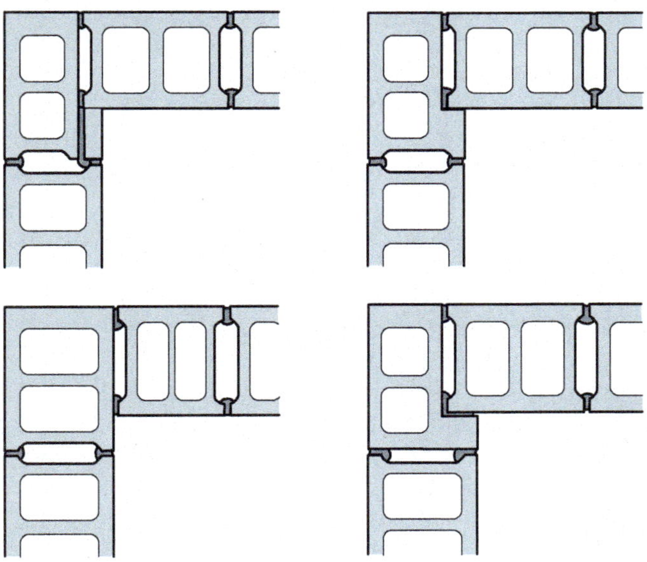

28105-13_F12.EPS

Figure 12 Block corner layouts.

1.2.6 Buttering Block

Block is larger and heavier than brick. Use two hands when lifting block to avoid strain. Block is more demanding than brick in that both units must be buttered to get a good head joint. Block is also more demanding than brick in that it calls for three different types of bed joints. There are two commonly used methods for buttering block. The method that you end up using will depend on a variety of factors such as the standard practice in your area and your level of experience as a mason.

One common method for buttering block is to prepare a full head joint. First, spread a bed joint. Position the first block in the mortar. Then stand two or three units on end next to their bed. Since block is wider at the top than at the bottom, stand the block so that the top sides will be on top when the block is placed. This makes it quick to butter several units at once.

You do not need to fill the trowel with mortar because one block does not take as much mortar as one brick. Butter the ear ends of the standing units. Wrap the mortar around the inside of each ear to help hold it in place. Then butter the ear end of the laid block. Lift the standing block by grasping the webs, or ends, as shown in *Figure 13*. Use two hands if needed. Do not jerk the block, or the mortar will fall off.

Place the block against the buttered, laid block. Tilt the block slightly toward you as you lay it into place, so that you can see the alignment of the cores and edges. Visually check that the edge of the block aligns with the block directly below.

To seat the block, gently press down and forward, so that the mortar squeezes out at the joints. Do not drop the block, but ease it into place. Continue laying the prebuttered block, being sure to butter the ear end of the laid block each time.

28105-13_F13.EPS

Figure 13 Placing a block.

Another method for buttering block is to butter one end, lift the block by the webs with one hand, and butter the other end. This method is not recommended for beginning masons.

After you place the block, cut off the excess mortar with the edge of your trowel. Check for level and plumb with your mason's level. Use the handle of your trowel to gently tap the block into place. After you place the block, the mortar joint spacing should be the standard ⅜ inch for both the bed and head joints.

Do not move the block after it is pushed against its neighbor. If you must move the block, take it off and remortar the bed joint and the head joint. Unlike brick with its solid and complete mortaring, block mortaring is fragile. Because its webs are so small in area compared to its size, block mortar joints are easily disturbed by movement. Do not take the chance of a weakened mortar joint developing a leak in the wall.

1.2.7 Block Bed Joints

Block can use one of three types of bed joint, depending on the purpose. Check the specifications before laying a block wall to confirm which type of bed joint to use. Consider the following:

- If the block is laid as the first course on a footing, it takes a full bed joint. Both face shells and all webbing, including cross webs, require a mortar bed.
- If the block is not to be in a reinforced wall, the bed joint has mortar on the face shells only. *Figure 14* shows this type of mortaring.
- If the block is part of a reinforced wall that will have reinforcing grout in some cores, the block needs a full-block bed joint. This has mortar on the face shells and on the webs, as shown in the detail in *Figure 14*. Mortaring the webs around the cells to be grouted will keep the grout from oozing out of the cores.

Sometimes, the specifications will call for an unreinforced wall to be laid with a full-block bed joint. Mortaring the webs as well as the shells increases the loadbearing strength of the wall. The architect or engineer may have calculated that a full-block bed joint will do the job instead of reinforcement. If you use only a shell bed joint, the wall will not have the calculated strength. This is another reason why it is important to read the specifications.

1.2.8 General Rules

These guidelines were covered in an earlier module, but they bear repeating. The way you work

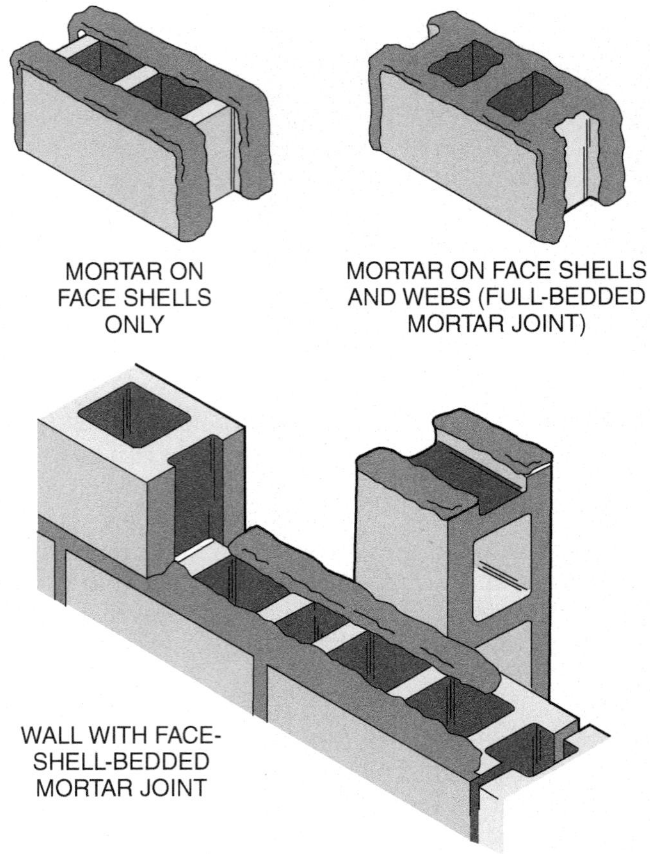

Figure 14 Buttering block.

the mortar determines the quality of the joints between the masonry units. The mortar and the joints form a vital part of the structural strength and water resistance of the wall. Learning these general rules and applying them as you spread mortar will help you build good walls:

- Use mortar with the consistency of mud, so it will cling to the masonry unit, creating a good bond.
- Butter the head joints thoroughly.
- When laying a unit on the bed joint, press down slightly and sideways, so the unit goes against the one next to it.
- If mortar falls off a moving unit, replace the mortar before placing the unit.
- Put down more mortar than the size of the final joint; remember that placing the unit will compress the mortar.
- Adjust the length of the mortar spread to the conditions and materials. Longer spreads may get too stiff to bond properly as water evaporates from them.
- Do not move a unit once it is placed, leveled, plumbed, and aligned, or once the mortar has reached its initial set.

- If a unit must be moved after it is placed, remove all the mortar on it and rebutter it.
- After placing the unit, cut away excess mortar with your trowel, and put it back in the pan, or use it to butter the next joint.
- It is best to use mortar within two hours. After that, the mortar begins to set and does not give a good bond.

1.2.9 Bonding Concrete Masonry Units

Block has its own set of commonly used bond patterns that are distinct from the patterns used for brick. *Figure 15* shows common block bonds, some of which are also seen in brickwork. As with brick, the stack bonds do not provide any structural strength. With block, however, it is simple to reinforce stack bond with grout and steel or grout in the cores.

Pattern variations, such as the coursed ashlar, can be made by using different sizes of block. Modern block walls can also add visual interest through texture and surface designs. Many designers use colors, textures, sizes, and surface designs, alone or in combination with bond patterns, to enhance block walls.

1.3.0 Laying and Tooling Concrete Masonry Units

Laying masonry units is a multistage process. As discussed in previous sections, the first step in any masonry job is reading the specifications. This is followed by planning the layout of the job. The next tasks are to locate and lay out the wall, then do the dry bonding. Dry bonding ensures that the layout will be correct and that the minimum number of cut block will be needed. Then, calculate the number of units to cut, and cut them. For the purpose of this module, assume all work is done in running bond on the modular grid system. Now you are ready to mix the mortar and start the actual laying.

The next tasks are to spread mortar, lay masonry units in place, and check their positioning.

An earlier module gave detailed procedures for spreading bed joints and buttering head joints. The following sections give procedures for positioning individual masonry units, laying to the line, and building corners and leads.

Placing block is similar to placing brick (*Figure 16*). You are more likely to have full head joints if you butter the units on both sides of head joints. Do not drop it, but move it slowly down and forward so it butts against the adjacent unit. Slightly delaying release allows the block to absorb moisture from the mortar, which provides good strength of bond. If the mortar does not ooze from the joints, you are not using enough mortar. There will be voids in the joints, and the wall will eventually leak.

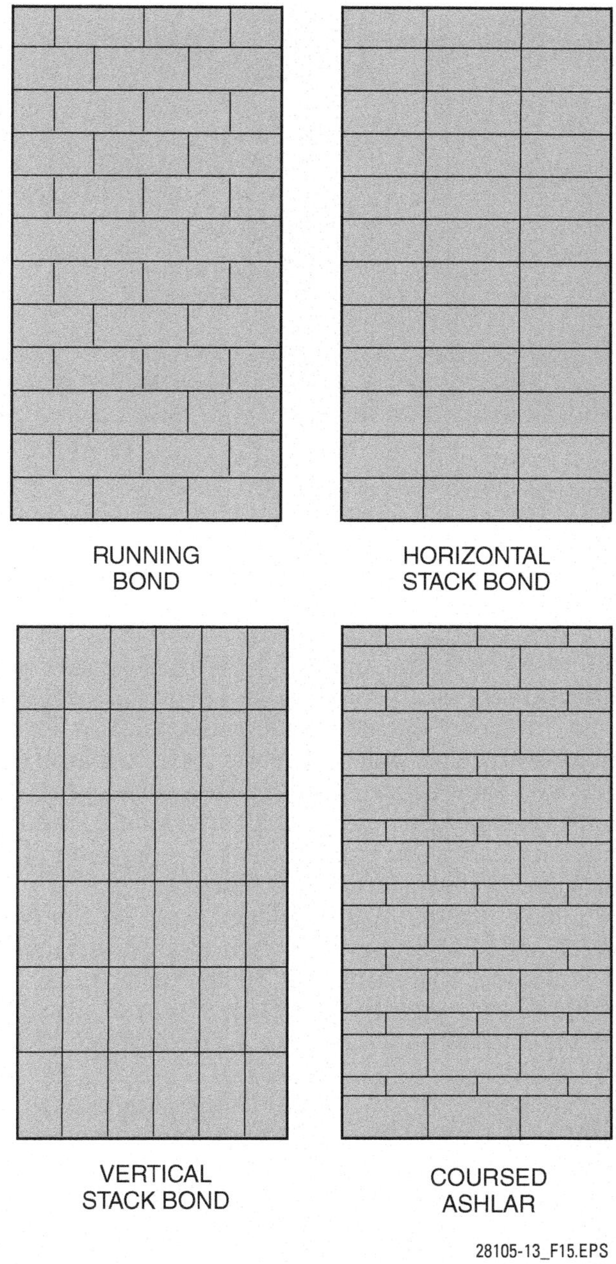

Figure 15 Block bonds.

Tilt the block toward you as you position it. Look down over the edge and into the cores to check alignment with the block underneath. If the block requires adjustment, lightly tap it into place. Check each block for alignment. If the block cannot be adjusted, take it up, remortar it, and reset it.

28105-13_F16.EPS

Figure 16 Placing a block.

1.3.1 Laying to the Line

To keep masonry courses level over a long wall, masons lay the units to a line. Working to a line allows several masons to work on the same wall without the wall moving in several directions. The line is set up between corner poles or corner lead units. The poles or leads must be carefully checked for location, plumb, level, and height. The mason usually works on the same side as the line but can also work from the other side depending on job conditions and experience.

Each masonry unit is placed with its top edge level with the line and 1⁄16 inch away from it. The distance is the same for all types of masonry units. Your eye will get trained to measure that distance automatically after some practice laying to the line.

A mason's line needs to be tied to something that will not move. It must be tied taut at a height that can be measured precisely. The mason's line is attached to corner poles or corner leads by means of line stretchers, line blocks, or line pins.

1.3.2 Setting Up the Line Using Corner Poles

Corner poles allow masons to lay to the line without laying the corners first. Attach the corner pole securely. It must not move as you pull a mason's line from it. For brick veneer walls, the corner pole can be braced against the frame or backing wall. You must check the placement of the corner poles before you string the line. If the pole has course markings, check that they are the correct distance from the footing. If the pole has no markings, transfer markings from your course rule. Make sure you start the measures from the footing.

Step 1 Attach the line to the left pole to start. If the pole has no clamps or fasteners, attach the line with a hitch or half-hitch knot. Stretch it to the right pole and gradually tighten it until it is stretched. Use a hitch or half hitch to secure it, tightening it as you tie. Check that it is at the proper height before you start laying.

Step 2 After laying each course, move the line up to the next course level. Stretch and measure it again. It is critical to make sure the line is at the proper height for each course.

Step 3 Use a modular rule or course spacing rule to check the line height at each end for every course.

1.3.3 Setting Up the Line Using Line Blocks and Stretchers

A mason's line can be set between corner leads or corners laid to mark the ends of the wall. The line can be attached by line blocks, line stretchers, or line pins.

Line blocks (*Figure 17*) have a slot cut in the center to allow the line to pass through. It takes two sets of hands to set up line blocks. The procedure is as follows:

Masonry as Art

Look around, and you will see many examples of the creative side of masonry, like the one in this photo. More than any other construction craft, masonry provides the opportunity to design eye-catching structures.

28105-13_SA05.EPS

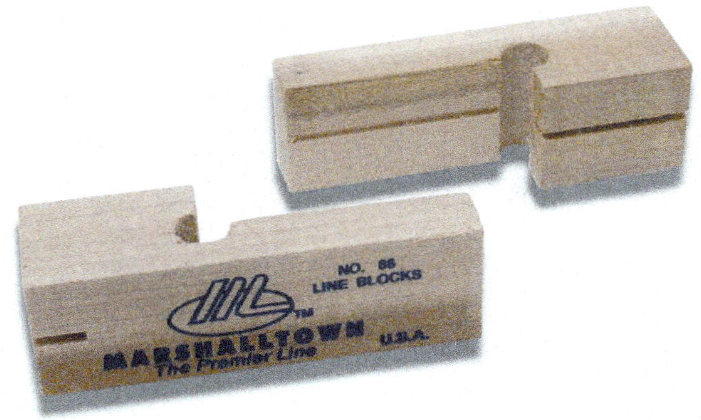

28105-13_F17.EPS

Figure 17 Wood line blocks.

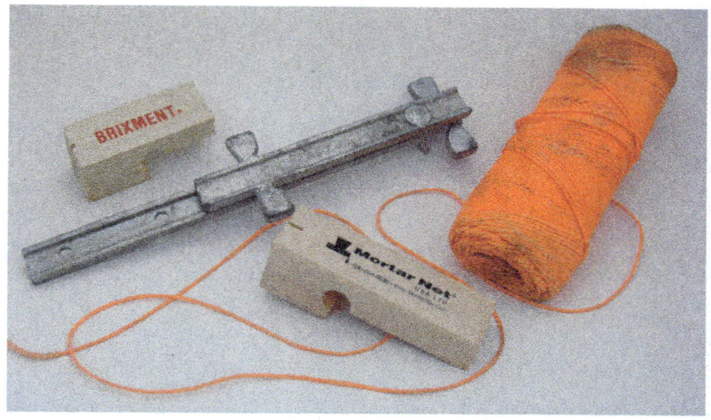

28105-13_F18.EPS

Figure 18 Line stretcher with line and wood line blocks.

Step 1 Pass the line through the slot of the block. Wrap the line around the block several times to keep it from passing through the slot.

Step 2 Have one person hold the line block aligned with the top of the course to be laid. Mason's lines are often first attached to the left side as you face the wall.

Step 3 Place the line block so that it hooks over the edge of the masonry unit, and hold it snug.

Step 4 The second person will walk the line to the right end of the wall.

Step 5 The second person then pulls the line sufficiently tight and wraps it three or four turns around the middle of the line block.

Step 6 The second person hooks the tensioned line block over the edge of the corner.

Step 7 Both parties check that the line is at the correct height.

Line stretchers (*Figure 18*) are put in place following the same steps. The line stretcher slips over the top of the block, rather than the edges. Line stretchers are useful when the corner lead is not higher than the course to be laid.

1.3.4 Setting Up the Line Using Line Pins

Steel line pins hold a mason's line in place. The line pin is less likely to pull out of the wall because of its shape. The peg end of the line, or the starting end, is traditionally started at the left as the mason faces the wall.

Step 1 Drive the line pin securely into the lead joint (*Figure 19*). Make sure that the top of the pin is level with the top of the course to be laid. Place the pin at a 45-degree downward angle, several units away from the corner. This will prevent the pin from coming loose as the line is pulled.

Step 2 Tie the line securely to the pin using the notches on the pin. Give the line a few very sharp, strong tugs. This tests whether the pin will come out as the line is tightened and helps to prevent injuries caused by flying line pins.

Step 3 Walk the line to the other lead. Drive the second line pin securely into the lead joint even with the top of the course. Check and measure that the pin is secure and in the correct position before applying the line.

Step 4 Wrap the line around the pin, and start tensioning. Pull the line with your left hand, and wrap it around the line pin with your right hand. Use a clove-hitch or half-hitch knot to secure the taut line to the pin. Be careful not to pull the line so tight that it breaks.

Step 5 When you move the line up for another course, immediately fill the pinholes with fresh mortar. If you wait until later to fill the line-pin holes, you will need to mix another batch of mortar. Taking care of the holes as you move the pins saves many steps at the end of the project.

1.3.5 Setting Up the Line Using Line Trigs

To keep a long line from sagging, set trigs to support the line midstring.

Step 1 Set the trig support unit in mortar in position on the wall. Be sure that this unit is set with the bond pattern of the wall, close to the middle of the wall.

28105-13_F19.EPS

Figure 19 Using a line pin.

Step 2 Check that the unit is level and plumb with the face of the wall. Check that the unit is at the proper height with a course rule or course pole.

Step 3 Sight down the wall to be sure that the trig unit is aligned with the wall and is set the proper distance from the line. The trig support unit is a permanent part of the wall, so place it carefully.

Step 4 After the trig support unit is in the proper place, slip a trig or clip over the taut line. Check that the line is still in position. Lay the trig on the top of the support unit with the line holder on the bottom side. Place another masonry unit on top of the trig to hold it in place. *Figure 20* shows the use of a trig.

Step 5 Check the line for accuracy once more. The line should just be level with and slightly off the corner of the trig support unit and the standard 1/16-inch away from it.

The difference in laying block and brick to the line is the difference in handling the units.

Block should be kept dry at all times, as moisture will cause them to expand. If they are used wet, they will shrink when they dry and cause cracks in the wall joints. To cut down on handling, stack them close to the work sites with the bottom (smaller) shells and webs down.

Practicing laying block will let you discover the easiest methods for yourself. Find a way to hold the buttered block that is comfortable for you. Lift the block firmly by grabbing the web at each end of it, and lay it on the mortar joint. Keep the trowel in your hand when laying block, to save time.

As you place the block, tip it toward you a little. You can look down the face to align the block with the top of the block in the course below. Then, roll the block back slightly so that the top is in correct alignment to the line. At the same time, press the block toward the last block laid. Moving the block slowly is key to this process. Do not release the block quickly, or you will have to remortar and reposition it.

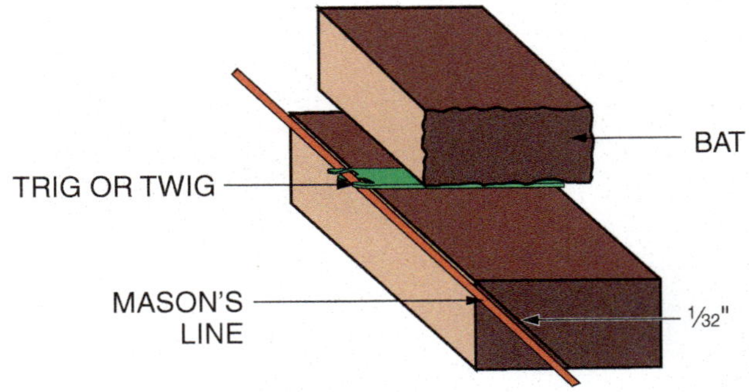

28105-13_F20.EPS

Figure 20 Using a line trig.

You can adjust the block by tapping. Be sure to tap in the middle of the block, away from the edges. Block face shells may chip if you tap on them. Using your trowel handle on the block may be necessary to adjust the unit into place.

1.3.6 Building Corners and Leads

Corners (*Figure 21*) are called leads because they lead the laying of the wall. They set the position, alignment, and elevation of the wall by serving as guides for the courses that fill the space between them. Building corners requires care as well as accurate leveling and plumbing to ensure that the corner is true.

As you learn to build corners, practice technique and good workmanship. Speed will follow. Be certain that each course is properly positioned before going on to the next. Once a corner is out of alignment, it is difficult to straighten it.

28105-13_F21.EPS

Figure 21 Building a corner lead with block.

Building a lead starts with marking the exact place in line with the corners and properly located for the bond pattern. Use a chalkline between the corners to locate the place. Lay block in the corner lead by following standard techniques. As each course is laid, check the course spacing on each course with the modular or spacing rule. Then check the course level, plumb, and **range** with your mason's level.

When the lead is complete, it needs to be checked for diagonal alignment. The diagonal is called the **tail**. Measure the alignment of the tail by holding the mason's level at an angle along the side edges of the end block. As shown in *Figure 22*, hold the level in line with the corner of each block. This lets you check that no block is protruding.

1.3.7 Block Corner Leads

Block requires special handling. Block chips easily when moved or tapped down in place; therefore, it must be eased slowly into position. Typically, the larger the block, the easier it is to keep level and plumb. Each unit needs to be checked for position in all dimensions. Even with these challenges, block does save time and money.

The procedure for laying a block outside-corner lead follows the same steps as for brick. The main difference is that block requires a different mortar spreading technique. The following steps do not repeat location material previously covered:

Step 1 Check the construction drawings for the location of the corner. Determine how high the corner should be. The corner is usually built to the full height of the lift for block. Subsequent corners are built as the work progresses.

28105-13_F22.EPS

Figure 22 Checking alignment of the tail.

Step 2 Determine the number of courses the corner will need. Use your rule or measure to calculate the courses in a given height; then calculate the number of block in each course. The sum of the stretchers in the first course must equal the number of courses high the corner will reach.

Step 3 Clean the foundation. Locate the point of the corner. Snap a chalkline from this point across the wall location to the opposite corner. Repeat the procedure on the other side of the corner. This aligns the corner with the other corners. Check the squareness of the chalkline with a framing square before laying any block.

Step 4 Check that the footing is level. Use your rule or tape, or measure the footing, to establish bond. Lay the first course as a dry bond. Because actual sizes of block may vary, space out the dry bond with bits of wood for the joints. Check that the dry bond is plumb to find any irregularities in the footing. If the footing is too high in places, make an adjustment by cutting off some of the bottom of the block.

Step 5 Lay the corner block first. Continue with the leg of the corner. Line up each block with the chalkline.

Step 6 Check the placement of each block for height, level, plumb, and range. When both legs are finished, check for alignment again.

Step 7 Do not remove excess mortar immediately, because this could cause the block to settle unevenly. Remove excess mortar from the first course after the second course is laid.

Step 8 For subsequent courses, apply mortar in a face-shell bedding on top of the previously laid course. Check each block for height, level, plumb, and range.

Step 9 Check for diagonal alignment; then tail the rack ends. If the edges of all units do not touch the level, the head joints are not properly sized. Adjust the block if the mortar is still plastic enough, or rebuild the courses, as required.

Step 10 After placing the block, check height, level, plumb, and range. Check across the diagonal as well. To train your eye, sight down the outermost point of the corner block from above to check plumb. Remove excess mortar from the outside and inside of the corner and from each exposed block edge.

Step 11 Continue until you have reached the required number of courses. Use one less block in each course. Lay and align each course. Remove excess mortar from each course, inside and outside. Be sure that aligning the block does not disturb the mortar bond. If the mortar bond is disturbed, take up the block, clean it, rebutter it, and replace it.

Step 12 If the corner does not measure up at each course, take the course up, and do it again. This is easier than taking the wall up and doing that again.

Step 13 Check the mortar, and tool the joints. Cut away any excess mortar from the top of the block. Check the height of the corner with the modular spacing rule. Recheck the corner for plumb, level, and range. If it does not measure up, take the corner down, and start over.

When measuring block, each one must touch and be completely flush with the mason's level. They must also be completely in line with the chalk marks on the footing. Repeat all measurements often to prevent bulges or depressions in the wall and to keep the courses in line.

The corner lead can now be used to anchor a line as detailed previously. Learning to build corner leads will teach you three-quarters of what you need to know about building corners.

1.3.8 Mortar Joint Finishes

Mortar joints between masonry units serve the following functions:

- Bonding units together
- Compensating for differences in the size of the units
- Bonding metal reinforcements, grids, and anchor bolts
- Increasing weather resistance
- Creating a neat, uniform appearance

Mortar joints are made by buttering masonry units with mortar and laying the units. The mason controls the amount of mortar buttered so that it fills a standard space between the units. Excess mortar oozes out between the units, and the mason trims it off. But this is not the last stage in making a mortar joint. When the joint is thumbprint-hard, the mortar left between the masonry units must be tooled to a proper finish. It is this last step, the tooling, that gives the mortar joints their uniform appearance and weather-resistant qualities.

Mortar joints can be finished in a number of ways. *Figure 23* shows some standard joint finishes. Usually, the joint finish will be part of the detailed specifications on a project. The process of tooling the joint compresses the mortar and thereby increases its weather resistance. Tooling also closes any hairline cracks that open as the mortar dries. Joints are tooled by shaped jointers. Raked joints are made by rakers. Struck, weathered, and flush joints are tooled by a trowel.

The mortar in a raked joint is partially removed, or raked, not compressed. It does not get the extra water resistance, so it is not recommended for exterior walls in wet climates. The struck joint collects dirt and water on the ledge, so it is not recommended for exterior walls. Flush joints are not compressed, only struck off. They are recommended for walls that will be plastered or parged.

The extruded joint, or weeping joint, shown in *Figure 23* is made when the masonry unit is laid. When the unit is placed, the excess mortar is not trimmed off. The mortar is left to harden to become the extruded joint. Since the mortar is not compressed in any way, this joint is not recommended for exterior walls.

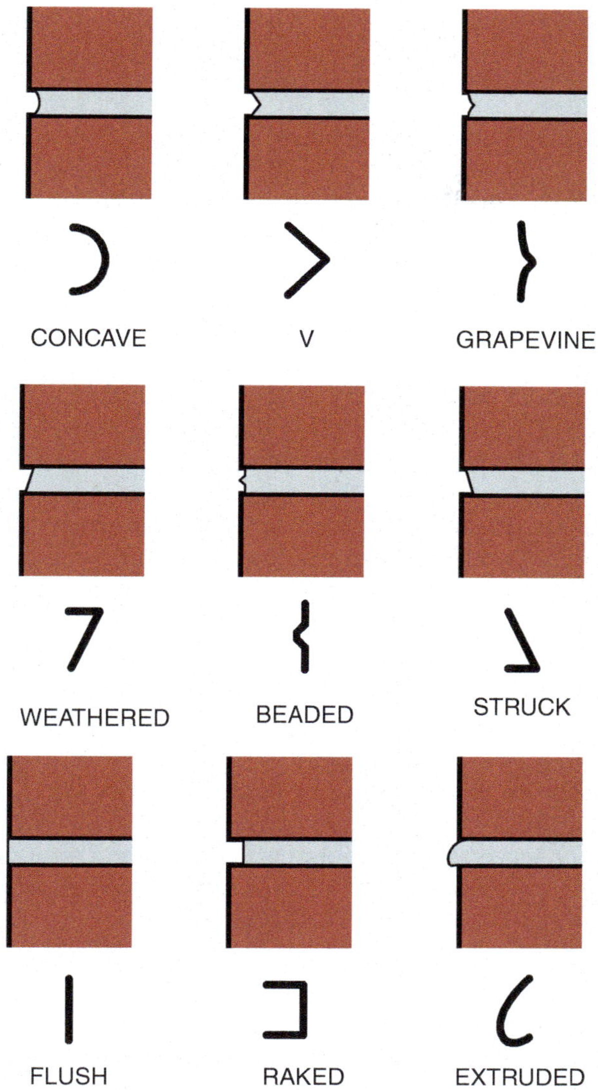

Figure 23 Mortar joint finishes.

1.3.9 Checking the Mortar

After you have laid the masonry units, the mortar must be at the proper consistency before it can be tooled. The test equipment for checking proper consistency is the mason's thumb. Press your thumb firmly into the mortar joint:

- If your thumb makes an impression, but mortar does not stick to it, the joint is ready.
- If the mortar sticks to your thumb, it is too soft. The mortar is still runny, and the joint will not hold the imprint of the jointing tool.
- If your thumb does not make an impression, the mortar is too stiff. Working the steel tool will burn black marks on the joints.

Working the joints with the jointer, raker, or trowel is called tooling the joints. Whichever tool you use to tool the joint, the procedure is the same. The first step is to test the mortar. The best time for tooling the joints will vary because weather affects mortar drying time. Test the mortar repeatedly to find the right window of time to do the finishing work.

1.3.10 Tooling

When the mortar is ready, the next step is the tooling. The tool should be slightly larger than the mortar joint to get the proper impression.

- Hold the tool with your thumb on the handle, so it does not scrape on the masonry unit.
- Apply enough pressure so that the runner fits snugly against the edges of the masonry units. Keep the runner pressed against the unit edges all the way through the tooling.
- Tool the head joints first (*Figure 24*). Tool head joints upward for a cleaner finish.
- Tool the bed joints last. The convex sled-runner tool (*Figure 25*) is most commonly used. To keep the joints smooth, walk the jointer along the wall as you tool. The joints should be straight and unbroken from one end of the wall to the other. If the head joints are struck last, they will leave ridges on the bed joints.

If you are making a raked joint, follow the same order of work. Some joint rakers, or skate rakers (*Figure 26*), have adjustable setscrews that set the depth of the rake-out. Do not rake out more than ½ inch, or you will weaken the joint and possibly expose ties or reinforcements in the joint. Be sure you leave no mortar on the ledge of the raked unit.

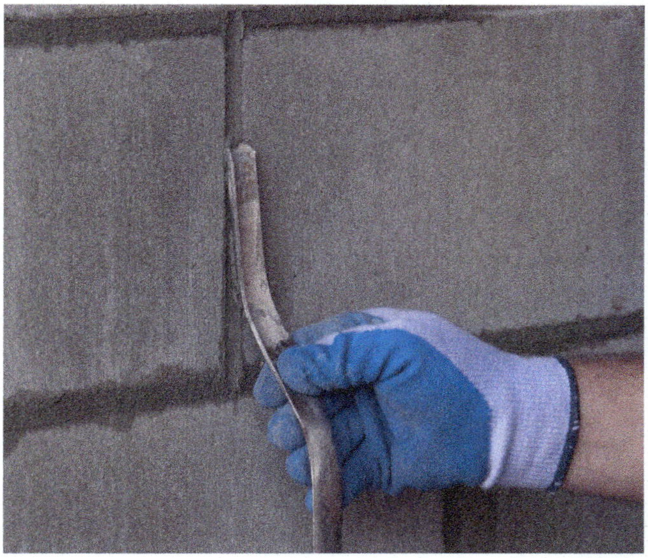

Figure 24 Using a jointer.

28105-13_F25.EPS

Figure 25 Using the convex sled.

28105-13_F26.EPS

Figure 26 Skate raker.

If you are making a troweled joint, follow the same order of work. Ensure that the angle of the struck or weathered joint faces the same way on all the head joints. If you are tooling flush joints with your trowel, tool up rather than down.

1.3.11 Cleaning Up Excess Mortar

After you tool the joints, you must clean up the excess mortar. Dried mortar sticks to masonry and is difficult to clean. Cleaning is much easier when you do it immediately after tooling. Follow these steps to clean up excess mortar:

Step 1 Trim off mortar burrs by using a trowel. Hold the trowel angled slightly away from the wall, as shown in *Figure 27A*. As you trim the burrs, flick them away so they do not stick to the units.

Step 2 Dress the wall after trimming off burrs. Dressing can be done with a soft brush, as shown in *Figure 27B*.

Step 3 Brush the head joints diagonally across the joints to avoid digging out the fresh mortar. Retool the joints after brushing to get a sharp, neat joint.

Step 4 After you finish cleaning the wall, clean the floor at the foot of the wall.

1.4.0 Cleaning Concrete Masonry Units

The hardest material to clean off masonry units is dried, smeared mortar that has worked its way into the surface of the masonry unit. This seriously affects the appearance of the finished structure. Your best approach is to avoid smearing and dropping mortar during construction. These guidelines will help you clean as you work:

- When mortar drops, do not rub it in. Trying to remove wet mortar causes smears. Let it dry to a mostly hardened state.
- Remove it with a trowel, putty knife, or chisel. Try to work the point under the mortar drop, and flick it off the masonry.
- The remaining spots can usually be removed by rubbing them with a piece of broken block or brick, then with a stiff brush.

You should spend some time cleaning every day, removing stray mortar from the wall sections as you complete them.

In addition to cleaning dropped mortar, there are other things to do. To keep masonry clean during construction, practice these good work habits:

- In order to prevent splashes, avoid stacking mortar pans and mortarboards against the wall.
- Temper the mortar with small amounts of water, so it will not drip or smear on the units.
- After laying units, cut off excess mortar carefully with the trowel.
- Wait until mortar is thumbprint-hard, to avoid smearing wet mortar on masonry units.
- After tooling joints, scrape off mortar burrs with your trowel before brushing.
- Avoid any motion that rubs or presses wet mortar into the face of the masonry unit.
- Keep materials clean, covered, and stored out of the way of concrete, tar, and other staining agents. Do not store materials under the scaffold.

(A) (B)

28105-13_F27.EPS

Figure 27 Trimming and dressing mortar burrs.

- Turn scaffold boards on edge with the clean side to the wall at the end of the day. This will prevent rain from splashing dirt and mortar onto the wall.
- Always cover the tops of walls at the end of the day to keep them dry and clean.

Implementing these good work habits should reduce the amount of time you spend cleaning the masonry units after construction is complete. It is especially important to follow these guidelines when working with block. Because of the rougher surface texture of block, mortar spilled on block is harder to clean.

CAUTION

Never use muriatic acid to clean masonry units. It will damage the surfaces of the units.

Cleaning block is difficult because the surface on standard block is very porous. If block is stained with mortar at the end of a job, rub it with a piece of block.

For further cleaning, it is important to check the manufacturer's safety data sheets (SDSs, formerly called material safety data sheets, or MSDSs) for recommended chemical cleaners and cleaning procedures. Proprietary cleaners can be very destructive to block if not used properly, and cannot be used without protective countermeasures. Read the block manufacturer's SDSs for recommended cleaning solutions. Read the SDSs, and follow manufacturer's directions for mixing, using, and storing any chemical solution. Detergents and proprietary cleaners are often recommended for use on block. If no cleaning procedures are given, follow the bucket-and-brush procedures given in the section "Cleaning Brick" in this module.

Any cleaner used should first be tested on an inconspicuous 4' × 5' section of wall. Sometimes, minerals in the block may react with some chemicals and cause stains.

WARNING!

Wear appropriate personal protective equipment when using chemical solutions.

High-pressure water washing involves the use of water and a detergent to clean the surface of block by causing dirt to absorb water and swell, thereby loosening it so that it can be washed away. To prevent efflorescence, plan to use the least amount of water and detergent that is required to clean the wall completely. Begin by removing clay and dirt with a dry fiber or nylon brush. Do not use steel brushes, as they can leave behind metal particles that rust and discolor the block. Use heated water on greasy surfaces or when cleaning during cold weather.

The NCMA recommends that masons follow these guidelines when using pressure-washing equipment:

- The water pressure should be limited to 400 to 600 pounds per square inch (psi).
- Use a wide-flange tip only, not a pointed tip.
- Keep the tip at least 12 inches away from the masonry while cleaning.
- Direct the spray at a 45-degree angle to the wall, never perpendicular to the wall.

Additional Resources

ASTM C90, Standard Specification for Loadbearing Concrete Masonry Units. 2012. West Conshohocken, PA: ASTM International.

TEK 2-3A, *Architectural Concrete Masonry Units.* 2001. Herndon, VA: National Concrete Masonry Association. **secure.ncma.org/source/Orders/ProductDetail.cfm?pc=TEK02-03**

TEK 8-4A, *Cleaning Concrete Masonry.* 2005. Herndon, VA: National Concrete Masonry Association. **www.ncma.org/etek/Pages/Manualviewer.aspx?filename=TEK%2008-04A.pdf**

1.0.0 Section Review

1. The compressive strength of block is measured in _____.
   a. parts per million
   b. pounds per square foot
   c. pounds per square inch
   d. foot pounds

2. A purely decorative bond formed by masonry units and mortar joints on the face of a surface is called a _____.
   a. simple bond
   b. structural pattern bond
   c. structural bond
   d. pattern bond

3. When setting up a line using line pins, the line pin should be driven securely into the lead joint _____.
   a. at a 45-degree downward angle, several units away from the corner
   b. at a 45-degree upward angle, several units away from the corner
   c. at a 45-degree downward angle, precisely at the corner
   d. at a 45-degree upward angle, precisely at the corner

4. When using pressure-washing equipment to clean block, the NCMA recommends that the distance from the tip to the masonry should be at least _____.
   a. 3 inches
   b. 6 inches
   c. 9 inches
   d. 12 inches

SECTION TWO

2.0.0 BRICK MATERIALS AND INSTALLATION

Objective

Describe how to install brick.

a. Identify the characteristics of brick.
b. Explain how to set up, lay out, and bond brick.
d. Explain how to lay and tool brick.
e. Explain how to clean brick.

Performance Tasks 5 through 8

Lay a dry bond for brick.

Tool a bed joint for brick.

Lay brick to the line in courses that are true for height, level, plumb, and range.

Build a brick lead.

Trade Terms

Closure unit: The last brick to fill a course.

Crowd the line: To touch the mason's line or a masonry unit too close to the line.

Racking: Shortening each course of masonry by half a unit so it is shorter than the course below it.

Return: A corner in a structure or lead.

Slack to the line: Masonry units set too far away from the mason's line.

As you learned in the module *Introduction to Masonry*, clay masonry units, collectively referred to as brick, are the second-oldest building material. The following sections review the characteristics of brick and introduce you to the steps for setting up, laying out, and bonding brick, as well as for cutting, laying, and tooling brick.

2.1.0 Identifying the Characteristics of Brick

The types of brick include the following:

- Uncored brick
- Cored brick and tile
- Architectural terra-cotta units

Brick is classified as solid if no more than 25 percent of its surface is cored, which means that the brick has holes extending through it to reduce weight. If the brick has no holes at all, it is called uncored. ASTM standards cover all types of masonry units, loadbearing and nonbearing. Brick comes in modular and nonmodular sizes, in a wide range of colors, textures, and finishes. This section will focus on brick and laying brick.

Brick can be installed in any of six positions. *Figure 28* shows each of these six positions and their names. The dark edge of the brick is the named part and the part that shows when the brick is laid in a pattern bond.

An important characteristic of brick is its initial rate of absorption, or the amount of water it can soak up in a fixed length of time. The percentage of water present in brick affects the hardening of the mortar around the brick. If the brick has absorbed too much moisture prior to being placed, the mortar will set more slowly than usual. On the other hand, if the brick is extremely dry prior to being placed, it may absorb too much moisture from the mortar. This will prevent the mortar from hardening properly because there will not be enough water left for good hydration.

Hard-surfaced brick usually need to be covered on the job site so they do not get wet. Soft-surfaced brick is usually very absorbent and may sometimes need to be wetted down before it is used.

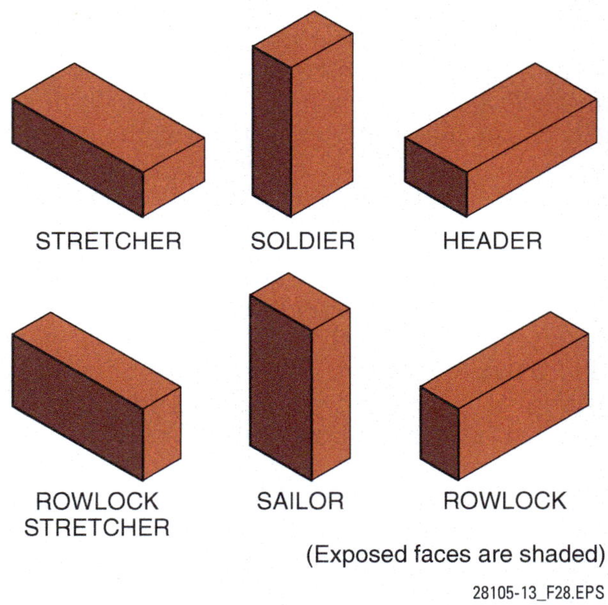

Figure 28 Brick positions.

The mason needs to determine whether the brick is too dry for a good bond with the mortar. The following test can be used to measure the absorption rate of brick:

Step 1 Draw a circle about the size of a quarter on the surface of the brick with a crayon or wax marker.

Step 2 With a medicine dropper, place 20 drops of water inside the circle.

Step 3 Using a watch with a second hand, note the time required for the water to be absorbed.

If the time for absorption exceeds 1½ minutes, the brick does not need to be wetted. If the brick absorbs the water in less than 1½ minutes, the brick should be wetted.

Wet brick with a hose directed on the brick pile until water runs from all sides. Let the surface of the brick dry before laying them in the wall.

2.2.0 Setting Up, Laying Out, and Bonding Brick

The procedures for setting up, laying out, and bonding brick are mostly similar to those you have already learned for block. The job setup and structure layout must be complete before you can start to lay brick.

2.2.1 Setting Up

Once the masonry contractor has read the contract, construction drawings, and specifications, and reviewed the schedule and any referenced standards and codes, the contractor can estimate the workers, materials, and equipment needed for the job.

As with setting up a block project, follow these steps:

- Check that all materials are stored close to workstations, protected from moisture, and not in direct contact with earth.
- Place mortar pans and mortarboards by workstations. If you are using scaffold, place the pans at intervals on the scaffold near the point of final use.
- Stockpile units on each side of the mortar pans and at intervals along the wall line. If you are using scaffold, place units along the top of the scaffold near the point of final use. Stockpiles should allow the mason to move units as little as possible once laying starts.
- Stack all units face up and in a way that does not requires the mason to move or turn them excessively.
- Check all scaffold for proper assembly and position. Ensure that braces are attached and planks are secured at each end. Scaffold should be level and close to the wall.
- Check all mechanical equipment, power tools, and hand tools. Make sure they are clean, in good condition, and the right size for the job.

2.2.2 Job Layout, Planning, and Locating

Review the plans and specifications before laying out the structure. The first steps are to plan out the work, locate where it will go, and then to establish bond, such as a dry bond (see *Figure 29*). It is especially important to dry-bond brick across openings to ensure proper layout and make any necessary adjustments in the mortar joints. Then check the plans for wall lengths, heights, door dimensions, and window openings. Identify the specified pattern or bond, the nominal size of the masonry unit, and the openings. Then draw a rough layout of the wall and lay out the bond pattern in order to count and calculate how many brick to cut. Remember that the dimensions provided on plans are nominal dimensions, while masons need to use actual dimensions when laying out courses.

The starting point of the pattern determines how many masonry units will need to be cut. By adjusting the starting unit of a nonmodular bond pattern, the mason can come up with a layout that calls for cutting the least amount of units. The mason will check these calculations by laying a dry bond before cutting any units.

28105-13_F29.EPS

Figure 29 Dry-bonded brick.

Locate the footing and clean it off to ensure a good bond between the footing and the first course. Ensure the footing is level and fix it if it is not. Then locate the walls, being sure not to confuse the measurements for the interior and exterior walls. Then establish your points and run a chalkline between them, remembering to mark key points along the chalkline with a marking pencil, nail, or screwdriver. Then mark the entire foundation for walls, openings, and control joints.

2.2.3 Bonding Brick

As you have already learned, masons deal with four types of bonds: simple bonds, pattern bonds, structural bonds, and structural pattern bonds. Remember, the distinction between a structural bond and a pattern bond is not always easy to make. By overlapping or interlocking masonry to create a structural bond, you are also creating a pattern. The definition of a particular pattern as a structural bond or a structural pattern bond varies from location to location.

Pattern bonds add design but not strength to masonry walls. The stack bond (*Figure 30*) is only a pattern bond. If the stack bond pattern is used in a loadbearing wall, the wythe must be bonded to its backing with rigid steel ties. In loadbearing construction, this patterned wall should be reinforced with steel joint-reinforcement ties.

Pattern and structural pattern bonding calls for placing brick in different positions in the wythes. *Figure 31* shows different ways of placing brick in order to make different kinds of patterns.

The stretcher is the everyday workhorse. Headers are used primarily for tying wythes together, capping walls, windowsills, and pattern bonds. Soldiers are used over doors, windows, or other openings, and in pattern bonds. Rowlock stretchers are used in pattern bonds, in brick walks. Rowlocks are found in capping walls, windowsills, ornamental cornices, and pattern bonds.

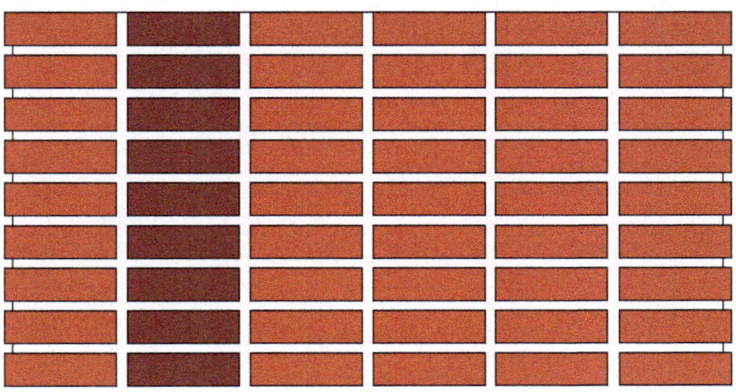

STACK BOND

28105-13_F30.EPS

Figure 30 Stack bond.

Sailors are rarely seen, except in pattern bonds and brick walks.

Wythes can be structurally bonded by using metal ties, joint reinforcements, anchors, grout, and steel rods. These engineering methods are used to increase strength and loadbearing by firmly tying masonry units and wythes together.

Another way, older than the preceding, to structurally bond a wythe is to lap masonry units. Lapping one unit halfway over the one under it provides the best distribution of weight and stress.

In a single-wythe wall, a structural bond is made by staggering the placement of the brick. This results in the brick in one course overlapping the brick underneath. The structural pattern bond resulting from this simple overlap is the running bond, as shown in *Figure 32*. Common overlaps are the half lap and the one-third lap. Changing the proportion of the overlap changes the look of the pattern.

In two-wythe walls, a structural bond is made between the wythes. This can be made by rigid steel ties that equalize loadbearing. It can also be made by overlapping a brick from the face wythe to the backup wythe. The overlap brick is turned into the header or the rowlock position. This results in a complex structural bond that is also a structural pattern bond, with different sizes of brick facing out. The results are the traditional English bond and Flemish bond shown in *Figure 33*.

The Flemish bond consists of alternating headers and stretchers in every course. The English bond consists of alternating courses of headers and stretchers. If the headers are not needed for structural bonding, cut brick is used. Brick can be laid to show different faces and cut in different ways.

The combination of the Flemish and English bonds with the running bond results in the common or American bond. As shown in *Figure 34*, the common bond is a running bond with headers every sixth course. The headers are in the Flemish or English pattern, according to the specifications.

2.2.4 Buttering Brick

The preferred way to butter brick is to butter it before placing it on the wall. Project specifications and some model building codes require that you incorporate full head joints. As you have already learned, the head joint is the vertical mortar joint between masonry units. The term *laying forward* means to butter a brick before placing it. One way to make this process more efficient is to butter the front edge of a brick and push it into the rear of

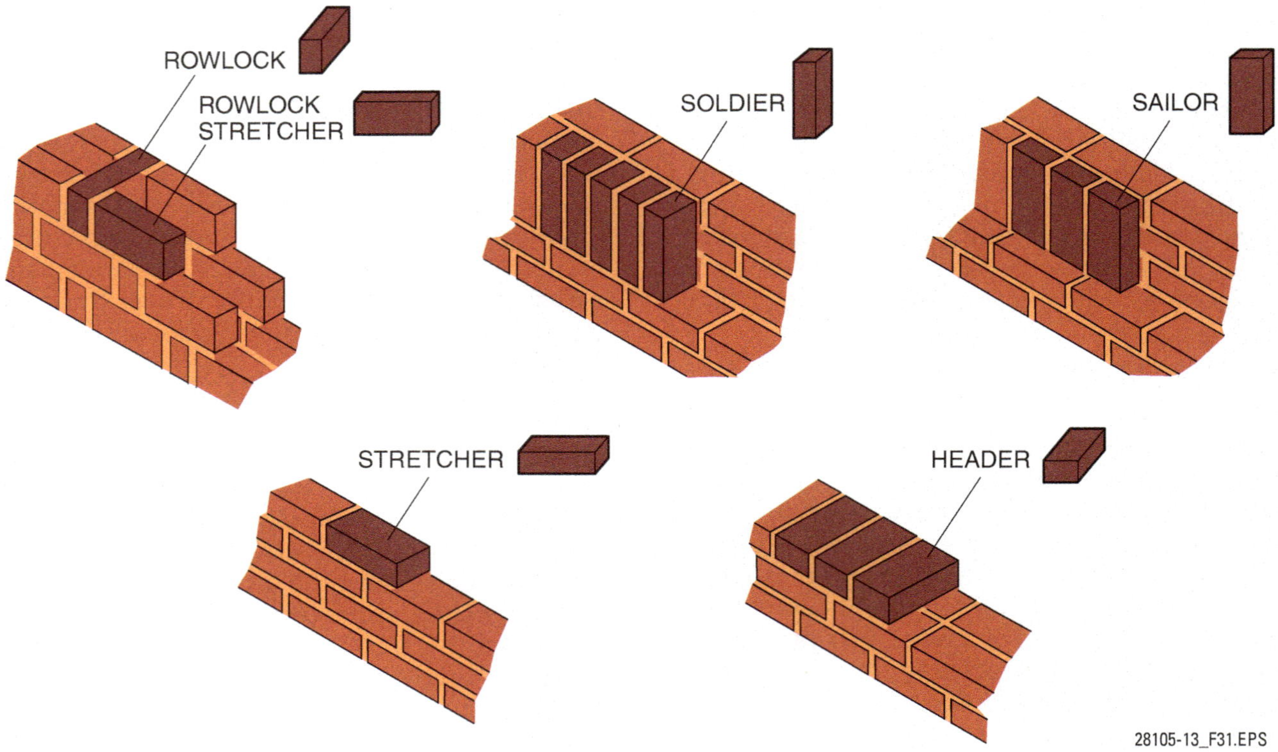

Figure 31 Brick positions in walls.

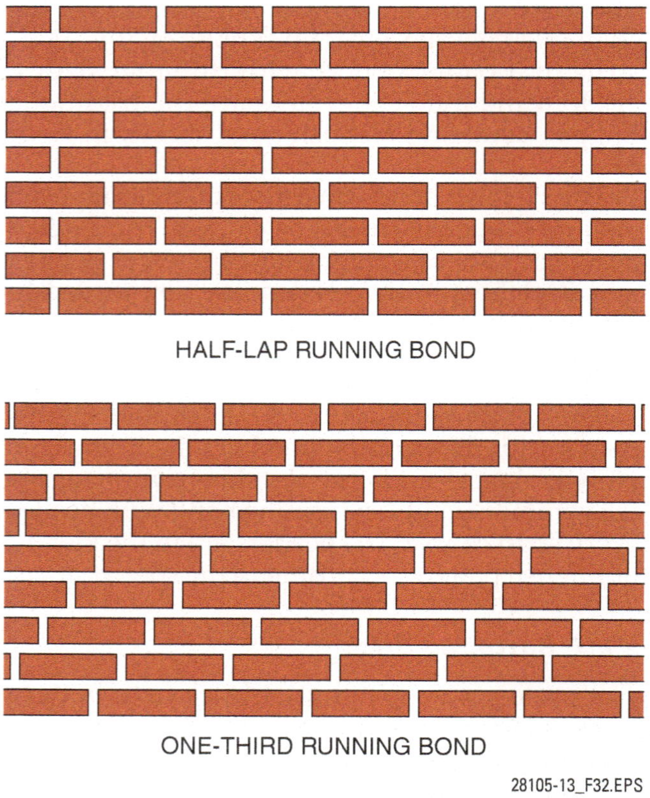

Figure 32 Running bonds.

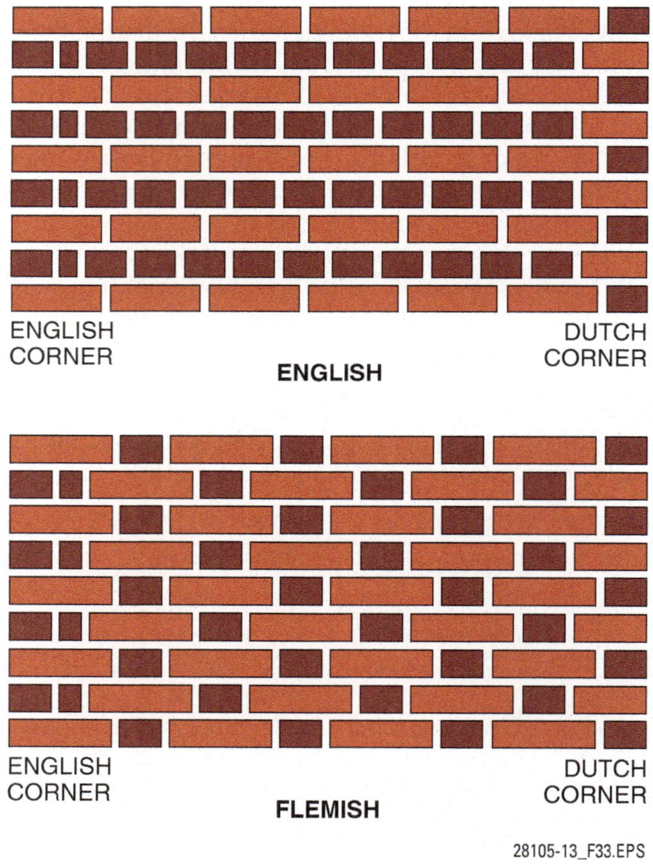

Figure 33 English and Flemish bonds.

28105-13_F34.EPS

Figure 34 Common bond.

the last brick (see *Figure 35*). Buttering a brick in this way is easier than locking and throwing. The process for buttering a brick is as follows:

Step 1 Throw a small amount of mortar on the front edge of a brick.

Step 2 Butter the bottom (wide) edge of the brick.

Step 3 Place mortar on the front and rear (narrow) edges of the brick.

Step 4 Push the brick to within ⅜ inch of the rear of the last brick and slice the oozing mortar in the direction of the last brick.

2.3.0 Laying and Tooling Brick

Using both hands to lay brick is an efficient practice. Keep your trowel in one hand, and use the other for picking, holding, and placing brick. This will make the work easier and faster. Use your fingers efficiently as well. When you pick up a brick, hold it plumb. Pick it up so that your thumb is on the face of the brick. Let your fingers and thumb curl down over the top edges of the brick, slightly away from the face. In this position, your fingers will not interfere with the line as you place the brick on the wall.

2.3.1 Placing Brick

The most important placing rule is to place gently. Do not drop the brick onto the bed joint; lower it down gently. Press the brick forward at the same time so that it will butt against the unit next to it. Mortar should ooze out slightly on both head and bed joints to show that there has been full contact.

Align the latest brick with the brick next to it and below it as you place it. Line it up with the mason's line if you are using one. By standing slightly to one side, you will be able to sight down the wall so as not to crowd the line or to be slack to the line. This will help maintain plumb head joints by sighting the brick below the newly laid unit.

28105-13_F35.EPS

Figure 35 Placing a buttered brick.

You may need to slightly adjust the unit in its bed. First try pressing downward on the brick with the heel of your hand. Keep part of the heel of your hand on the brick next to the one you are adjusting. If this is not sufficient, you may need to tap the brick with the handle of your trowel. After adjusting the brick, cut off the extruded mortar with your trowel, and lay the next unit. Cutting mortar as you go will help you to keep the masonry clean.

2.3.2 Checking the Height

The first check is always course height. If this is off, there is no use checking anything else. Use your modular or standard brick spacing rule to check the height of the brick. Follow these steps:

Step 1 After the brick is laid on the wall, unfold the rule (*Figure 36*), and place it on the base or footing used for the mortar and brick.

Step 2 Hold the rule vertically. Check that the end of the rule is flat on the base, so the reading is accurate. If you are using modular brick, the first course should be even with number 6 on the modular rule. If you are using a different size of brick, check the appropriate scale on the modular or standard rules.

Moving Blended Stacks of Brick

When using a brick cart to move a stack of masonry units that has already been blended, it is important to preserve the pattern while loading and unloading the cart.

The height, or vertical course spacing, depends on the thickness of the mortar joints. Practicing laying full bed and head joints is the fastest way to learn to make standard-size joints.

If more than one course is laid, always set the modular rule on the top of the first course to measure. The base may have been irregular, and a large joint may have been used to level the first course.

2.3.3 Checking Level

After checking the height for your line of six brick units, check with your mason's level for levelness using the following steps:

Step 1 Remove any excess mortar on top of the brick.

Step 2 Place your mason's level lengthwise on the center width of the six brick units to be checked.

Step 3 Use your trowel handle to gently tap down any units that are high with relation to the mason's level (*Figure 37*). Do not tap them so hard they sink too low.

Step 4 If individual brick units are low, pick them up. Clean and mortar the bed and the head joint again, and reposition the brick. Reposition the mason's level again, and get it level.

Commonly Used Bonds

The English cross, or Dutch bond, uses a structural pattern bond that repeats every four courses. The pattern courses are all stretcher, all header, and a course of three stretchers and one header. The last pattern course is all header again. The Dutch bond is shown in on the left (*Figure SA06*).

The Dutch bond may seem complicated until you look at a traditional garden wall bond. The figure on the right (*Figure SA07*) shows two variations on the garden wall structural pattern bond. The double-stretcher garden wall pattern shown in A (*Figure SA07A*) repeats every five courses. The dovetail garden wall pattern shown in B (*Figure SA07B*) repeats every 14 courses. More variations are possible. The only limit on patterning is the skill and ingenuity of the mason.

ENGLISH CROSS OR DUTCH BOND

28105-13_SA06.EPS

(A) DOUBLE STRETCHER WITH UNITS IN DIAGONAL LINES

(B) DOVETAIL

GARDEN WALL BONDS

28105-13_SA07.EPS

2.3.4 Checking Plumb

The next step after leveling is to check for plumb, or vertical range. Follow these steps:

Step 1 Hold the level in a vertical position against the end of the last brick laid (*Figure 38*).

Step 2 Tap the brick with the trowel handle to adjust the brick face either in or out.

Step 3 Move the level to the end of the first brick laid, and repeat the process.

Figure 39 gives profiles and names for units that are plumb and out of plumb. The large black dot represents the mason's line. By looking and touching, you can train your hand and eye to know when brick is plumb and when brick is not plumb.

Other Common Brick Bonds

Other commonly used bonds include the patterns shown here. Note that the running bond, horizontal and vertical stack bonds, and the coursed ashlar bonds are also commonly used block bonds.

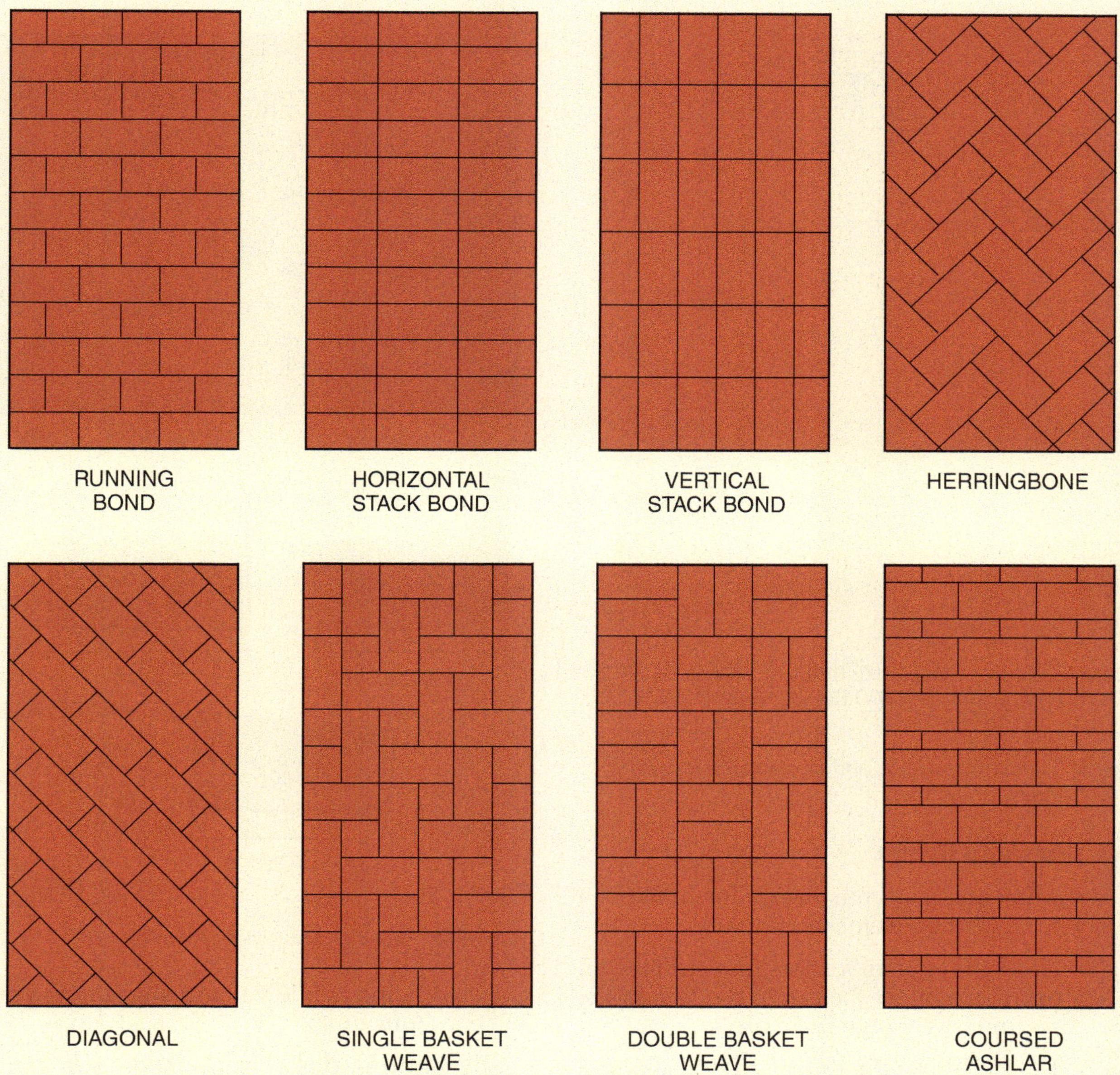

OTHER COMMON BRICK BONDS

28105-13_SA08.EPS

28105-13_F36.EPS

Figure 36 Standard, oversize, and modular brick spacing rules.

2.3.5 Checking Range

After the first and last units in the line have been plumbed, check the rest for range:

Step 1 Hold the mason's level in a horizontal position against the top of the face of the six units, as shown in *Figure 40*.

Step 2 Tap the brick either forward or back until it is all aligned against the mason's level. Be careful not to move the end plumb points while you are aligning the middle four units.

28105-13_F37.EPS

Figure 37 Leveling the course.

28105-13_F38.EPS

Figure 38 Checking plumb.

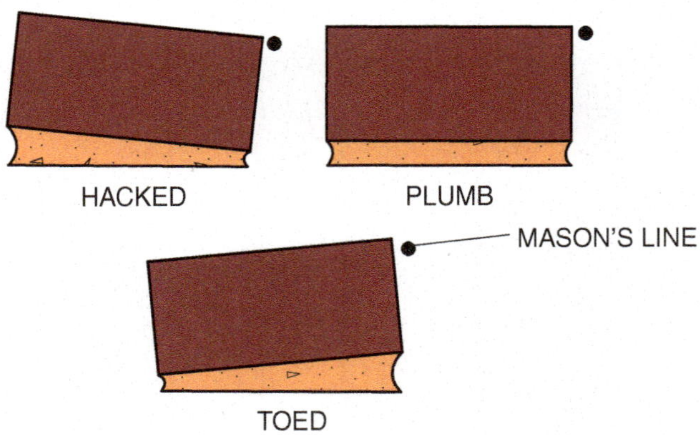

28105-13_F39.EPS

Figure 39 Plumb and out-of-plumb brick.

By sighting down from above, you can train your eye to know when brick is straight and when brick is not straight.

28105-13_F40.EPS

Figure 40 Checking for range.

2.3.6 Laying the Closure Unit

The last unit in a course is called the closure unit. Masons lay corners of a wall first then work from each corner toward the middle. The last unit, or closure unit, must fit in the gap between the masonry units that have already been laid (*Figure 41*). The closure unit should fall toward the middle of a wall. The space left for it should be large enough for the unit and its two head joints.

The process for laying the closure unit is as follows:

Step 1 Butter the closure unit on both head joints.

Step 2 Butter the adjacent units on their open head joints.

Step 3 Gently ease the unit into the space.

Step 4 If the mortar falls completely out of a closure unit joint, remove the unit, and reset it in fresh mortar. Check to ensure that all joints are full.

If the head joints have been properly spaced, the closure unit will slide in with the specified joint spacing. Otherwise, the closure unit will have head joints that are too large or too small. If this is the case, remove the last three or four units that were laid on either side of the closure unit. Remortar them, and re-lay them to correct for the closure head-joint size. The objective is to avoid a sudden jump in the size of a head joint. A big change in joint size will catch the eye and can also skew the pattern bond. If you must move brick, be sure to check them again for height, level, plumb, and range.

2.3.7 Laying Brick to the Line

After you set the line, you can begin to lay masonry units to the line. The advantage of laying to the line is that it cuts down on the need for the mason's level.

You must lay to the line without disturbing the line. This is to crowd the line. If the line is hit, other masons on the line have to wait for the line to stop moving. Even experienced masons crowd the line occasionally. To avoid crowding the line, hold your brick from the top, as shown in *Figure 42*. As you release the brick, roll your fingers or thumb away from the line, then press the brick into place.

The brick must come to sit $\frac{1}{16}$ inch inside the line. The top of the brick must be even with the top of the line. Looking at the brick from above, you should be able to see a sliver of daylight between the line and the brick. Another way to think about this is to lay the brick so that it is one line width away from the line.

28105-13_F41.EPS

Figure 41 Placing a closure unit in a brick wall.

28105-13_F42.EPS

Figure 42 Laying brick to the line.

> **WARNING!**
>
> When tightening a line, the line can snap under tension and cause injury. Before tightening a line, always inform your crew so that they can take appropriate precautions.

While brick set too close is said to be crowding the line, brick set too far from the line is said to be slack to the line. When laid correctly, the bottom edge of the brick should be in line with the top of the course under it, and the top edge of the brick should be 1⁄16 inch back and even with the top edge of the line.

Adjust the brick to the line by pressing down with the heel of your hand. Check that the brick is not hacked, which means having its bottom edge set back from the plane of the wall, or toed, which means having its bottom edge protruding out from the plane of the wall. While you are learning to lay to the line, it is a good idea to check your placement with the mason's level. After you have gained some skill in working to the line, you will find that you will not need to use the mason's level so often.

Much of the mason's time can be spent laying brick to the line. Practice will improve your ability to lay precisely without disturbing the line constantly. As you learn to do this, there are some additional habits you should pick up to save yourself time and energy:

- Always pick up a brick with the face out so that it is in the same position in which you will lay it. Limit turning brick in your hand, as this slows and tires you.
- Pick up frogged brick with the frog down because this is the way it will be laid.
- Fill head and bed joints completely. This cuts down the time you will need to tool joints later. This also ensures stronger, water-resistant walls.
- Stock your brick within arm's length, or approximately 2 feet away. When working on a veneer or cavity wall, stock your brick on the backing wythe.

2.3.8 Building Corners and Leads

As you have already learned, corners are called leads because they lead the laying of the wall (*Figure 43*). They set the position, alignment, and elevation of the wall by serving as guides for the courses that fill the space between them. Building corners requires care as well as accurate leveling and plumbing to ensure that the corner is true. The techniques for building corners and leads with brick are similar to those that you have already learned for building them with block.

2.3.9 Brick Corner Leads

Sometimes it is necessary to build a lead or guide between corners on a long wall. This is a lead without corner angles, or **returns**. It is merely a number of brick courses laid to a given point. A corner lead requires **racking**, or stepping back a half brick on each end with each new course. This means that the lead is laid in a half-lap running bond with one less brick in each course.

The first course is usually six brick units long, the length of the mason's level. Each course is one brick less until the sixth course has only a single brick, pyramid fashion.

A corner lead has return or bend in it. The return must be a 90-degree angle unless the specifications say otherwise. Placement and alignment of the corner are crucial because the corner will

28105-13_F43.EPS

Figure 43 Building a corner lead with brick.

Making Adjustments

If brick is not in line, some adjustment can be made as long as the mortar has not set. Moving a brick after the initial set has started will disturb the line or smear the wall. If it is necessary to move a brick after the initial set has begun, the brick and mortar must be removed and new mortar laid. Do not lift a brick or pull it from the head joint because this will destroy the bond.

Source: Brick Institute of America

set the location of the remainder of the wall. Laying a corner can be intricate, demanding work if there is a pattern bond to follow. The following steps are for building an unreinforced outside-corner lead in a half-lap running bond pattern.

Step 1 Check the specifications for the location of the corner. Determine how high the corner should be. The corner should reach halfway to the top of a wall built with no scaffold, or halfway to the bottom of the scaffold. Subsequent corners are built as the work progresses.

Step 2 Determine the number of courses the corner will need. Use your rule or measure to calculate the courses in a given height; then calculate the number of brick in each course. The sum of the stretchers in the first course must equal the number of courses high the corner will reach.

Step 3 Locate the building line and the position of the corner on the footing or foundation. Clean off the footing, and check that it is level. Lay out the corner with a square. Mark the location directly on the footing. Check the plan or specifications to determine which face of the corner gets the full stretcher and which face gets the header. Some plans have detailed drawings of corners.

Step 4 Establish bond using your rule or measure. If adjustments are necessary, dry-bond the units along the first course in each leg of the corner. Mark the spacing along the footing.

Step 5 Lay the first course of one leg in mortar, and check the height, level, and plumb. Be sure to level the corner brick and the end brick before the units in the center. Check height, level, plumb, and range for the second leg of the corner.

Step 6 Check level on the diagonal, as in *Figure 44*. Lay the mason's level across each diagonal pair of brick. This will let you make sure that the corner continues level across the angle. This may only be necessary in a classroom project setting.

Step 7 Remove excess mortar along the bed and head joints and from the leg ends. Also remove excess mortar from the inside of the corner.

Step 8 Range the brick. To range is to sight along a string to check horizontal alignment. A range is performed after the first course is laid.

Step 9 Lay the second course, reversing the placement of the brick at the corner. The leg that had the full stretcher before now gets the header. Use one less brick in the second course to rack the ends. Since each course is racked, stop spreading mortar half a brick from the end of each course.

Step 10 After placing the brick, check height, level, plumb, and range. To train your eye, sight down the outermost point of the corner unit from above to check plumb. Remove excess mortar from the outside and inside of the corner and from each exposed edge brick.

Step 11 Continue until you have reached the required number of courses. Use one less brick in each course. Lay and align each course. Remove excess mortar from each course, inside and outside. Be sure that aligning the brick does not disturb the mortar bond. If the mortar bond is disturbed, take up the brick, clean it, rebutter it, and replace it.

Step 12 If the corner does not measure up at each course, take the course up, and do it again. This is easier than taking the wall up and doing that again.

Step 13 Check the mortar, and tool the joints. Brush the loose mortar carefully from the brick. Check the height of the corner with the modular spacing rule. Recheck the corner for plumb, level, and range. If it does not measure up, take the corner down, and start over.

28105-13_F44.EPS

Figure 44 Leveling the diagonal.

Because the corner is so important to the wall, speed is not half as important as accuracy. Learn to be accurate, and the speed will follow.

2.4.0 Cleaning Brick

The best method of cleaning any new brick masonry is the least severe method. If the daily cleaning practices listed in the section "Cleaning Concrete Masonry Units" are not enough, the next step is bucket-and-brush hand cleaning. This may include using a proprietary cleaning compound. *Table 2* lists cleaning methods developed by the Brick Industry Association (BIA) for different types of brick.

Any chemical compound you use should first be tested on an inconspicuous 4' × 5' section of wall. Sometimes, minerals in the brick may react with some chemicals and cause stains. Read the brick manufacturer's SDS for recommended cleaning solutions, and follow the manufacturer's directions for mixing, using, and storing any chemical solution.

WARNING!

Wear appropriate personal protective equipment when using chemical solutions.

When cleaning, you will need a hose, bucket, wooden scraper, chisel, and stiff brush. Follow these guidelines:

- Clean new masonry as soon as it has hardened, which is typically seven days. This gives the mortar time to cure and set. Do not wait longer than six months because the mortar will be almost impossible to remove.
- Before wetting, protect any metal, glass, wood, limestone, and cast-stone surfaces. Mask or cover windows, doors, and fancy trim work.
- Prepare the chemical cleaning solution. Follow manufacturer's directions. Remember to pour chemicals into water, not water into chemicals.
- Wet the wall with the hose until the water runs off, to remove loose particles or dirt.
- Start working from the top. Also keep the area wet immediately below the space you are scrubbing, to prevent the chemicals from drying into the wall.
- Scrub a small area with the chemical applied on a stiff brush. Keep the scrub area small enough so that the solution does not dry on the wall as you are working.
- As you finish cleaning each area, rinse the wall thoroughly. Rinse the surrounding wall area above and below, all the way to the bottom of the wall, to keep chemicals from staining the wall.
- Flush the entire wall for 10 minutes after you finish cleaning. This will dilute any remaining proprietary cleaner and prevent burns.

High-pressure water washing, steam cleaning, and sandblasting are also used to clean new and old masonry. Because these techniques can damage masonry surfaces, they require trained operators. If you have not been trained, do not use this equipment. Sandblasting is rarely used, and only under special circumstances, for example, to remove paint from brick.

High-pressure water washing of brick involves the use of a potable water tank and compressor to spray water on the brick. Effective cleaning that

Make Sure the Corner Is Square

You can use a framing square to make sure the corner is square. Each brick in the lead should touch the framing square.

28105-13_SA09.EPS

Table 2 Cleaning Guide for New Masonry

Brick Category	Cleaning Method	Remarks
Red and Red Flashed	Bucket and Brush Hand Cleaning Pressurized Water Abrasive Blasting	Water, detergents, emulsifying agents, or suitable proprietary compounds may be used.
White, Tan, Buff, Gray, Pink, Brown, Black, Specks and Spots	Bucket and Brush Hand Cleaning Pressurized Water Abrasive Blasting	Clean with water, detergents, emulsifying agents, or suitable proprietary compounds. Unbuffered muriatic acid solutions tend to cause stains in brick containing manganese and vanadium. Light-colored brick are more susceptible to "acid burn" and stains, compared to darker units.
Sand Finish or Surface Coating	Bucket and Brush Hand Cleaning	Clean with water and scrub brush using light pressure. Stubborn mortar stains may require use of cleaning solutions. Abrasive blasting is not recommended. Cleaning may affect appearance. See **Brick Category** for additional remarks based on brick color.
Glazed Brick	Bucket and Brush Hand Cleaning Pressurized Water	Wipe glazed surface with soft cloth within a few minutes of laying units. Use a soft sponge or brush plus ample water supply for final washing. Use detergents where necessary and proprietary cleaners only for very difficult mortar stain. Consult brick and cleaner manufacturer before use of proprietary cleaners on salt-glazed or metallic-glazed brick. Do not use abrasive powders. Do not use metal cleaning tools or brushes.
Colored Mortars	Method is generally controlled by **Brick Category**	Many manufacturers of colored mortars do not recommend chemical cleaning solutions. Unbuffered acids and some proprietary cleaners tend to bleach colored mortars. Mild detergent solutions are generally recommended.

28105-13_T02.EPS

does not damage the brick involves maintaining a consistent pressure, water flow, distance, and spray angle. It also requires the operator of the cleaning equipment to use uniform horizontal strokes. Otherwise, pressure washing may damage brick (*Figure 45*). This method is less labor intensive than cleaning by bucket and brush, because it allows masons to clean larger areas more quickly. It is not recommended for sand-finished or surface-coated brick.

Refer to the brick manufacturer's guidelines to ensure that pressurized water cleaning is suitable for the type of brick being used. The same goes for mortar. The BIA recommends that nozzle pressures not exceed 300 psi. The BIA has developed the following steps for performing high-pressure water washing of brick:

Step 1 Clean new masonry as soon as it has hardened, which is typically a minimum seven days. Colored mortar may require more time before cleaning.

Step 2 Remove large mortar clumps using nonmetallic tools prior to pressure washing.

Step 3 Select a proprietary cleaner that is suitable for the type of brick, mortar, and adjacent materials. Follow the cleaner manufacturer's instructions when preparing the cleaner and be sure to verify its compatibility with the cleaning equipment to be used.

Step 4 Protect adjacent windows, doors, and materials such as sealants, metal, glass, wood, limestone, cast stone, concrete masonry and ornamental trim from the cleaning solution. Take similar precautions for grass, plants, and soil.

Step 5 Use a maximum pressure of 30 to 50 psi with a 25- to 50-degree fan-shaped nozzle to saturate the brick masonry to be cleaned by flushing with water from the top down to ensure that it does not absorb the cleaning solution or dissolved mortar particles. Ensure that areas below the section being cleaned are also kept saturated until after the final rinse, to prevent streaking and to prevent the brick from absorbing the runoff.

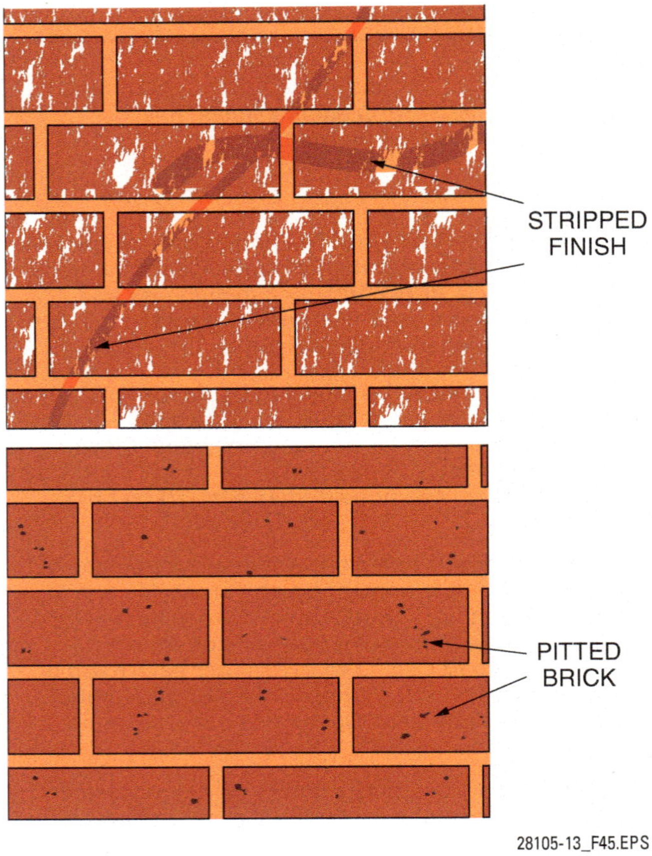

Figure 45 Results of improper pressure washing.

Step 6 Use a low-pressure sprayer of 30 to 50 psi and a 50-degree fan-shaped sprayer nozzle to apply the cleaning solution according to the manufacturer's instructions. A brush may also be used to apply the cleaning solution. Using a high-pressure sprayer to apply cleaning solution can cause the solution to be driven into the masonry, causing staining.

Step 7 After applying the cleaning solution, use a 25- to 50-degree fan-shaped nozzle and a maximum water pressure of 200 to 300 psi to rinse the wall thoroughly by flushing it with large amounts of potable water from top to bottom.

Brick Cleaning Solutions

When selecting a brick cleaning solution, choose the weakest solution available that is compatible with the range of available types of brick.

Additional Resources

Good Practice for Cleaning New Brickwork. 2003. Charlotte, NC: Brick SouthEast.

Technical Note TN20, *Cleaning Brickwork.* 2006. Reston, VA: The Brick Industry Association. **www.gobrick.com/Portals/25/docs/Technical%20Notes/TN20.pdf**

2.0.0 Section Review

1. When measuring the absorption rate of brick using the medicine-dropper test, the brick does not need to be wetted if the time for absorption _____.
   a. exceeds 1½ minutes
   b. exceeds ½ minute
   c. is less than 1½ minutes
   d. is less than ½ minute

2. On a loadbearing wall, the wythe must be bonded to its backing with rigid steel ties when using the _____.
   a. running bond pattern
   b. stack bond pattern
   c. English bond pattern
   d. Flemish bond pattern

3. After laying a brick, the first check is always _____.
   a. course height
   b. range
   c. plumb
   d. level

4. When using high-pressure water washing to clean brick, the BIA recommends that nozzle pressures not exceed _____.
   a. 600 psi
   b. 500 psi
   c. 400 psi
   d. 300 psi

SECTION THREE

3.0.0 CUTTING CONCRETE MASONRY UNITS AND BRICK

Objective

Describe how to cut concrete masonry units and brick.

a. Explain how to cut with chisels and hammers.
b. Explain how to cut with masonry hammers.
c. Explain how to cut with saws and splitters.
d. Explain how to check units and cuts.

Performance Tasks 9 and 10

Accurately cut block using the following tools:

- A masonry hammer
- A brick set
- A power saw
- A splitter

Accurately cut brick using the following tools:

- A masonry hammer
- A brick set
- A power saw
- A splitter

Trade Term

Lintel: The support beam over an opening such as a window or door. Also called a *header*.

Masonry units often need to be cut to fit a specific space. Even when building on a modular grid, structural bond patterns, door and window openings, and corners usually call for some cut masonry units. English and Dutch corners specifically call for cut masonry units as part of the patterning.

On a large job, the masonry contractor or supervisor will figure the pattern layouts and calculate the number of masonry units to be cut. Someone will be assigned to cut the units with a masonry saw or a splitter before they are needed. Sometimes, masons need to cut a few more units or cut to a slightly different size. This is when you need to know how to cut masonry with hand tools.

Block is usually cut in several standard ways. It can be cut across the stretcher face, both horizontally and vertically, as shown in *Figure 46*. You may easily make these cuts by hand. Block cut across the face horizontally is called split or rip. If the block is cut exactly in half, it is called a half-high rip. Rip block is often used under windows. They act as a filler to reach a height of 8 inches, so normal coursing can continue.

Block units also get their webs cut out. Taking one end off a block makes an opening easily slipped over a pipe. Fitting the block to its location may take more cuts. You might make these cuts using a masonry saw. The cuts have their own names to save time and confusion.

The following four types of cuts are shown in *Figure 47* on a two-cell block:

- The open-end cut has the end of one cell cut out, leaving one cell.
- The center-web or web-out cut has the internal web cut out.
- The double open-end or H-block cut has both ends cut out.
- The clothespin or hairpin cut has one end cut out. This leaves only one end to hold the block together.

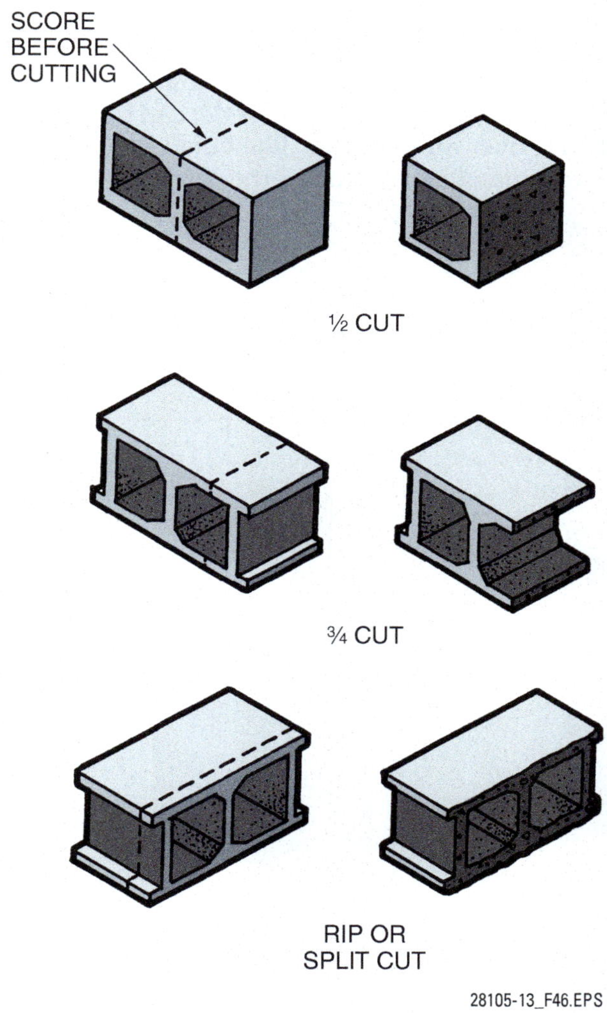

Figure 46 Horizontal and vertical face cuts.

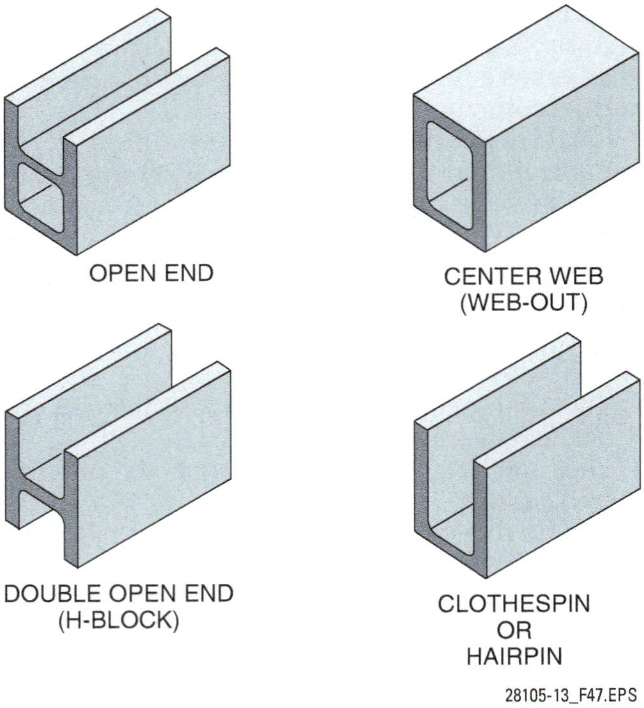

Figure 47 End and web block cuts.

The bond-beam block has the ends and inside web of the block cut down one-quarter and one-half of the way. This cut (*Figure 48*) can be used for a lintel over an opening. The cuts give room for the reinforcement on top of the opening.

> **WARNING!**
>
> Remember to wear a hard hat and eye protection when cutting with hand tools. Never cut masonry over the mortar pan or near other workers. Chips may fly off, causing injury.

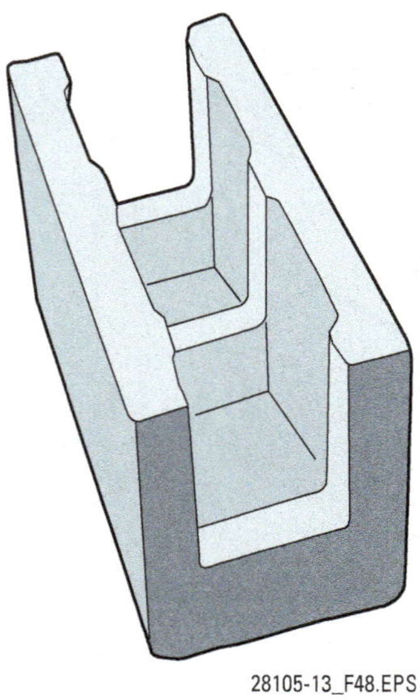

Figure 48 Bond-beam cut.

Block can be cut with chisels or a brick hammer. The procedures are detailed in the following sections. Block can also be cut easily by hand tool, masonry saw, or splitter. Sometimes you will need to cut brick for finishing corner patterns, bonds, and around openings. *Figure 49* shows the common cut-brick shapes and names. The soap closure is used for cornering.

3.1.0 Cutting with Chisels and Hammers

Using the chisel and hammer can result in a smooth cut for block. This procedure also works well for brick:

Step 1 Check the tools you will use. Cutting edges should be sharp, and the hammer handle should be firmly attached.

Step 2 Put on your hard hat and safety goggles or other eye protection.

Step 3 Put the block or brick on a bag of sand, a board, or the ground to make a safe cutting surface. Make sure it is resting flat and plumb on a surface with some give to it.

Step 4 Use a square and a pencil to mark the cut all the way around the masonry unit.

Step 5 Hold the chisel (for block) or the brick set (for brick) vertically on the marked line. The flat side of the chisel should face the finished cut, or the part you want to keep.

Step 6 Give the chisel end several light taps with the striking end of the hammer to score

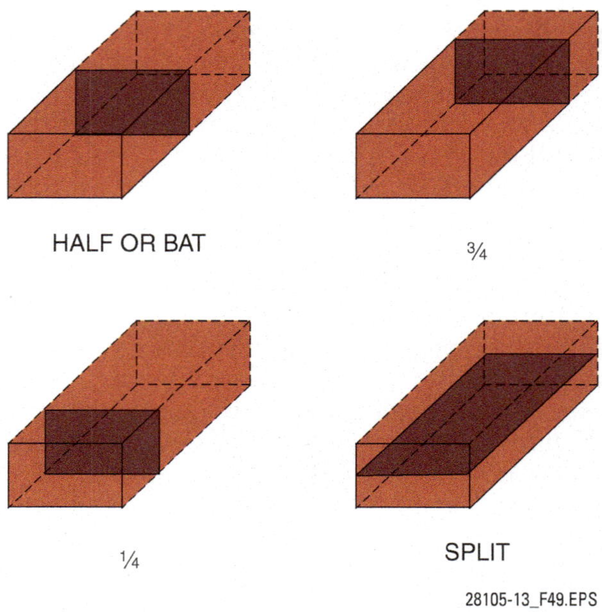

Figure 49 Common brick cuts.

the masonry unit. Move the hammer and chisel all around the unit, scoring all along the cut mark. Be sure to keep your fingers above the cutting edge of the chisel. *Figure 50* shows this step.

Step 7 Place the chisel on the scored line, with the flat side facing the finished cut. Deliver a hard blow to the chisel head with the hammer. Sometimes two blows are needed.

Cutting in this way gives an accurate and clean cut. You can also make an accurate cut by using a hammer, instead of a chisel, for the final step. If you are cutting with the mason's hammer, follow Steps 1 through 7 just listed, then continue with these steps:

Step 1 Place the scored block on top of another block so that the waste part hangs free. You can hold the wanted part secure with your foot.

28105-13_F50.EPS

Figure 50 Cutting with a hammer and chisel.

Step 2 Strike the waste end of the block with the striking end of the hammer. This knocks off the waste end, leaving a clean, finished cut.

NOTE

Hammer and chisel cutting may not be permitted on some commercial jobs. Check the project specifications.

3.2.0 Cutting with Masonry Hammers

Cutting with the chisel end of the masonry hammer gives a rougher cut. The steps for cutting brick in this way are as follows:

Step 1 Check the tool you will use. Cutting edges should be sharp, and the hammer handle should be firmly attached.

Step 2 Put on your hard hat and safety goggles or other eye protection.

Step 3 Use a square and a pencil to mark the cut all the way around the masonry unit.

Step 4 Hold the brick in one hand and your hammer in the other. Hold the part of the brick you want to keep, with the waste part down.

Step 5 Strike the brick lightly with the chisel end of the hammer to score it along the marks on all sides. As you turn the brick, be sure to keep your fingers and thumb off the side of the brick being scored. *Figure 51* shows this step.

Step 6 Strike the face of the brick sharply with the chisel end of the hammer. Let the waste part fall to the ground.

Step 7 If necessary, use the striking end of the hammer to dress out any small, rough edges left by the cut.

The same procedure can be used for block, except that block should not be held in your hand. Set the block on sand, the ground, or a board for a safe cutting surface. Follow Steps 1, 2, 3, and 5. Then tilt the block face away and prop it with another block. Hold it with your foot and apply a sharp blow with the hammer. Block may need to be struck on both faces. Finish by dressing out any rough edges.

NOTE

Never use a trowel to cut block or brick.

28105-13_F51.EPS

Figure 51 Cutting with a brick hammer.

3.3.0 Saws and Splitters

Cutting masonry units with power saws or splitters takes two kinds of awareness: you must be aware of how to operate the machinery, and also be aware of the masonry units and cuts.

Masonry saws are available in freestanding and portable models (*Figure 52*). They use either diamond or Carborundum™ blades. Diamond blades are often irrigated to control dust and to prolong the life of the blade. The water wets the masonry unit, which must dry out before it can be laid. Carborundum™ blades are not irrigated, but they make clouds of dust. Dust must be controlled according to requirements in the local applicable code.

WARNING!

Silicosis is a serious lung disease that is caused by inhaling sand dust. Silica is a major component of sand and is therefore present in concrete products and mortar. Silica dust is released when cutting brick and cement, especially when dry-cutting with a power saw. Any time you are involved in the cutting or demolition of concrete or masonry materials, be sure to wear approved respiratory equipment.

Splitters (*Figure 53*) do not use water or generate dust. They do, however, exert tremendous force through gearing and hydraulic power.

As with any potentially dangerous equipment, follow these general safety rules:

- Do not operate any saw or splitter until you have had specific instructions in handling that equipment.
- Check the condition of the equipment before using it.

28105-13_F52.EPS

Figure 52 Portable masonry saw.

28105-13_F53.EPS

Figure 53 Brick splitter.

- Wear appropriate personal protective equipment, including hearing protection, hard hat, respiratory protection, eye protection, and gloves as needed.
- Never force the equipment.
- For handheld saws, secure and brace the unit before cutting it.
- Do not operate equipment when you are feeling ill or are taking any medication that may slow your reaction time.

Review the safety rules as well as the operating instructions before operating any equipment.

3.4.0 Checking Units and Cuts

After you have checked out the equipment, the safety procedures, and the operating procedures, check out the masonry units:

- Know what the finished item should look like.
- Mark all cutting lines before the blade starts running.
- Mark cutting lines in grease pencil for wet-cut saws.
- Do not cut a cracked masonry unit.

If you are not clear about the cuts to be made, ask your supervisor for more direction.

Additional Resources

Technical Note TN3A, *Brick Masonry Material Properties*. 1992. Reston, VA: The Brick Industry Association. **www.gobrick.com/Portals/25/docs/Technical%20Notes/TN3A.pdf**

3.0.0 Section Review

1. When cutting with the mason's hammer, once the masonry unit has been scored, _____.
   a. place the scored block on top of another block so that the waste part hangs free
   b. hold the blocking chisel or the brick set vertically on the marked line
   c. use a square and a pencil to mark the cut all the way around the masonry unit
   d. tool the face of the brick sharply with the chisel end of the hammer

2. When cutting a concrete masonry unit with a masonry hammer, the block should *not* be placed _____.
   a. on sand
   b. on the ground
   c. on a board
   d. in your hand

3. The device used to cut masonry units by exerting tremendous force through gearing and hydraulic power is called a _____.
   a. masonry saw
   b. splitter
   c. pile driver
   d. breaker

4. When using a wet-cut saw to cut masonry units, mark the cutting lines using a _____.
   a. felt-tip marker
   b. grease pencil
   c. lead pencil
   d. laser etcher

SECTION FOUR

4.0.0 MASONRY REINFORCEMENT AND ACCESSORIES

Objective

Describe how to install masonry reinforcement and accessories.

a. Describe how to install masonry reinforcements.
b. Describe how to install masonry accessories.

Trade Terms

Hurricane strap: A special kind of anchor that is used to tie the frame to the foundation.

Veneer tie: An anchor that is used to tie veneer to masonry structures.

Masonry walls usually need material in addition to mortar to make the wall stronger, to hold it in place, or to handle moisture. These additional materials include reinforcements and accessories. The masonry contractor or general contractor will buy these materials and supply them as part of the job. Plans and specifications typically detail the locations and types of the reinforcements and accessories to be used.

4.1.0 Installing Masonry Reinforcements

Reinforcements that are commonly used in masonry include anchors for brick veneer over all types of structural backing, metal and **veneer ties**, reinforcement bars, and joint reinforcement ties. Each of these will be discussed separately below. Be sure to follow the manufacturer's instructions when installing reinforcements. Be sure to use reinforcements that are appropriate for the type of masonry unit that you are installing.

4.1.1 Anchors

Anchors are used to tie brick veneer to steel and wood stud walls. They are also used to anchor a wall that meets another wall at a 90-degree angle. Anchors must be installed according to the specifications, as they affect the loadbearing of the wall. *Figure 54* shows two types of shaped anchors. Anchors should be spaced no more than 18 inches on center vertically and no more than 32 inches on center horizontally. Anchors are attached to the studs, not to the sheathing. A special kind of anchor called a **hurricane strap** is used to tie the frame to the foundation (*Figure 55*). Hurricane straps do not anchor veneer.

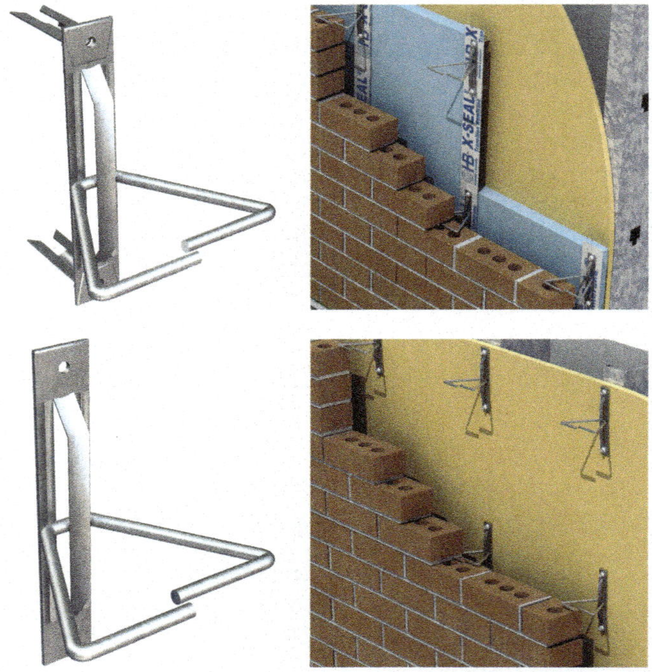

28105-13_F54.EPS

Figure 54 Anchors for brick veneer over steel stud walls.

4.1.2 Metal and Veneer Ties

Metal ties are sized and spaced according to project specifications. The ties keep the walls from separating when weight is placed on them by the other parts of the structure. Metal ties are also used for composite walls. The ties equalize the loadbearing and also tie the two wythes together.

Veneer ties (*Figure 56*) are used to tie a masonry veneer wall to a backing wall. Veneer ties can equalize loadbearing and help keep the veneer wall from moving away from its backing. They are made of corrugated galvanized steel or wire.

4.1.3 Reinforcement Bars

Steel reinforcement bars come in different thicknesses and lengths. They are inserted in block cores, and then the cores are filled with grout. They add strength and weight-bearing capacity to block walls. Sometimes they are placed in the middle of cavity walls where the cavity is to be grouted.

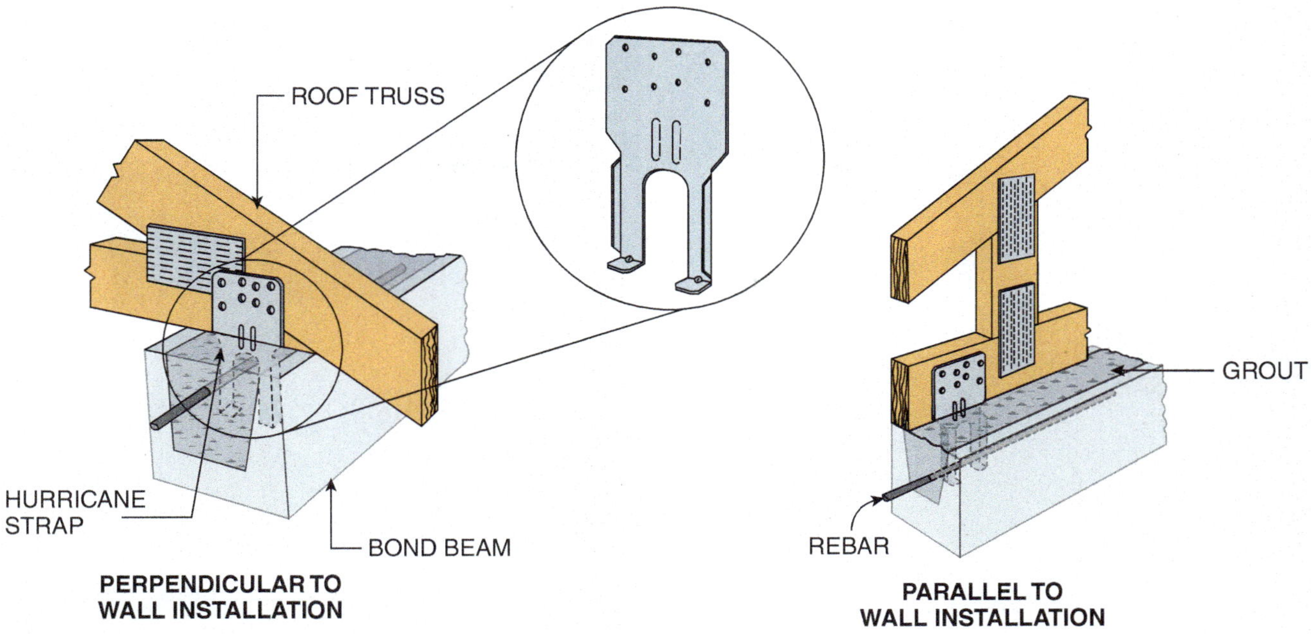

Figure 55 Hurricane strap.

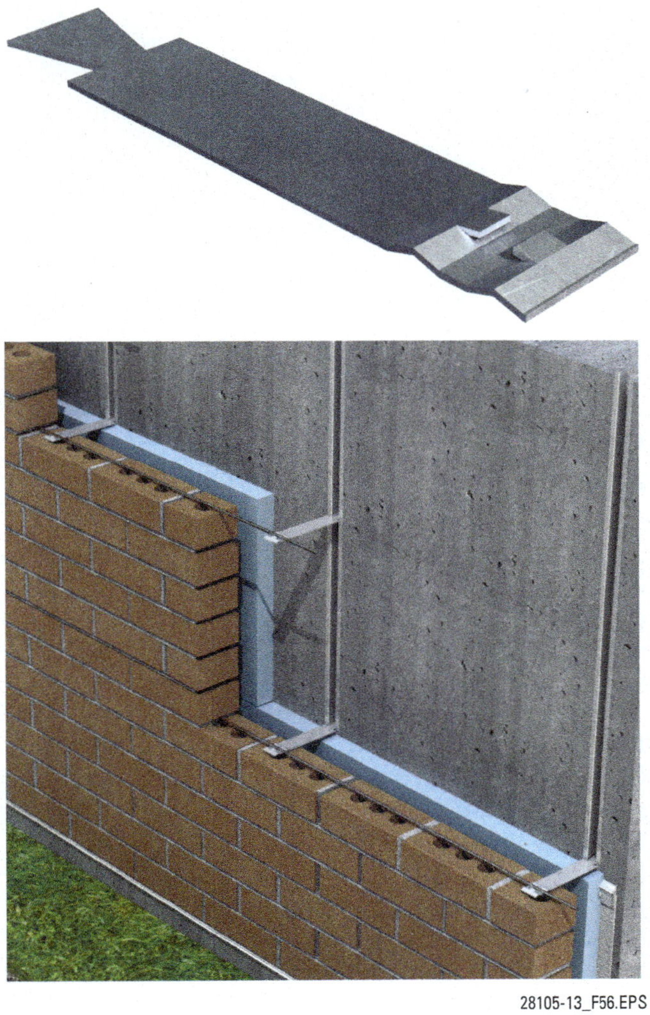

Figure 56 Veneer ties.

4.1.4 Horizontal Joint Reinforcement

Horizontal joint reinforcement is made of two 10-foot lengths of steel bars welded together by rectangular or triangular cross bracing. *Figure 57* shows ladder (rectangular) and truss (triangular) versions of horizontal joint reinforcement. They are used in horizontal joints every second or third course, as specified.

4.2.0 Installing Masonry Accessories

Accessories that are commonly used in masonry include flashing and joint fillers. Each of these accessories is discussed below. Always follow the manufacturer's instructions when installing accessories, and ensure that the accessories you are using are appropriate for the type of masonry unit being installed.

4.2.1 Flashing

Flashing within a wall allows moisture trapped in a masonry wall to exit through a weep system to the outside face of the wall. Flashing is placed under masonry lintels, sills, copings, and spandrels. The most common flashing is made of copper, stainless, or galvanized metal. Bituminous flashing is made of fabric saturated with asphalt. Newer types of flashing are made of plastics. They are cheaper and easier to work with. *Figure 58* shows flashing in position under a sill and a lintel.

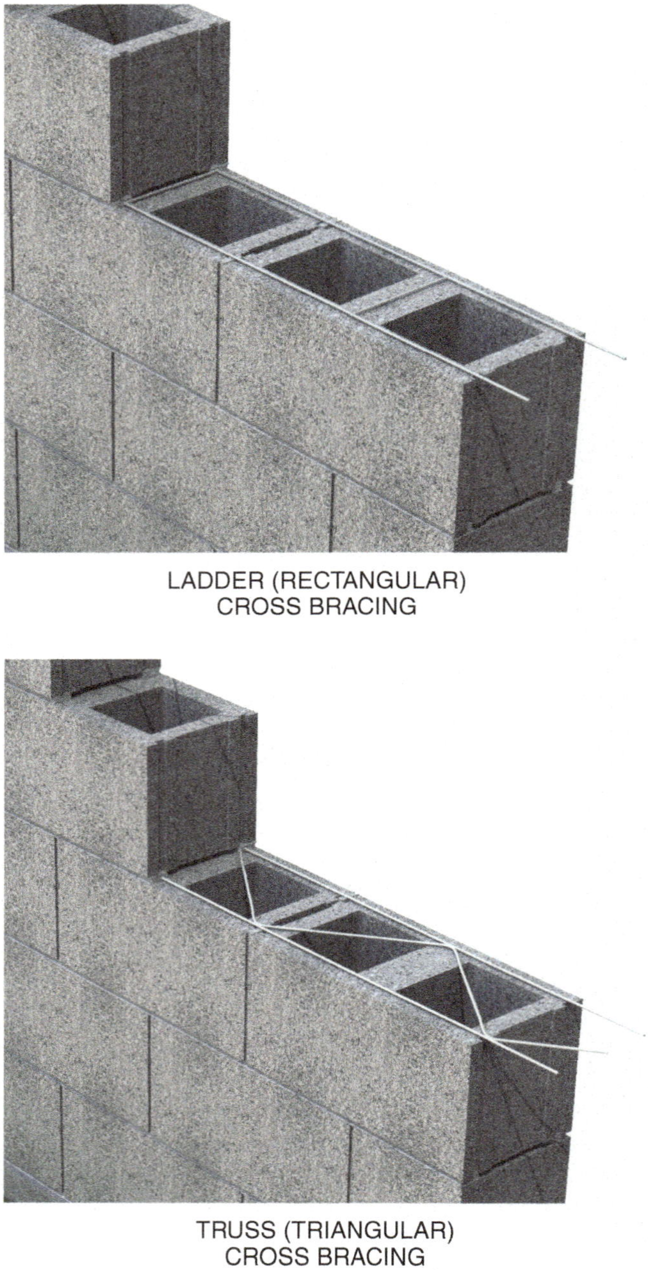

28105-13_F57.EPS

Figure 57 Joint reinforcements.

Some manufacturers produce complete flashing systems that include a sheet metal or flexible membrane, metal drip edge, mortar collection device, and a termination bar. End dams and corner boots can complete the system. *Figure 59* shows a flashing system under construction and the completed system. Unitized flashing systems are also available. These systems are assembled in a manufacturing facility, allowing the contractor to simply remove a complete product from the box and install it in the wall. *Figure 60* shows an example of a unitized system.

28105-13_F58.EPS

Figure 58 Flashing applications.

4.2.2 Joint Fillers

Plastic or rubber joint fillers are used to replace mortar in expansion or contraction joints in block walls. They break the bond between adjacent block and allow expansion and contraction of the wall. They fill the control joints in order to keep moisture out of the space. *Figure 61* shows molded joint fillers used for block.

When control joints are not used, block construction is prone to cracking from shrinkage, while brick construction is prone to cracking from expansion. Shrinkage cracking occurs as the concrete slowly finishes drying. As the concrete shrinks, it moves slightly. The movement causes cracks in a rigid slab or wall. The shrinkage cracking in a block's structure is controlled in the same way as for concrete slabs. This is done by using reinforcement in combination with contraction or control joints.

28105-13_F59.EPS

Figure 59 A flashing system.

28105-13_F60.EPS

Figure 60 Unitized flashing system.

28105-13_F61.EPS

Figure 61 Joint fillers.

Figure 62 shows some typical locations for contraction joints. Walls are likely to crack at abrupt changes in wall thickness or heights, at openings, over windows, and over doors. Control joints are also used at intersections between loadbearing walls and partition walls.

Block walls use two kinds of reinforcement: grout and steel with grout. Either the grout is poured into block cores, or steel rods are inserted in block cores, and grout is poured around the steel. The reinforcement gives rigidity and strength to the wall. Block walls need reinforcement of either kind on both sides of a control joint.

Control joints establish movement points in the structure. Cracks occur at the control joint instead

Chimney Flashing

Proper flashing is extremely important around chimneys, especially in areas subject to snow buildup. On a new building, roofers will apply the flashing. If existing flashing is disturbed during a chimney repair, however, it is up to the mason to make sure the flashing is secure.

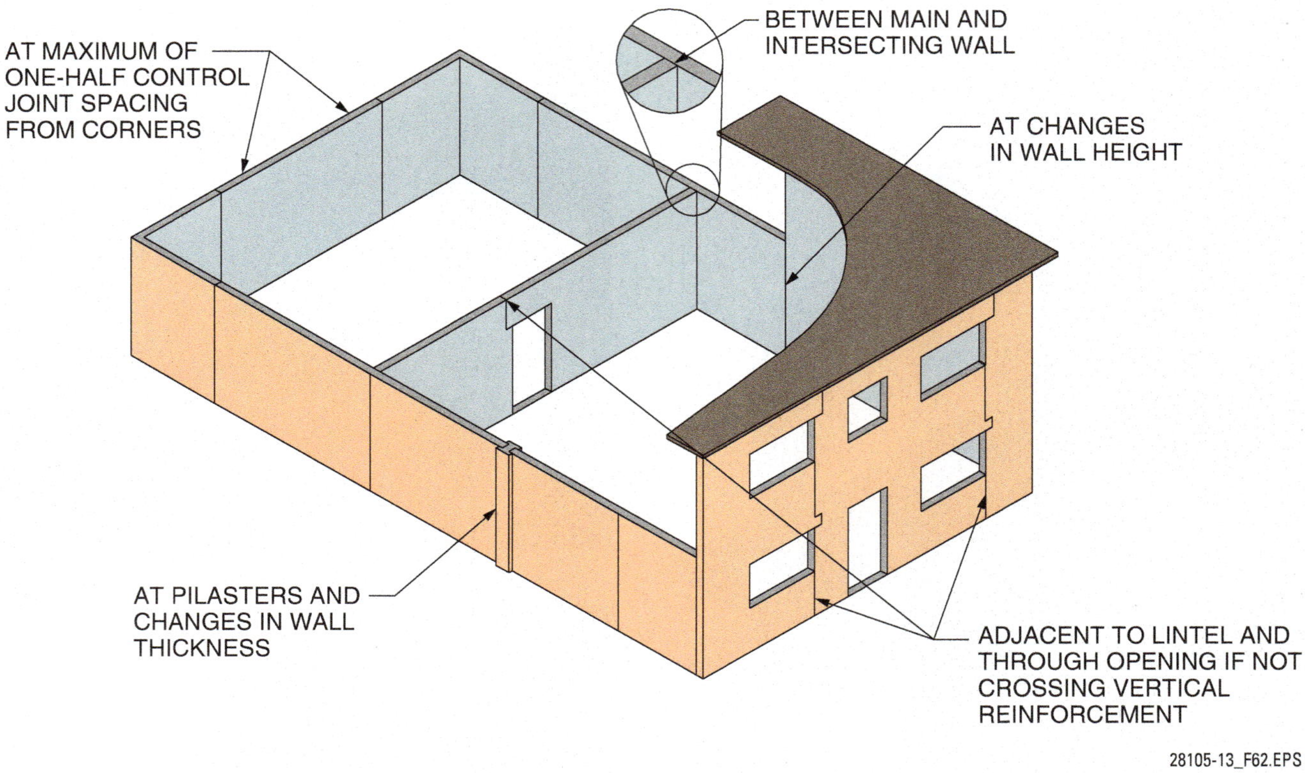

Figure 62 Locations of control joints in a block structure.

of randomly. In a concrete slab, these control joints are made by cutting grooves. In a block wall, this is done by breaking the contact between two columns of units, as shown in *Figure 63*.

The control joints replace a standard mortar joint every 20 feet or so. Control joints must be no more than 30 feet apart. The control joints in block walls have no mortar; they are filled with silicone or another flexible material. The control-joint filler keeps the rain out but allows slight movement. The reinforcement keeps the edges of the contraction joints aligned as they move.

Clay masonry units do not contract and shrink as they age, but they do change size very slightly with temperature and moisture changes in the air. Clay masonry structures need expansion joints to handle this type of movement. Clay masonry walls must include expansion joints. These are usually soft, mortarless joints filled with foam and covered with a layer of silicon paste. As with contraction joints, the filler keeps the rain out and allows slight movement of the masonry.

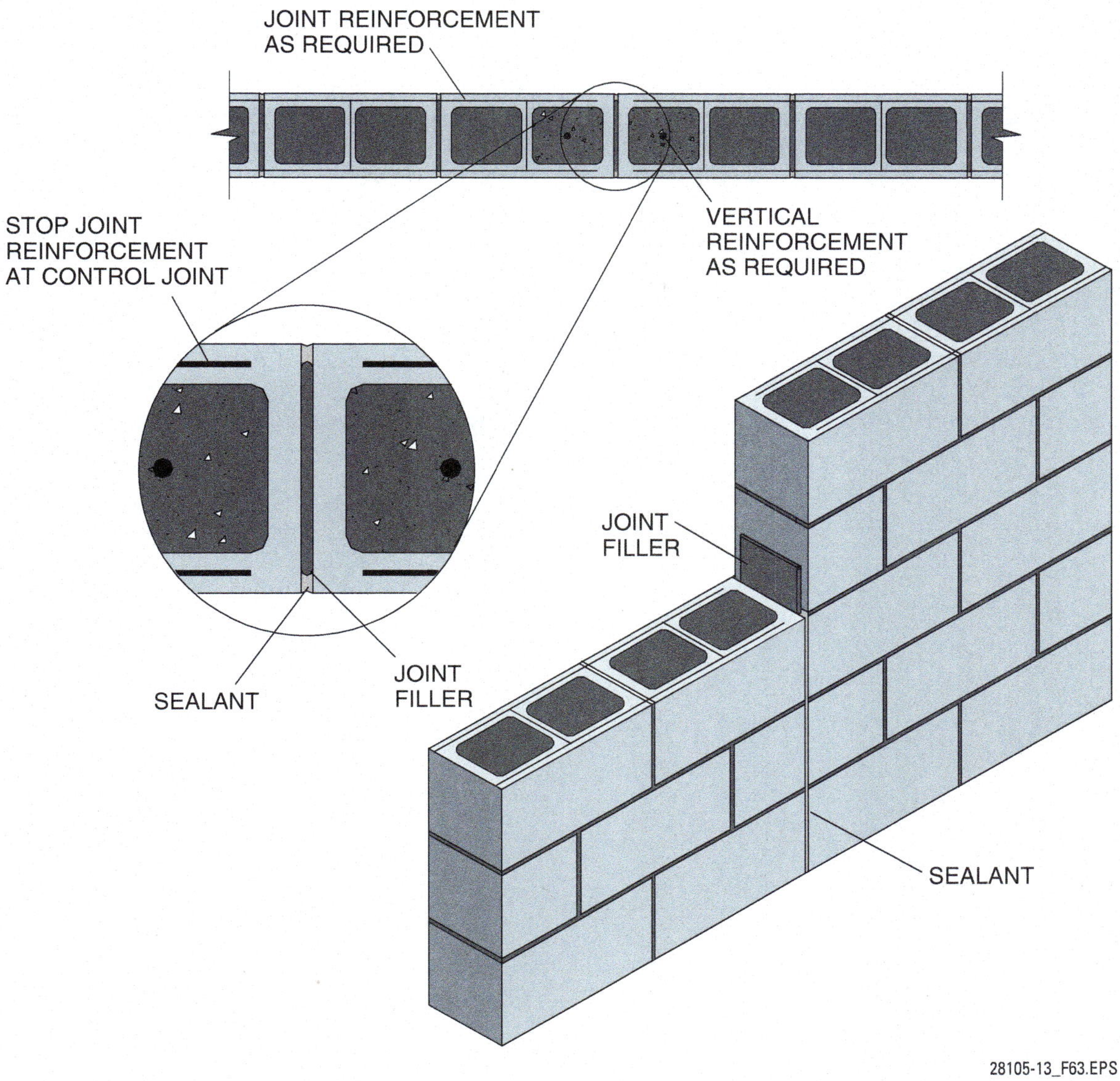

Figure 63 Vertical control joint with plastic filler.

Additional Resources

The ABCs of Concrete Masonry Construction. Skokie, IL: Portland Cement Association (Video, 13 min. 34 sec.).

Bricklaying: Brick and Block Masonry. Reston, VA: Brick Institute of America.

Building Block Walls: A Basic Guide for Students in Masonry Vocational Training. 1988. Herndon, VA: National Concrete Masonry Association.

Concrete Masonry Handbook. Skokie, IL: Portland Cement Association.

4.0.0 Section Review

1. The maximum distance that anchors should be spaced on center vertically is _____.
   a. 12 inches
   b. 18 inches
   c. 24 inches
   d. 30 inches

2. A control joint in concrete masonry walls replaces a standard mortar joint approximately every _____.
   a. 10 feet
   b. 15 feet
   c. 20 feet
   d. 25 feet

SUMMARY

Masons are skilled craft professionals whose work often stands not just for many years, but also for many decades after completion. It is up to each individual mason to learn the skills necessary to create strong, properly built structures. These skills take practice and patience to master.

In order to have a long, productive career as a mason, you must work hard toward mastering your craft. You must learn all of the relevant specifications and standards, and keep yourself informed of changes in the regulations governing the masonry industry. As well, you need to be familiar with the specific challenges and special project types common to your particular region.

Learning the basic skills in this module is essential to becoming a skilled mason. However, being a successful mason is about much more than understanding how to choose masonry units, lay a true course, and properly mix mortar. It is about having pride in your workmanship and about taking the time to do each job right. By mastering the craft, having real pride in your work, and always continuing to learn, you can ensure yourself a long, productive career in masonry.

Review Questions

1. The number of types that ASTM classifies concrete masonry units (block) into is _____.
   a. two
   b. four
   c. six
   d. eight

2. The amount of weight a block can support without collapsing is its _____.
   a. tensile strength
   b. structural strength
   c. integral strength
   d. compressive strength

3. Uncored loadbearing block is available in grade(s) _____.
   a. N only
   b. N and S
   c. S only
   d. M and S

4. The concrete block shape shown in *Review Question Figure 1* is a(n) _____.
   a. stretcher unit
   b. lintel unit
   c. control joint unit
   d. double open-end unit

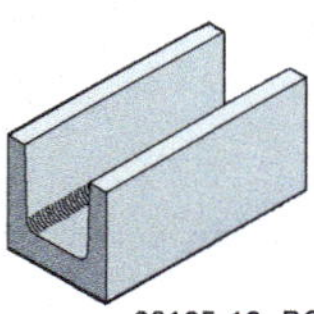

28105-13_RQ01.EPS

Figure 1

5. Nonstructural concrete block has a minimum compressive strength of _____.
   a. 500 psi
   b. 750 psi
   c. 1,000 psi
   d. 1,500 psi

6. Compared to Grade N concrete block, Grade S concrete brick _____.
   a. is more resistant to moisture
   b. is more readily available
   c. is less resistant to frost penetration
   d. provides better insulation

7. Allowing block to become exposed to moisture can result in _____.
   a. difficulty in aligning units
   b. shrinkage upon drying
   c. efflorescence
   d. expansion of mortar joints

8. When laying out courses, a mason must use _____.
   a. control dimensions
   b. modular units
   c. nominal dimensions
   d. variable joint spacing

9. The concrete block bond pattern shown in *Review Question Figure 2* is a(n) _____.
   a. English bond
   b. horizontal stack bond
   c. Flemish bond
   d. common bond

28105-13_RQ02.EPS

Figure 2

10. To keep a long line from sagging, masons use a support device called a _____.
    a. line stretcher
    b. line pin
    c. line block
    d. line trig

11. Soft-surfaced brick _____.
    a. may need to be wetted down before use
    b. should never be dried
    c. may need to be dried before use
    d. should never be wetted down

12. To identify the top side of a concrete masonry unit, look for _____.
    a. a stamped marking
    b. a larger shell and web
    c. a grooved indentation
    d. a thinner shell and web

13. The brick position that is used most in laying walls is the _____.
    a. soldier
    b. header
    c. rowlock
    d. stretcher

14. The vertical mortar joint between masonry units is called the _____.
    a. filler joint
    b. stretcher joint
    c. head joint
    d. tail joint

15. A buttered brick is placed and then adjusted to achieve a mortar joint that is approximately _____.
    a. ¼-inch wide
    b. ⅜-inch wide
    c. ½-inch wide
    d. ¾-inch wide

16. An out-of-plumb brick that tilts toward the mason's line is referred to as _____.
    a. toed
    b. slanted up
    c. slanted down
    d. hacked

17. To keep walls from separating when weight is placed on them by the other parts of the structure, you would use _____.
    a. reinforcement bars
    b. hurricane straps
    c. metal ties
    d. veneer ties

18. When masons want to break the bond between adjacent block and allow expansion and contraction of a block wall, an alternative to mortar in the expansion or contraction joints is _____.
    a. grout
    b. plastic or rubber joint filler
    c. expanded rigid foam
    d. flashing

19. The type of block cut in which the end of one cell is cut out, leaving one cell, is called the _____.
    a. center-web cut
    b. open-end cut
    c. clothespin cut
    d. double open-end cut

20. The brick type used for cornering is the _____.
    a. split closure
    b. ¾ closure
    c. half or bat closure
    d. soap closure

Trade Terms Quiz

Fill in the blank with the correct term that you learned from your study of this module.

1. A corner in a structure or lead is called a(n) _______.
2. To _______ is to lay a bed of mortar in a line for a course of masonry units.
3. When a person touches the mason's line, or a masonry unit is too close to the line, the person or masonry unit is said to _______.
4. A lead built at a corner in which each course of masonry is shorter than the course below it is called a(n) _______.
5. A(n) _______ is a special kind of anchor that is used to tie the frame to the foundation.
6. Masonry units set too far away from the mason's line are said to be _______.
7. A(n) _______ is the support beam over an opening such as a window or door.
8. Laying out masonry units without mortar to establish spacing is called a(n) _______.
9. To _______ is to check the spacing of head joints by checking the diagonal edges of the courses on a lead or corner.
10. The last brick to fill a course is called the _______.
11. _______ is to align a masonry unit or wall along the plane or face of the wall.
12. Shortening each course of masonry by half a unit so it is shorter than the course below it is called _______.
13. _______ is a style of brick in which a depression in the bedding surface of a brick lightens its weight.
14. An anchor that is used to tie veneer to masonry structures is called a(n) _______.
15. _______ is laying out masonry units to establish spacing.

Trade Terms

Bond
Closure unit
Corner lead
Crowd the line
Dry bond
Frogged
Hurricane strap
Lintel
Racking
Range
Return
Slack to the line
Spread
Tail
Veneer tie

Cornerstone of Craftsmanship

Bryan Light

Technical Services Manager
Brick Industry Association Southeast Manager

Bryan Light learned from an early age that hard work and dedication are key ingredients for career success. Today, when he talks to designers and inspects masonry installations, he is reminded how important those characteristics are for ensuring quality—not only of a particular building, but of the entire masonry industry.

Describe your job.
It is my job to promote the use of masonry on all levels. If the public has a false conception about masonry, if installers are not knowledgeable, if designers are not up to speed, or if our workforce cannot meet consumer demand, then lesser claddings will be employed on future building projects. I try to make masonry desirable in as many applications as possible so that it will continue be the premier product that it has been for millennia.

How did you get started in the construction industry?
I have two older sisters who married masons, so from the time I was 14 years old I had the perfect opportunity for summer jobs in the family business! Masonry started out as a job, but quickly turned into a passion. I started my own company at the age of 23.

What do you think it takes to become a success?
There is no easy way to succeed; it takes hard work and dedication to do what means most to you. The concept of success in business is not the same for all people. Many miserable people appear to be successful.

Masonry is not portrayed as the most glamorous of careers. However, when you have the knowledge and ability to erect structures that can be used for many different purposes, and that can last for several lifetimes, then that fosters a very rewarding and successful feeling in a person.

What do you enjoy most about your career?
The journey and the legacy. I am happy to know that my introduction to masonry was as a laborer, but that I was able to move into decades of being the owner of a masonry business, and then on to Technical Services Manager for the Brick Industry Association. As for the legacy, I continue to be very proud when I drive by homes, banks, schools, religious structures, and other buildings on which I installed the masonry that still are just as beautiful as they were originally.

Why do you think training and education are important in construction?
The need for a trained and educated workforce is perhaps the most pressing issue facing the masonry industry today in order to ensure that masonry remains the preferred building material in construction. When I encourage designers to specify more masonry in their projects, the first thing they ask is, "Will we have a well-trained workforce to install it?" And when I conduct an inspection to determine why a particular masonry installation fell short of expectations, some owners have said to me, "I wish I had used another cladding material." Both of these issues must be addressed with proper training.

Why do you think credentials are important in construction?
Credentials facilitate the smooth movement of members of our workforce from one location to another. Well-trained individuals have long been referred to as journeymen. When masons move, credentialing makes it easy for them to carry along proof of their level of expertise. Thus, they can get back to work quickly in the new location, and their new employers will feel more confident about their abilities.

How has training/construction impacted your life and career?
Masonry construction has given me a level of pride that I may not have been able to achieve in another profession. It has rewarded me handsomely financially as well, and as a result I have been able to provide everyone in my family a very comfortable life.

Would you recommend construction as a career to others?
Looking ahead to what will be needed in this country as well as around the world, I see that masonry will continue to be highly sought after in institutional, commercial, and residential markets. For this reason, a career in masonry installation will continue to be lucrative as well as rewarding for a very long time. So yes, I highly recommend it!

What does craftsmanship mean to you?
To be called a craftsman is special. A famous philosopher of old said, "Have you seen a man skillful in his work? Before kings is where he will station himself; he will not station himself before commonplace men."

Trade Terms Introduced in This Module

Bond: Laying out masonry units to establish spacing.

Closure unit: The last brick to fill a course.

Corner lead: A lead built at a corner in which each course of masonry is shorter than the course below it.

Crowd the line: To touch the mason's line or to place a masonry unit too close to the line.

Dry bond: Laying out masonry units without mortar to establish spacing.

Frogged: A style of brick in which a depression in the bedding surface of a brick lightens its weight.

Hurricane strap: A special kind of anchor that is used to tie the frame to the foundation.

Lintel: The support beam over an opening such as a window or door. Also called a header.

Racking: Shortening each course of masonry by half a unit so it is shorter than the course below it.

Range: To align the masonry unit or wall along the plane or face of the wall. Walls should be ranged from one corner to another.

Return: A corner in a structure or lead.

Slack to the line: Masonry units set too far away from the mason's line.

Spread: To lay a bed of mortar in a line for a course of masonry units.

Tail: To check the spacing of head joints by checking the diagonal edges of the courses on a lead or corner.

Veneer tie: An anchor that is used to tie veneer to masonry structures.

Additional Resources

This module presents thorough resources for task training. The following resource material is suggested for further study.

The ABCs of Concrete Masonry Construction. Skokie, IL: Portland Cement Association (Video, 13 min. 34 sec.).

ASTM C90, Standard Specification for Loadbearing Concrete Masonry Units. 2012. West Conshohocken, PA: ASTM International.

Bricklaying: Brick and Block Masonry. Reston, VA: Brick Institute of America.

Building Block Walls: A Basic Guide for Students in Masonry Vocational Training. 1988. Herndon, VA: National Concrete Masonry Association.

Concrete Masonry Handbook. Skokie, IL: Portland Cement Association.

Good Practice for Cleaning New Brickwork. 2003. Charlotte, NC: Brick SouthEast.

Technical Note TN3A, *Brick Masonry Material Properties*. 1992. Reston, VA: The Brick Industry Association. **www.gobrick.com/Portals/25/docs/Technical%20Notes/TN3A.pdf**

Technical Note TN20, *Cleaning Brickwork*. 2006. Reston, VA: The Brick Industry Association. **www.gobrick.com/Portals/25/docs/Technical%20Notes/TN20.pdf**

TEK 2-3A, *Architectural Concrete Masonry Units*. 2001. Herndon, VA: National Concrete Masonry Association. **secure.ncma.org/source/Orders/ProductDetail.cfm?pc=TEK02-03**

TEK 8-4A, *Cleaning Concrete Masonry*. 2005. Herndon, VA: National Concrete Masonry Association. **www.ncma.org/etek/Pages/Manualviewer.aspx?filename=TEK%2008-04A.pdf**

Figure Credits

Courtesy of National Concrete Masonry Association, Figure 1, Figure 2, Figure 5, Table 1, Figure 62, RQ01

ChemGrout, Inc., SA01

Courtesy of Portland Cement Association, Figure 3, Figure 9, Figure 20 (top)

Courtesy of Air Vol Block, Inc., Figure 4

Architect: USA Architects, Planners + Interior Designers. Project: William Allen High School, Allentown PA. We are: Oldcastle®. Product: Trenwyth™ Astra-Glaze®, Figure 7

Courtesy of the Brick Industry Association, SA04, Figure 28, Figure 30, Figure 31, Figure 33, Figure 34, SA06, SA07, Table 2, Figure 49, RQ02, E01

Courtesy of Haskell, Figure 10

Courtesy of Dennis Neal, FMA&EF, SA05, Figure 40

Marshalltown Company, Figure 17

Courtesy of iQ Power Tools, Figure 52

Granite City Tools, Figure 53

Courtesy of Hohmann & Barnard, Inc., Figure 54, Figure 56, Figure 57, Figure 58 (top), Figure 61

Courtesy Of WIRE-BOND®, Figure 58 (bottom)

Courtesy of Mortar Net Solutions, Figure 59, Figure 60

Section Review Answer Key

Answer	Section Reference	Objective
Section One		
1. c	1.1.1	1a
2. d	1.2.0	1b
3. a	1.3.4	1c
4. d	1.4.0	1d
Section Two		
1. a	2.1.0	2a
2. b	2.2.3	2b
3. a	2.3.2	2c
4. d	2.4.0	2d
Section Three		
1. a	3.1.0	3a
2. d	3.2.0	3b
3. b	3.3.0	3c
4. b	3.4.0	3d
Section Four		
1. b	4.1.1	4a
2. c	4.2.2	4b

NCCER CURRICULA — USER UPDATE

NCCER makes every effort to keep its textbooks up-to-date and free of technical errors. We appreciate your help in this process. If you find an error, a typographical mistake, or an inaccuracy in NCCER's curricula, please fill out this form (or a photocopy), or complete the online form at **www.nccer.org/olf**. Be sure to include the exact module ID number, page number, a detailed description, and your recommended correction. Your input will be brought to the attention of the Authoring Team. Thank you for your assistance.

Instructors – If you have an idea for improving this textbook, or have found that additional materials were necessary to teach this module effectively, please let us know so that we may present your suggestions to the Authoring Team.

NCCER Product Development and Revision
13614 Progress Blvd., Alachua, FL 32615
Email: curriculum@nccer.org
Online: www.nccer.org/olf

❑ Trainee Guide ❑ Lesson Plans ❑ Exam ❑ PowerPoints Other ____________________

Craft / Level: Copyright Date:

Module ID Number / Title:

Section Number(s):

Description:

Recommended Correction:

Your Name:

Address:

Email: Phone:

Glossary

Admixture: A chemical or mineral other than water, cement, or aggregate added to mortar immediately before or during mixing to change its setting time or curing time; to reduce water; or to change the overall properties of the mortar.

Adobe: Sun-dried, molded clay brick.

Aggregate: Materials such as crushed stone or gravel used as a filler in concrete and concrete block.

Air-entraining: A type of admixture added to mortar to increase microscopic air bubbles in mixed mortar. The air bubbles increase resistance to freeze-thaw damage.

American Society for Testing and Materials (ASTM) International: The publisher of masonry standards.

Arresting force: The force needed to stop a person from falling. The greater the free-fall distance, the more force is needed to stop or arrest the fall.

Ashlar: A squared or rectangular cut stone masonry unit; or, a flat-faced surface having sawed or dressed bed and joint surfaces.

Bed joint: A horizontal joint between two masonry units; the horizontal joint between two masonry units in separate courses.

Bond: Laying out masonry units to establish spacing.

Bundled: To be wrapped or bound in a cube shape.

Butter: To apply mortar to a masonry unit prior to laying it.

Capital: The top part of an architectural column.

Carabiner: A coupling link fitted with a safety closure.

Caustic: Capable of causing chemical burns.

Closure unit: The last brick to fill a course.

Competent person: An individual who is capable of identifying existing and predictable hazards or working conditions that are hazardous, unsanitary, or dangerous to employees, and who has authorization to take prompt, corrective measures to eliminate or control these hazards and conditions.

Concrete masonry unit (CMU): A hollow or solid block made from portland cement and aggregates.

Converting: The process of changing from one form of measure to another, for example, from feet to inches or from inches to feet.

Cored: Brick that has holes extending through it to reduce weight.

Corner lead: A lead built at a corner in which each course of masonry is shorter than the course below it.

Corner pole: Any type of post braced into a plumb position so that a line can be fastened to it. Also called a *deadman*.

Cornice: The horizontal projection crowning the wall of a building.

Course: A row or horizontal layer of masonry units.

Crowd the line: To touch the mason's line or to place a masonry unit too close to the line.

Cube: A strapped bundle of approximately 500 standard brick, or 90 standard block. The number of units in a cube will vary according to the manufacturer.

Deceleration device: A device such as a shock-absorbing lanyard or self-retracting lifeline that brings a falling person to a stop without injury.

Deceleration distance: The distance it takes before a person comes to a stop when falling. The required deceleration distance for a fall arrest system is a maximum of 3½ feet.

Dry bond: Laying out masonry units without mortar to establish spacing.

Extent of bond: The adhesion of mortar to masonry units, as determined by its lime and moisture content.

Facing: That part of a masonry unit or wall that shows after construction; the finished side of a masonry unit.

Flammable: Capable of easily igniting and rapidly burning; used to describe a fuel with a flash point below 100°F.

Footing: The base for a masonry unit wall, or concrete foundation, that distributes the weight of the structural member resting on it.

Free-fall distance: The vertical distance a worker moves after a fall before a deceleration device is activated.

Frogged: A style of brick in which a depression in the bedding surface of a brick lightens its weight.

Grout: A mixture of portland cement, lime, and water, with or without fine aggregate, with a high-enough water content that it can be poured into spaces between masonry units and voids in a wall.

Head joint: The vertical joint between two masonry units.

Hurricane strap: A special kind of anchor that is used to tie the frame to the foundation.

Hydration: A chemical reaction between cement and water that hardens the mortar. Hydration requires the presence of water and optimal air temperatures of between 40°F and 80°F.

Hygroscopic: Having a high initial rate of moisture absorption.

Joints: The area between each brick or block that is filled with mortar.

Kickback: A reaction caused by a pinched, misaligned, or snagged tuckpoint grinder wheel that causes the wheel to stop momentarily, propelling the grinder away from the surface and toward the operator.

Lead: The two corners of a structural unit or wall, built first and used as a position marker and measuring guide for the entire wall.

Legend: A table, list, or chart used in construction drawings to explain the meanings of the various lines, symbols, and abbreviations used in that particular set of drawings.

Lintel: The support beam over an opening such as a window or door. Also called a *header.*

Management system: The organization of a company's management, including reporting procedures, supervisory responsibility, and administration.

Manufactured stone veneer: A premade veneer consisting of cast cementitious material with pigments and other added materials to give the appearance of natural stone. Also called *adhered concrete masonry veneer* (ACMV).

Mason: A person who assembles masonry units by hand, using mortar, dry stacking, or mechanical connectors.

Masonry cement: Cement that has been modified by adding lime and other materials.

Masonry unit: Any building block made of brick, cement, ashlar, clay, adobe, rubble, glass, tile, or any other material that can be assembled into a structural unit.

MasterFormat™: A standard indexing system for construction specifications in the United States and Canada, developed by the Construction Specifications Institute (CSI) and Construction Specifications Canada (CSC) and used by project planners to prepare project specifications.

Mortar: A mixture of portland cement, lime, fine aggregate, and water, plastic or stiff enough to hold its shape between masonry units.

Nitrile: A synthetic rubberlike material used in masonry gloves to protect hands while permitting a tactile response.

Nominal dimension: The size of the masonry unit plus the thickness of one standard (⅜ to ½ inch) mortar joint, used in laying out courses.

Nonstructural: Not bearing weight other than its own.

Occupational Safety and Health Administration (OSHA): The division of the US Department of Labor mandated to ensure a safe and healthy environment in the workplace.

Parapet: A low wall or railing.

Parge: A thin coat of mortar or grout on the outside surface of a wall. Parging prepares a masonry surface for attaching veneer or tile, or parging can waterproof the back of a masonry wall.

Personal protective equipment (PPE): Equipment or clothing designed to prevent or reduce injuries.

Pilaster: A square or rectangular pillar projecting from a wall.

Plasticity: The ability of mortar to flow like a liquid and not form cracks or break apart.

Pointing: Troweling mortar or a mortar-repairing material, such as epoxy, into a joint after masonry is laid.

Potable: Water that is safe for cooking and drinking.

Racking: Shortening each course of masonry by half a unit so it is shorter than the course below it.

Range: To align the masonry unit or wall along the plane or face of the wall. Walls should be ranged from one corner to another.

Respirator: A device that provides clean, filtered air for breathing, no matter what is in the surrounding air.

Return: A corner in a structure or lead.

Safety data sheet (SDS): A document that must accompany any hazardous substance. The SDS identifies the substance and gives the exposure limits, the physical and chemical characteristics, the kind of hazard it presents, precautions for safe handling and use, and specific control measures. Formerly known as material safety data sheet (MSDS).

Scaffold: An elevated platform for workers and materials.

Silicosis: A respiratory disease caused by the inhalation of silica dust.

Slack to the line: Masonry units set too far away from the mason's line.

Slaked lime: Lime reduced by mixing with water to a safe form that can be used in the production of mortar.

Spread: A row of mortar placed into a bed joint; to lay a bed of mortar in a line for a course of masonry units.

Stringing: Spreading mortar with a trowel on a wall or footing for a bed joint.

Structural: Bearing weight in addition to its own.

Tail: To check the spacing of head joints by checking the diagonal edges of the courses on a lead or corner.

Tempering: Adding water to mortar to replace evaporated moisture and restore proper consistency. Any tempering must be done within the first 2 hours after mixing, as mortar begins to harden after 2 ½ hours.

Tuckpointing: Filling fresh mortar into cut-out or defective joints in masonry.

Uncored: Brick that has no holes extending through it.

Veneer tie: An anchor that is used to tie veneer to masonry structures.

Water retention: The ability of mortar to keep sufficient water in the mix to enhance plasticity and workability and to decrease the need to temper.

Weephole: A small opening in mortar joints or faces to allow the escape of moisture.

Workability: The property of mortar to remain soft and plastic long enough to allow the mason to place and align masonry units and tool off the mortar joints before the mortar hardens completely.

Working stack: A stack of brick that has been removed from a bundle and set up near where the brick will be used.

Wythe: A continuous section of masonry wall, one masonry unit in thickness, or that part of a wall that is one masonry unit in thickness.

Index

A

B

C

G

H

I

J

N

O

P

Q

R

T

U

V

W

Y